S0-BLS-358

Global Tropospheric Chemistry

A PLAN FOR ACTION

Global Tropospheric Chemistry

A PLAN FOR ACTION

Global Tropospheric Chemistry Panel
Board of Atmospheric Sciences and Climate
Commission on Physical Sciences, Mathematics, and Resources
National Research Council

NATIONAL ACADEMY PRESS
Washington, D.C. 1984

NATIONAL ACADEMY PRESS 2101 CONSTITUTION AVE., NW WASHINGTON, DC 20418

This material is based upon work supported jointly by the National Science Foundation and the National Aeronautics and Space Administration under Grant Number ATM 80-24257.

Library of Congress Catalog Card Number 84-61498

International Standard Book Number 0-309-03481-7

Printed in the United States of America

Global Tropospheric Chemistry Panel

ROBERT A. DUCE, University of Rhode Island, *Chairman*
RALPH CICERONE, National Center for Atmospheric Research, *Vice Chairman*
DOUGLAS DAVIS, Georgia Institute of Technology
C. C. DELWICHE, University of California, Davis
ROBERT DICKINSON, National Center for Atmospheric Research
ROBERT HARRISS, National Aeronautics and Space Administration
BRUCE HICKS, National Oceanic and Atmospheric Administration
DONALD LENSCHOW, National Center for Atmospheric Research
HIRAM LEVY II, National Oceanic and Atmospheric Administration
SHAW LIU, National Oceanic and Atmospheric Administration
MICHAEL McELROY, Harvard University
VOLKER MOHNEN, State University of New York, Albany
HIROMI NIKI, Ford Motor Company
JOSEPH PROSPERO, University of Miami

Board on Atmospheric Sciences and Climate

Commission on Physical Sciences, Mathematics, and Resources

Foreword

As the world approaches the threshold of the twenty-first century, higher levels of understanding of the physical environment are becoming attainable and more necessary. Just as science and technology have permitted world human population to grow and life expectancy to increase through modern industry and agriculture, so they permit more rigorous investigations into how the earth's planetary life support system works. Prudent management will become imperative if the general health and stability of human life on this planet are to be assured. Effective management will require a good understanding of the complex physical, chemical, and biological processes in that system that enables it to combine solar radiant energy with the cycling of chemical nutrients through the biosphere to sustain plant, animal, and human life.

The important role of chemical and physical processes in the troposphere in the planetary life support system has been brought into sharp focus in recent years not only by research discoveries, but also by a disturbing, recurring sequence of problem identification and response, e.g., impacts of smog on health, of acid rain on lakes, forests, and agriculture, of increasing carbon dioxide and other trace gases on climate, and of chemicals moving upward through the troposphere to the stratosphere. It has become clear that the troposphere is an integral component of the planetary life support system—receiving, transporting, transforming, and depositing substances that either contribute to the efficiency of the system or deleteriously perturb it. Yet relatively little effort has been expended on obtaining a fundamental understanding of the global troposphere and its dynamical behavior and cycles. Perturbations can be expected to increase in frequency and variety during the next several decades, and their significant economic impact will grow. Because the atmosphere is a moving and restless continuum enveloping the planet, the issues are international; since physical, chemical, and biological processes are inextricably intertwined, the effort to understand them must be interdisciplinary.

Accordingly, it was timely that a panel of atmospheric chemists and meteorologists be convened to develop the conceptual framework and propose a scientific strategy for a study of the chemistry of the global troposphere. Rapid advances in the theoretical understanding of chemical reactions in the troposphere, field-measurement capabilities, laboratory techniques, data handling, and numerical modeling capacities strongly support the conviction that a coordinated international effort can lead, before the end of the century, to the kind of understanding that would provide the predictive capability necessary to anticipate the impact on the planetary ecosystem of conscious or inadvertent changes in the chemistry of the lower atmosphere.

The institutional framework for such a study exists in the International Council of Scientific Unions. Atmospheric scientists in the United States and around the world are enthusiastic. Now is a propitious time to act.

The panelists have presented a challenging but tractable scientific endeavor, with highly attractive societal benefits. It might well constitute an important element of an international program dedicated to understanding the behavior of the geosphere and biosphere as an integrated system.

THOMAS F. MALONE, *Chairman*
Board on Atmospheric Sciences and Climate

Preface

Prompted by an increasing awareness of the influence of human activity on the chemistry of the global troposphere, a meeting of 10 atmospheric chemists and meteorologists was held at the National Center for Atmospheric Research in the spring of 1981. In a letter report to the National Science Foundation (NSF) following this meeting, this group called for the development of a comprehensive plan for a coordinated study of tropospheric chemistry on a global scale. They also recognized the complexities of tropospheric biogeochemical cycles and the difficulties in predicting tropospheric responses to both natural and anthropogenic perturbations but expressed confidence that the necessary research is feasible. In response, the NSF asked the National Research Council to form a Panel on Global Tropospheric Chemistry. This panel was formed by the NRC's Committee on Atmospheric Sciences (now the Board on Atmospheric Sciences and Climate) during the summer of 1982; the panel's work has been supported by the NSF and by the National Aeronautics and Space Administration.

The panel was given the following charge:

1. assess the requirement for a global study of the chemistry of the troposphere;
2. develop a scientific strategy for a comprehensive plan taking into account the existing and projected programs of the government;
3. assess the requirements of a global study in terms of theoretical knowledge, numerical modeling, instrumentation, observing platforms, ground-level observational techniques, and other related needs; and
4. outline the appropriate sequence and coordination required to achieve the most effective utilization of available resources.

The entire panel held meetings at Kingston, Rhode Island, in September 1982; at Santa Monica, California, in December 1982; and at Boulder, Colorado, in April 1983. Subgroups of the panel met at other times. During these meetings a scientific framework was developed for the proposed program based on the fundamental processes controlling biogeochemical cycles in the troposphere: sources, transport, transformations, and removal.

Efforts were made to keep both the U.S. and the international atmospheric chemistry community aware of the panel's deliberations, and comments and suggestions from this community were solicited. Many thoughtful and helpful responses were received and used by the panel in preparing its report.

Chapters 1 through 4 (Part I) of this report present the details of the rationale and framework for the proposed *Global Tropospheric Chemistry Program*. The overall program includes intensive field, laboratory, and modeling investigations in four areas: biological sources for tropospheric constituents, global distribution and long-range transport of trace species, fast-photochemical cycles and transformations, and wet and dry removal processes. Instrument and platform development requirements are assessed, and the need for strong international cooperation is stressed.

Chapters 5 through 9 (Part II) present the background information from which the proposed program was developed. Brief reviews are given of current understanding and gaps in knowledge concerning sources, transport, transformation, and removal of trace species in the troposphere. Reviews of the primary chemical cycles in the troposphere and the role of modeling in understanding tropospheric chemical processes are presented along with community surveys and reviews of currently available chemical instrumentation techniques used in atmospheric studies as well as aircraft, ship, and spaceborne sampling platforms. Current research programs in tropospheric chemistry in the United States are also reviewed in an appendix.

Development of this program was a joint effort involving every panel member. Without this cooperative involvement, it would not have been possible to complete this task. The panel expresses its gratitude to a number of individuals who contributed time, effort, and enthusiasm to the development of this report. The panel thanks Dieter Ehhalt, William Chameides, and Dan Albritton for their detailed reviews of early sections of the report, and the anonymous National Research Council reviewers of this report for their constructive criticism and suggestions. Robert Charlson, Paul Crutzen, Leonard Newman, Frank Allario, and Roger Tanner contributed significantly to the panel's efforts through participation in some panel or subpanel meetings or extensive discussions with panel members.

Global Tropospheric Chemistry Panel

ROBERT A. DUCE, *Chairman*
RALPH CICERONE, *Vice Chairman*
DOUGLAS DAVIS
C. C. DELWICHE
ROBERT DICKINSON
ROBERT HARRISS
BRUCE HICKS
DONALD LENSCHOW
HIRAM LEVY II
SHAW LIU
MICHAEL McELROY
VOLKER MOHNEN
HIROMI NIKI
JOSEPH PROSPERO

Contents

PART I

A PLAN FOR ACTION

Executive Summary

The lower atmosphere, or troposphere, is that part of the atmosphere in closest contact with the earth's surface. The weather phenomena that are experienced daily are manifestations of tropospheric processes, and many chemicals, both natural and synthetic, flow through the troposphere. These substances undergo chemical transformations as they travel. They are eventually removed from the troposphere by rain or by uptake at the earth's surface (often by vegetation) or by upward transport into the overlying stratosphere. Because humans and much of the biosphere reside and respire in the lower troposphere, chemical processes and cycles in this region are a critical component of the atmosphere-ocean-soil life support system of the planet.

Much of the current understanding of the earth's troposphere has arisen from a decade or more of intensive investigation of urban air pollution, the chemistry of the stratosphere, and the physics of the climate and from some exploratory studies of the nonurban troposphere. Yet, fundamental questions remain unanswered about the basic state and dynamical response patterns of the global tropospheric chemical system. Considering the state of this science and the importance of the questions, we believe the time has arrived to initiate a major international research program aimed at understanding the fundamental processes that control the chemical composition and cycles of the global troposphere and how these processes and properties affect the physical behavior of the atmosphere. Highly skilled individuals and instrumentation are available to begin the research. Accordingly, **we recommend that the United States assume a major role in initiating a comprehensive investigation of the chemistry of the global troposphere.**

The long-term goals of this *Global Tropospheric Chemistry Program* should be as follows:

1. **To understand the basic chemical cycles in the troposphere through field investigations, theory aided by numerical modeling, and laboratory studies.**
2. **To predict tropospheric responses to perturbations, both natural and human-induced, of these cycles.**
3. **To provide the information required for the maintenance and effective future management of the atmospheric component of the global life support system.**

The chemical composition of the unperturbed troposphere is dynamically balanced; it is controlled to a large extent by the terrestrial and marine biosphere. Indeed, the present tropospheric composition could not persist in the absence of biological activity. The interaction is bidirectional, however. The productivity of the biosphere is dependent upon the troposphere for oxygen and carbon dioxide, for the fixation and transport of nutrients such as nitrogen and sulfur and other trace species, and for the maintenance of a stable climate and solar radiation environment.

Changes in tropospheric composition induced by human activity can alter this balance over a range of spatial and temporal scales. This realization has stimulated an explosive growth in interest in the atmosphere as an integrated chemical system. It is now known that such changes can be relatively long lasting and that they may have serious consequences for the overall function of the planetary life support system, including food production, the quality of air and water, the integrity of the natural biosphere, the global chemical cycles essential to life, and the stability of terrestrial climate.

Impacts can be subtle. Release of synthetic chlorofluorocarbon gases can alter the abundance of halogens in the stratosphere, leading to an enhanced rate of removal for ozone and, consequently, to greater fluxes of ultraviolet radiation at the planetary surface. A slow, long-term buildup of trace gases that absorb infrared radiation can significantly alter the earth's climate, lead to a change in the water cycle, and result in a shift in the photochemistry of the lower atmosphere that can affect the biogeochemical cycling of nutrient elements. There are also relatively direct effects of concern. Emission of nitrogen oxides and hydrocarbons in urban areas can have immediate and readily observable effects on ambient levels of ozone, with potentially deleterious consequences for agriculture and the health of local populations. The release of sulfur and nitrogen oxides by combustion and industrial processes can lead to increased deposition of acidifying substances on ecologically sensitive regions and the formation of climatically significant aerosol particles.

Some human activities have already had a dramatic impact on certain chemical cycles in the troposphere. As a consequence of the combustion of fossil fuel and biomass burning, nitrogen oxides are introduced into the global troposphere at rates believed to exceed production rates in the natural environment. A similar conclusion holds for sulfur, primarily because of the combustion of sulfur-rich fuels and the smelting of ores. The manufacture of chemical fertilizers, agricultural practices, and changes in land use may significantly alter the production of long-lived tropospheric and stratospheric gases such as carbon dioxide, methane, and nitrous oxide. Concentrations of these radiatively active gases are already increasing at a disturbing rate on a global scale. The capability for observing these changes has outstripped the ability of scientists to understand their causes and to evaluate their impact. Estimating the extent and effects of pollution on the natural global environment, including the stratosphere, is difficult, if not impossible, in the absence of a comprehensive understanding that embraces the entire hierarchy of important chemical cycles in the global troposphere.

Historically, the atmospheric chemistry community has operated by responding to crises. Examples include acid precipitation with its regional and hemispheric scale impacts; unexpected disturbances to the stratospheric ozone layer resulting from ground-level emissions of several trace gases; and the potential effects of several trace gases, including carbon dioxide, on the earth's climate. Time and money have not been available for the development of the systematic measurement and modeling programs required to anticipate possible future perturbations of the global troposphere and to respond knowledgeably to such events when they occur. If tropospheric chemical cycles are to be understood and a predictive capability is to be attained, a long-term commitment must be made to the development of comprehensive models of the global tropospheric chemical system. **This in turn requires coordinated research efforts aimed at attaining a thorough understanding of the fundamental processes controlling global tropospheric biogeochemical cycles. These processes include the input of trace species into the troposphere, their long-range transport and distribution, their chemical transformations, and their removal from the troposphere.** For these reasons **we recommend that major research efforts in the *Global Tropospheric Chemistry Program* be undertaken with the following specific scientific objectives**:

1. **To evaluate biological sources of chemical substances in the troposphere.** Primary emphasis should be placed on investigations of temperate and tropical forests and grasslands, intensely cultivated areas, coastal waters and salt marshes, open ocean regions, tundra regions, and biomass burning.
2. **To determine the global distribution of tropospheric trace gases and aerosol particles and to assess relevant physical properties.** This program calls for field measurements and analyses coordinated with the development and validation of tropospheric chemical transport models, the development of a regional and global data base for key species in chemical cycles, and the continuation and improvement of existing monitoring programs for the accurate measurement of long-term trends in environmentally important trace gases and aerosol particles.
3. **To test photochemical theory through field and laboratory investigations of photochemically driven transformation processes.** Particularly important tests will be investigations over tropical oceans and rain forests with additional studies in midlatitudes.
4. **To investigate wet and dry removal processes for**

trace gases and aerosol particles. Research should be directed not only toward evaluating chemical fluxes to land and water surfaces, but also toward a fundamental understanding of aqueous-phase reaction mechanisms and scavenging processes.

5. **To develop global tropospheric chemistry systems models (TCSMs) and the critical submodels required for the successful application of TCSMs.** A wide range of models of individual processes important for tropospheric chemistry as well as comprehensive global models that include the most important chemical and meteorological processes must be developed. Modeling and laboratory and field studies are necessarily symbiotic; progress in each area is dependent upon contributions from the others.

Projects to pursue these objectives should be initiated immediately. They should proceed simultaneously throughout the development of the proposed *Global Tropospheric Chemistry Program.*

The *Global Tropospheric Chemistry Program* would build on the significant insights already achieved through preliminary studies of global tropospheric chemistry, research on urban and regional air pollution, and on the last decade of studies of the stratosphere. Many well-posed questions are to be found in the numerous subtle interactions that couple the atmosphere to the global biosphere. The proposed program would contribute to a better understanding of the functioning of the natural troposphere and how its chemistry affects its physical behavior. The program would also allow a more comprehensive assessment of the impact of human activities on the global environment. Furthermore, it would provide knowledge that would enable mankind to use the resources of the planet in a manner that is most efficient and least disruptive to the harmony of nature and that is considerate of the long-term interests of life on earth.

To attain these objectives, sensitive instrumentation will be required for the measurement of chemical species in the remote troposphere. Much of this instrumentation is available now, and recent advances in instrumentation technology convince us that the additional required instrumentation can be developed. **We recommend that a vigorous program of instrument development, testing, and intercalibration be undertaken immediately and that it be continued throughout the *Global Tropospheric Chemistry Program.***

The *Global Tropospheric Chemistry Program* will also require the continuing commitment of laboratory chemical kineticists and the further development of laboratory systems for investigation of the mechanisms and kinetics of gas- and liquid-phase reactions. **We recommend that an increased effort be initiated on laboratory studies of the rates and pathways of fundamental chemical reactions in tropospheric chemical systems.**

The *Global Tropospheric Chemistry Program* should be international in scope. It must engage the resources and commitment of not just the national, but also the international scientific and political communities. **We recommend that the United States play a major role in a cooperative effort with other countries to attract these resources and that it do so with confidence that the international community is both ready and willing to join in this initiative.**

Currently available manpower in atmospheric chemistry is adequate to initiate this program, but manpower requirements are a concern over the longer term. A number of major U.S. and European universities have begun to intensify their graduate programs in atmospheric chemistry. To assure that highly trained personnel are available to conduct later portions of the *Global Tropospheric Chemistry Program*, **we recommend that a manpower survey for atmospheric chemistry be conducted through an appropriate scientific group.** This survey should be cognizant of existing channels from the basic physical and life sciences into atmospheric chemistry. Organizations that could be of assistance in such a survey include the American Chemical Society, the American Geophysical Union, the American Meteorological Society, and the University Corporation for Atmospheric Research.

The long-term vitality of tropospheric chemistry research and graduate education depends to a large degree on the undirected research of individuals—research that does not always result in immediate applied results. Strong support for the high-quality research of individual investigators at academic institutions must be maintained.

Although several federal agencies are now conducting and supporting research that addresses various aspects of global tropospheric chemistry, their current programs are not adequate to achieve the goals of the recommended *Global Tropospheric Chemistry Program.* Federal support is currently focused on urban or regional problems, such as air pollution or acid rain, or on questions related to climate, such as carbon dioxide buildup. Nevertheless, many of the research tasks encompassed by existing programs will contribute substantially to the goals of the *Global Tropospheric Chemistry Program* and can readily serve as a strong foundation for it. We estimate that the current size of the federal program in areas recommended by our panel is $10-20 million per year. We believe that the recommended program could be launched with the addition of an increment of approximately the same level in FY 1986. This increase would enable planning and exploratory observations to begin for the major field-oriented research programs and would support initial new thrusts in model development and chemical instrumentation. This funding would also be used for the improvement and required modifications of the necessary field platforms for the *Global Tropospheric Chemistry Program*, particularly research aircraft, ships, and ground-based platforms. It will also be necessary to develop some instrumentation for the meteorological measurements that are required for the chemistry field programs.

This budgetary augmentation should be sustained in subsequent years, and substantial additional increments will be required as the extensive field experiments get underway. Although a detailed cost analysis is not feasible at this time, we estimate that the annual funding required to conduct the U.S. component of a fully operational *Global Tropospheric Chemistry Program* will be in the tens of millions of dollars for a period of a decade or more.

We recommend that an appropriate U.S. scientific organization exercise general scientific oversight of the U.S. component of the *Global Tropospheric Chemistry Program*. This group should also be responsible for providing interaction with the international scientific community.

The recommended program involves a combination of small research projects involving individual principal investigators and large-scale, highly coordinated field projects. The latter will require extensive, shared facilities and innovative, effective project management. The individual investigator could be covered by the grant-support mechanism provided principally, at the present time, by the National Science Foundation (NSF) and the National Aeronautics and Space Administration (NASA) through their atmospheric chemistry programs. **We believe these agencies and others can work together to provide the necessary mechanisms and management to coordinate the efforts of U.S. scientists in universities and government and industrial laboratories.** The participation of other government agencies—such as the National Oceanic and Atmospheric Administration, the Department of Energy, the Environmental Protection Agency, and the Department of Defense—and of universities, private research organizations, and industries will also be essential for the success of the program. Because of the need for effective global observations, we suggest that a study be undertaken to examine the potential for satellite-based remote sensing and to define the role, if any, of satellite measurements for the *Global Tropospheric Chemistry Program*.

The *Global Tropospheric Chemistry Program* represents a fundamental step toward understanding and predicting effects of human society on the planetary life support system. Benefits from this program will transcend national boundaries and extend to all. The scientific community is ready to meet this exciting and demanding challenge.

1 The Need for a Program

The recent history of atmospheric chemistry research (i.e., since 1970) is characterized by a multitude of surprising discoveries. The perception of the atmosphere as a chemical system has changed dramatically. It is now known that the atmosphere is a dynamic system where many chemical reactions, physical transformations, and types of transport occur. There are intense inputs of raw materials from natural processes (often biological) and from human activities. The movement and reactions of chemicals in the atmosphere are now clearly seen as components and links in global biogeochemical cycles of the chemical elements. It is not a trivial task to appreciate the significance of these current concepts and the opportunities they present, largely because so many new ideas and facts have emerged so quickly.

When one reviews the knowledge of atmospheric chemical composition as it existed two or three decades ago, one is struck by the primitive state of the science. Quantitatively, only the atmospheric concentrations of nitrogen (N_2), oxygen (O_2), the noble gases, carbon dioxide (CO_2), water below the tropopause, and ozone (O_3) in the stratosphere were then known. By 1950, methane (CH_4), nitrous oxide (N_2O), carbon monoxide (CO), and hydrogen (H_2) had been detected, but measured only to about 50 percent accuracy. The existence of airborne particles was known, but little information, even on their bulk properties, was available. On the basis of extant data on the visible and ultraviolet light spectrum of the sun, one could speculate that stratospheric O_3 was important as an ultraviolet shield, but other possible absorbers were unexplored. Tropospheric O_3, although known to be a product of photochemical reactions in urban smog, was little more than a curiosity when detected in clean background air. The roles of CO_2, water, and O_3 in climate and atmospheric dynamics were identified qualitatively, but they were not well understood.

The atmosphere near the earth was viewed as a fluid in motion, transporting moisture and heat. It also transported pollutants arising from cities, factories, and fires. The chemical species in the air were regarded as essentially inert and for good reason—most of the components that were known were inert gases. A fair amount was known about radionuclides in the atmosphere from studies related to nuclear weapons testing. Indeed, the use of radiochemical techniques in atmospheric studies was more prevalent in 1960 than in 1980. Because of these studies, a great deal has been learned about stratospheric transport processes; this knowledge has been extremely valuable in the efforts to assess the global impact of anthropogenic chlorofluorocarbons and the exhaust from high-flying aircraft.

Since those early days, progress has been rapid and it is still accelerating. It is now understood that the troposphere is a reactive environment. Because of the complexity inherent in such an environment, new programs in which a broad spectrum of studies are coordinated, are needed. Indeed, much of the current understanding of tropospheric chemistry is the result of the coupling of new chemical data with new theoretical insights and models.

It has also become clear, both to atmospheric chemists and to laymen, that humans are increasingly capable of perturbing the atmosphere. Often inadvertent and unforeseen, these perturbations are sometimes direct and sometimes subtle, and they can extend to the earth's soils, waters, biota, and climate. Thus, perturbations to tropospheric processes can affect earth's biogeochemical cycles and the total life support system of the planet.

PUBLIC POLICY PROBLEMS AND ATMOSPHERIC CHEMISTRY

Over the past decade or so, human perturbations and influences on the chemistry of the atmosphere have been identified at a rate that exceeds the ability of scientists to predict the behavior of the perturbed system, even though knowledge has grown explosively. The array and scope of these perturbations are impressive, as are the related research and policy questions they raise. Notable examples are (1) the acid rain phenomenon[1,2] with its regional and hemispheric scale manifestations and its contributions from gas-phase, liquid-phase and solid-phase species, reactions, and deposition; (2) disturbances to the stratospheric O_3 layer and its photochemistry[3,4,5,6,7]—caused by ground-level emissions of chlorofluorocarbons, chlorocarbons, N_2O, and direct stratospheric injections of nitric oxide (NO) by high-altitude nuclear weapons testing and, potentially, from stratospheric aviation; and (3) the potential effects of the growing concentrations of carbon dioxide[8] and several radiatively active trace gases (whose sources are largely either directly or indirectly under human control) on the entire background tropospheric chemical system and the earth's climate. Indeed, the combined effects of CH_4, nitrogen oxides, chlorofluoromethanes, and other radiatively active trace gases could be equivalent to the doubling of CO_2 during the next 30 to 40 years. Further examples include (4) perturbations to global nutrient-element cycles that have significant atmospheric components and consequences, e.g., the carbon, nitrogen, and sulfur cycles; and (5) modification of the radiative properties of the atmosphere by aerosol particles.

These problems are real, and the areas of the world affected by pollution are large and they are growing; indeed, some of the problems are demonstrably and inherently global and are of concern to atmospheric chemists and public policymakers. Further, the recent history of technology is one of exponential growth, for example, in the variety and rates of production of manufactured chemicals and of their application to agricultural soils and their release to water supplies. Also, ever-increasing energy production leads to increased releases of combustion products to the atmosphere. The increased use of technology and the introduction of new technological processes will have further impacts on the atmosphere and thus more questions concerning pollution control and resource management will inevitably arise.

---

[1] *Acid Deposition: Atmospheric Processes in Eastern North America, A Review of Current Scientific Understanding,* Committee on Atmospheric Transport and Chemical Transformation in Acid Precipitation, National Academy Press, Washington, D.C., 1983, 375 pp.

[2] *Atmosphere-Biosphere Interactions: Toward a Better Understanding of the Ecological Consequences of Fossil Fuel Combustion,* Committee on the Atmosphere and the Biosphere, National Academy Press, Washington, D.C., 1981, 263 pp.

[3] *Stratospheric Ozone Depletion by Halocarbons: Chemistry and Transport,* Panel on Stratospheric Chemistry and Transport of the Committee on Impacts of Stratospheric Change, National Academy of Sciences, Washington, D.C., 1979, 238 pp.

[4] *Halocarbons: Effects on Stratospheric Ozone,* Panel on Atmospheric Chemistry of the Committee on Impacts of Stratospheric Change, National Academy of Sciences, Washington, D.C., 1976, 352 pp.

[5] *Environmental Impact of Stratospheric Flight: Biological and Climatic Effects of Aircraft Emissions in the Stratosphere,* National Academy of Sciences, Washington, D.C., 1975, 348 pp.

[6] *Causes and Effects of Stratospheric Ozone Reduction: An Update,* Committee on Chemistry and Physics of Ozone Depletion, and Committee on Biological Effects of Increased Solar Ultraviolet Radiation, National Academy Press, 1982, 339 pp.

[7] *Causes and Effects of Stratospheric Ozone Reduction: Update 1983,* Committee on Causes and Effects of Changes in Stratospheric Ozone: Update 1983, National Academy Press, Washington, D.C., 1983, 254 pp.

[8] *Changing Climate,* Carbon Dioxide Assessment Committee, National Academy Press, 1983, 496 pp.

ATMOSPHERIC CHEMISTRY: TOOL, SCIENCE, OR BOTH?

A detailed examination of the activities and achievements in the field of atmospheric chemistry over the past 15 to 20 years reveals a recurring pattern in which a potentially serious problem is identified and a crisis response is evoked from the scientific community. Research carried out in a crisis-response mode attempts to obtain, as quickly as possible, the minimum amount of information needed for policy formulation. In this manner, considerable progress was made in quantifying the atmospheric effects of many human activities. However, there has been too little time and insufficient support to carry out the systematic and exploratory research needed to go beyond the confines of the immediate problem, i.e., to achieve a more complete understanding of the global troposphere so as to be able to anticipate pollution problems and to establish a more reliable base from which to formulate a response.

The identification of an anthropogenic effect on the atmosphere is in itself progress. However, it is difficult, if not impossible, to assess accurately the extent and implications of human impacts on natural processes if their workings are not understood. Thus, it is necessary to obtain a more quantitative understanding of the dynamics of the perturbations and of the background state and dynamics of the unperturbed natural atmosphere and biogeochemical system. The formulation of effective strategies for pollution control and resource management require this. Because of the rapid increase in human population, technology, and consumption of resources and because of the limited knowledge available even in 1970, the relatively small community of atmospheric chemists could not foresee or keep pace with the need for more and more quantitative information on air chemistry. A further proliferation of problems has occurred since then.

One can draw upon the experiences of atmospheric chemists over the last 10 to 15 years to devise more effective research strategies. For example, major discoveries have arisen from isolated, undirected research, often conducted by individuals. Indeed, major problems in environmental chemistry have been so discovered. Examples are Lovelock's early detection of the chlorofluorocarbons CCl_3F and CCl_2F_2 in the atmosphere and the theoretical investigation of their atmospheric chemistry by Molina and Rowland. At the outset of this research, no specific goal demanded that it be conducted. Similarly, when Keeling began his high-precision CO_2 monitoring in 1957, only a few farsighted individuals recognized that such data would ever be of such practical relevance or scientific value. One concludes from these and other such important examples that there must always be a place for the undirected research of individuals, research that does not always offer immediate applied results. One suspects that the field of atmospheric chemistry supports too little of this research largely because of the more pressing demands for problem- or mission-oriented research that is often directed at specific, identified pollution problems.

The science of atmospheric chemistry has reached a much more robust state in the last few years. The present state can now permit more progress to be made from more systematic research. For example, the potential effects of human activities on the atmosphere can be better quantified and future problems can be anticipated on the basis of a growing understanding of the pathways of atmospheric chemical transformations and dynamics and from a growing data base on atmospheric chemical composition. With increased knowledge of the chemical composition of the atmosphere and of the chemical pathways in it, one can draw from relevant knowledge of chemistry, physics, and biology and perform laboratory experiments; in this way one can deduce what other chemical substances might be found in the air and how they might behave. General principles have been sought and applied with increasing success. A prominent example involves the mapping out of chain reaction schemes that achieve chemical transformations through chemical catalysis. Thereby, it has become clear that certain reactive species can have importance far beyond that suggested by their extremely small concentrations. For example, NO is apparently important even at levels of one part in 10^{11}, and the hydroxyl radical (OH) is important at much lower concentrations still. It is important to note that the very existence of odd-electron or radical species like the hydroxyl radical has been proposed and explored only in the last 20 years or so; their significance in atmospheric processes was firmly established only a decade ago.

TROPOSPHERIC CHEMISTRY: THE PROSPECTUS

Atmospheric chemistry is confronted by an array of identified environmental problems for which public policymakers are demanding research solutions. We suspect that even more problems that derive from human activities are on the horizon. Although dramatic progress has been made toward understanding the basic systems and implementing solutions to applied problems, it is clear that future environmental issues and policy decisions will require a deeper understanding of the entire global atmospheric chemical system and of biogeochemical cycles. A major research effort in tropospheric chemistry should be made with this objective in mind. In Chapters 1 through 4 (Part I), we propose and justify a coordinated program of tropospheric chemistry research whose scope extends to global scales. In Chapters 5 through 9 (Part II) of this report, the current status of the science is described in more detail.

The program described here is realistic and feasible. It is soundly based on the results of a decade of intensive investigation of the earth's stratosphere and climate and about two decades of research on urban air pollution. Instrumentation is available now for many of the proposed investigations, and it is at an advanced stage of development for others. The development of instrumentation and advanced numerical models is being conducted in a number of laboratories that have demonstrated excellence. Highly skilled individuals are available to perform the research, and many of these individuals are already mounting research activities very relevant to the proposed research. A number of U.S. and European universities are expanding their graduate programs in atmospheric chemistry. Several U.S. scientific institutions and agencies have begun to accept responsibilities and propose initiatives; similar activity is evident in Europe, Australia, and Canada and may reasonably be anticipated elsewhere. Further, the need to perform research that effectively couples chemistry and meteorology is now well recognized. Moreover, because of the inherent heterogeneity of the troposphere, the plethora of sources, and the complexity of biological processes and of surface effects, there will be tremendous challenges to be faced. Great care must be exercised in formulating scientific concepts and in implementing a coordinated program. Given these, society will reap the rewards of more effective and efficient management of resources, maintenance of human health, and a safer global environment.

2 A Framework

To date, much of the research in tropospheric chemistry has focused on isolated questions or on one of the many elemental cycles, for example, carbon, nitrogen, or sulfur. Because of the complexity of tropospheric chemistry and the poor state of knowledge at that time, narrowly focused studies and exploratory programs were justified. Indeed, such studies were successful in that they advanced the field of tropospheric chemistry to its present status where one can begin to discern the overall structure and interaction of the chemical systems in the troposphere. Because of these complex interactions and the dynamic nature of the chemistry and physics of the troposphere, we believe that future advances in many areas of tropospheric chemistry can be best achieved by fostering research within a unifying conceptual framework based on atmospheric chemical processes. This framework is that of geochemical and biogeochemical cycles.

Four major categories of processes dominate chemical cycles in the troposphere: those related to sources, to chemical reactions and transformations, to transport, and to removal. Within the context of tropospheric chemistry, a chemical cycle begins when a substance is emitted into the troposphere. Consequently, a knowledge of the strength and distribution of sources is critical. Materials injected into the troposphere can undergo chemical reactions, some of which are cyclic and some of which produce a wide range of species that can have chemical and physical properties very different from those of the reactants; such transformations can effectively remove the species from a specific chemical cycle. The distribution of a species in the troposphere will be dependent on source characteristics and on the controlling chemical reactions; however, distributions are also greatly affected by a variety of transport processes that range in scale from that of boundary layer turbulence to that of planetary flow. Often, physical interactions occur in which the composition can influence the radiative or physical properties of the atmosphere or the underlying surface. Finally, the tropospheric cycle is terminated by removal of the species from the reacting system, usually through deposition at the earth's surface.

In this chapter, we outline a research strategy for tropospheric chemistry that encompasses these four fundamental processes and their roles in mediating atmospheric physical processes. A major effort would also be directed toward the development of global tropospheric chemistry models that can satisfactorily incorporate and describe these processes. A process-oriented discussion of tropospheric chemistry cycles has heuristic advantages over one that focuses on individual cycles in that it incorporates in a coherent manner many interrelated aspects of tropospheric chemistry. The disadvantage of this approach is that important aspects of tropospheric chemistry might be ignored if they do not fit into the framework of the discussion. The fact that a specific area of research is not mentioned in this program is not meant to imply that such work necessarily has a low priority. Similarly, the fact that the program calls for an increased level of cooperation and coordination does not mean that it must be highly regimented. Indeed, throughout the report, we stress that this effort can succeed only if independent active research scientists participate fully in its design, implementation, and management. Such participation is essential because the program is an evolutionary one in which there must be an interchange among scientists working on various processes and a feedback between experimental scientists, theoreticians, and modelers. Finally, the research activities of independent individual scientists, unaffiliated with the proposed program, will continue to be needed for the health of this scientific field.

SOURCES

Primary sources for tropospheric substances fall into three major categories: surface emissions, in situ formation from other species, and, in some cases, injection from the stratosphere. In each of these categories there are natural and anthropogenic components.

Natural surface sources are affected by a wide range of factors such as season, temperature, nutrient level, organic content, and pH; consequently, source strengths are highly variable in space and time. The terrestrial and marine biospheres are of particular importance as natural sources for chemical species in all the element cycles, but very little quantitative information is available on these sources. Some measurements of local fluxes from biological sources have been made for methane (CH_4) and a few reduced sulfur compounds. Estimates of local source strengths in the ocean, as well as in rivers and lakes, have been made for methyl chloride (CH_3Cl), N_2O, and reduced sulfur compounds by measuring their degree of supersaturation in surface waters, a quantity that varies greatly. To estimate fluxes from measured supersaturations, theoretical models of gas exchange across the water surface are required. At this time, it is not possible to make accurate global estimates of the natural surface sources for such globally important trace gases as CH_4, N_2O, ammonia (NH_3), carbonyl sulfide (COS), carbon disulfide (CS_2), hydrogen sulfide (H_2S), dimethyl sulfide ($(CH_3)_2S$), and CH_3Cl. It is even more difficult to determine the natural source strengths of the many reactive hydrocarbons that are released by vegetation. Future research must not only develop equipment and experimental protocols that are capable of accurately determining the local fluxes from the biosphere, but it must also concentrate on understanding the biological, chemical, and meteorological factors that control these fluxes.

For the United States and Europe, there are relatively accurate inventories of the anthropogenic sources of combustion products, such as nitric oxide (NO), nitrogen dioxide (NO_2), carbon monoxide (CO), and sulfur dioxide (SO_2); considerably less accurate estimates are available for the rest of the globe. The same is true of a wide range of industrial emissions. Emission rates from low-technology combustion, such as wood fires and slash-and-burn agriculture, are much less accurately known, even on local or regional scales. There is growing evidence that biomass burning, particularly in the tropics, may be a significant source of many trace species in the troposphere. In general, existing estimates of fluxes to the troposphere from anthropogenic surface sources are much more accurate than those available for natural sources. We emphasize here the need for research on the latter, and indeed on a subset—biological natural sources.

Trace substances are introduced above the lowest layers of the troposphere by in situ sources, both human (e.g., aircraft and tall stacks) and natural (e.g., lightning and volcanoes). While in situ sources are much smaller than surface sources on a global scale, they are more effective than surface sources because they inject water-soluble and surface-reactive gases into the middle and upper troposphere, where gas lifetimes are considerably longer than near the surface. Lightning is possibly an important source of nitric oxide and nitrogen dioxide (both hereafter referred to as NO_x), and volcanoes can be major episodic sources of sulfur compounds and other materials. Aircraft and tall stacks are important in situ sources of anthropogenic NO_x and SO_2. Many species are produced in situ by various tropospheric photochemical processes; these processes are considered to be secondary sources and will be discussed in the sections dealing with transformations.

Stratospheric sources are the result of high-energy ultraviolet radiation that can dissociate species such as N_2O and oxygen, which are normally nonreactive in the troposphere. Materials are transported from the stratosphere through the tropopause into the troposphere; such stratospheric injections serve as a relatively uniform source of many species compared to emissions of the same species from surface sources. However, current theories, observational data, and dynamical model calculations all show that injections into the troposphere occur predominantly at midlatitudes and more in the northern than the southern hemisphere. Stratospheric injection of ozone (O_3), originally produced in the middle stratosphere, is a major and perhaps dominant source of tropospheric O_3. There is considerable controversy about the importance of stratospheric injection of O_3 relative to in situ photochemical production; this controversy extends to some other substances as well. Although the stratospheric injection of NO_x and nitric acid (HNO_3) is much smaller than the combustion surface source, it could be the dominant source for these substances in the upper troposphere.

TRANSPORT AND DISTRIBUTION

Transport processes involve a wide range of space and time scales. However, the discussion of the troposphere can be greatly simplified by separating it, conceptually, into three layers: the planetary boundary layer, the free troposphere, and the tropopause region.

The planetary boundary layer (PBL) is the layer of the atmosphere that interacts directly (on the order of hours) with the earth's surface. The PBL plays a critical role in transport and distribution processes because most surface sources, whether natural or anthropogenic, emit directly into this layer. The structure of the PBL is strongly dependent upon surface properties such as roughness, temperature, and quantity and type of vegetation. Typically, the daytime continental boundary layer is convectively mixed as a result of solar heating at the surface. Atmospheric constituents in the PBL are efficiently mixed throughout its depth, which can extend up to several kilometers. Because of the cooling of the earth's surface at night, a stably stratified boundary layer (a few hundred meters in depth) is formed over land; the internal mixing within this layer is less efficient than that in the daytime PBL. Chemicals emitted into the stably stratified nighttime boundary layer may be transported long distances horizontally without extensive vertical mixing. Over the ocean, the PBL is well mixed to a height ranging from 0.5 to 2.0 km. Where the surface layer of water is warm, e.g., in the tropics and the summertime midlatitudes, this mixing is primarily convective and driven by evaporation. Where the water is colder than the air, the mixing is caused by mechanical turbulence, and a much shallower layer is formed. In both cases there is no strong diurnal cycle similar to that observed over land.

The winds in the free troposphere, i.e., above the PBL, have a strong latitudinal component that is easterly in the tropics and westerly in midlatitudes. Consequently, materials that have a residence time in the atmosphere of greater than about one month will have a distribution that reflects the spatial distribution of sources, and there will be a stronger concentration gradient in the north-south direction than around the latitude circle. However, superimposed on the mean east-west horizontal flow are large-scale, wave-like oscillations, particularly in midlatitudes, and strong vertical and north-south, thermally driven regional flows such as the Indian monsoon in the tropics. Thus the distribution of relatively long-lived trace species in the atmosphere will depend primarily on the location of the source of the injected material relative to the features of the general circulation. The distribution of shorter-lived species will be more complicated because the transport patterns will be determined to a considerable degree by the nature of the local weather systems occurring at the time.

An especially distinctive feature of the global tropospheric circulation is the relatively restricted interhemispheric flow across the equatorial regions of the oceans. As a result, the transfer of materials between the northern and southern hemispheres proceeds at a relatively slow rate, leading to differences in the interhemispheric concentration distribution of some species. These differences are especially noticeable for certain anthropogenic materials. Their concentrations are significantly greater in the northern hemisphere where industrialization and energy utilization are greater than in the southern hemisphere.

The tropopause separates the relatively turbulent and well-mixed free troposphere from the relatively stable and stratified stratosphere. Net upward transport through the tropopause occurs predominantly in the tropics, and net downward transport occurs mostly at midlatitudes.

Vertical transport processes within the troposphere range in scale from that of deep cumulus convection through that of cyclone-scale interactions in the polar and subtropical jets to that of the global-scale Hadley circulation. These processes mix trace gases and particles throughout the free troposphere and provide a linkage between the PBL and lower stratosphere. Local convection is particularly important in tapping the PBL and vertically mixing the free troposphere. Synoptic-scale cyclones also mix the free troposphere vertically. Though less intense than individual convective clouds, they tap much larger regions of the PBL. Cyclones, in combination with upper tropospheric jet streams, also provide an effective mechanism for transporting materials downward from the upper troposphere and lower stratosphere. On the largest scale, thermally driven upward transport in the tropics mixes the troposphere as a whole in that region and transports trace gases from the subtropical and tropical PBL to the tropical lower stratosphere.

As a general rule, the smaller the ratio of the input (or removal) rate of any substance to the total mass of that substance in the troposphere, the more uniform is its tropospheric distribution. Long-lived gases—such as N_2O, COS, and chlorofluoromethanes, which have atmospheric lifetimes of decades or more—are well mixed throughout the troposphere. Gases such as CH_4 and CH_3Cl, with somewhat shorter tropospheric lifetimes of a few years, are generally well-mixed vertically, and there are only small hemispheric differences. Gases with tropospheric lifetimes of a few months or less (CO, O_3, SO_2, NO, NO_2, HNO_3, and reactive hydrocarbons, for example) are not well mixed, and their distributions show large vertical and latitudinal gradients that are generated by source and sink distributions, chemical transformations, and removal processes. As the tropospheric lifetime of a species decreases, transport has less influence on the distribution on the hemispheric and global scale. Highly reactive species, such as the hydroxyl (OH) and hydroperoxyl (HO_2) radicals, have very short lifetimes; thus their concentration distributions do not depend directly on transport. However, because of reaction pathways involving other cycles, their distributions may depend on other species whose distributions are influenced by transport processes (e.g., water, O_3, and CO).

TRANSFORMATION

A key process in all the biogeochemical cycles is the chemical transformation of tropospheric trace gases into species that are either nonreactive in the troposphere or easily removed by rain or surface deposition. The oxidation of CO to CO_2 by the OH radical is an example of the former, while the oxidation of NO by O_3 to NO_2 followed by further oxidation to HNO_3 by the OH radical is an example of the latter. All the chemical transformations can be grouped into three basic classes: homogeneous gas-phase reactions (reaction of one gaseous species with another), homogeneous aqueous-phase reactions (reaction of one dissolved species with another), and heterogeneous reactions (reactions of species at a phase interface).

Homogeneous Gas-Phase Transformations

Ozone plays a significant role in tropospheric gas-phase chemistry. It reacts directly with some compounds such as NO_x and unsaturated hydrocarbons, and, more importantly, the photodissociation of O_3 leads to the formation of OH radicals. Hydroxyl-radical-initiated oxidation is the major pathway for the transformation of a large variety of tropospheric compounds and determines their chemical lifetimes. The oxidation of NO_2 leads directly to HNO_3 vapor. In the case of CH_4 and more complex hydrocarbons, numerous reaction intermediates, including free radicals, are produced; ultimately, these are either converted to stable nonreactive products, such as CO_2 and water, or removed from the gas phase by heterogeneous processes. Chemical transformations of a number of trace gases are interrelated by reactions involving common reactive species. This leads to a strong chemical coupling of the various element cycles and, in many cases, to a chemistry that is cyclic in nature. At present, there is considerable uncertainty as to the identity and fate of compounds produced in situ in the troposphere during the oxidation processes. Many of the reaction steps following the initial chemical attack of the OH radical on CH_4, NH_3, SO_2, and the more complex hydrocarbons are not known for certain. A number of potential intermediate products such as aldehydes, peroxides, and organic nitrates have been found in the troposphere, but conclusive laboratory confirmation of the mechanisms leading to these products and their subsequent reactions is still needed. Furthermore, there is a need to determine the reaction rates under conditions similar to those found in the troposphere (1 percent water, 20 percent oxygen, and atmospheric pressure in the range 0.1 to 1.0 atm).

Homogeneous Aqueous-Phase Transformations

Homogeneous reactions in water droplets appear to have a significant impact on the cycles of sulfur, nitrogen, and perhaps other elements. This chemistry takes place in both the submicrometer aqueous aerosol particles and the larger, 2- to 40-μm droplets found in convective and stable stratiform clouds and in fog. The reactions that occur in these two aqueous environments can be quite complex because of the presence of many interacting species: neutral free radicals, free radical ions, nonfree radical ions, as well as neutral semistable species. Representative of this type of system is the H_xO_y/sulfur/halogen system shown in Figure 2.1. This system illustrates the important oxidative capacity of H_xO_y species (both in their ionic and neutral forms) in the aqueous phase.

There is still much to learn about aqueous-phase processes in the troposphere. A major question is the

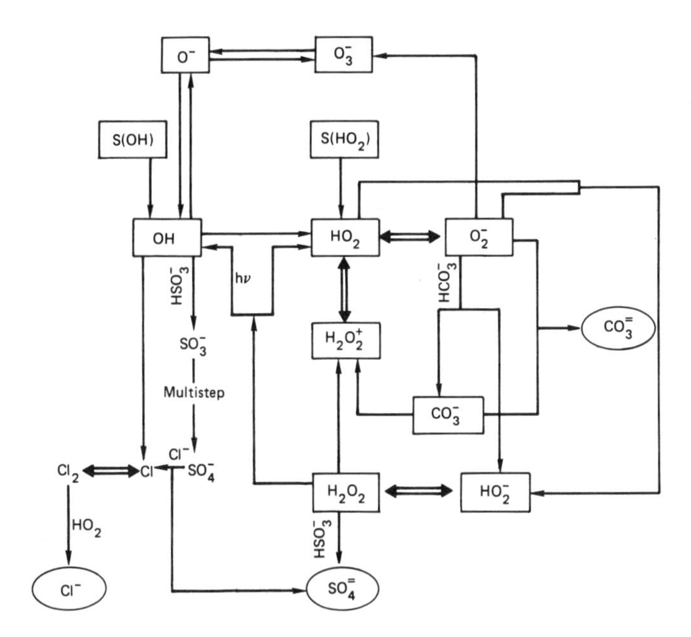

FIGURE 2.1 The proposed chemical pathways for $(OH)_{aq}$ and $(HO_2)_{aq}$ in a cloud droplet. S(OH) and $S(HO_2)$ represent the scavenging sources of these species to the droplet. The open double arrow indicates rapid chemical equilibrium, and the closed single arrow indicates an aqueous-phase reaction. End products are circled.

degree to which laboratory reaction rate constants apply to the actual environment of a water droplet. In addition, there are still many key reactions for which there are no data. Finally, one must recognize that there are few data on the concentrations of many critical species within aqueous aerosol particles and cloud droplets at different tropospheric locations and as a function of time.

Heterogeneous Transformations

Normally, a heterogeneous reaction occurs at an interface between two phases, although several interfaces can be involved in an overall process. Most reactions that occur on such surfaces are thought to be noncatalytic. These include chemical reactions in which both phases participate as consumable reactants and physical processes that involve either transport or growth, or both. Adsorption or absorption are, of course, heterogeneous processes. In heterogeneous catalytic processes the interfacial material or a species adsorbed on it is conserved. A special class of heterogeneous process, gas-to-particle conversion, converts a trace gas to a particle or a liquid droplet suspended in the atmosphere.

Important heterogeneous reactions in tropospheric chemistry include the conversion of gas-phase ammonia and nitric acid to ammonium nitrate and the conversion of gas-phase sulfur dioxide to sulfate in cloud droplets. There is still not much known about the role of heterogeneous processes in many trace gas and trace element cycles. However, it is clear that such processes are important in removal by deposition to surfaces.

REMOVAL

The final stage of any tropospheric elemental cycle is removal from the troposphere. The process can involve the episodic collection of particles and gases in water drops and ice crystals that fall as rain and snow (wet deposition), the more continuous direct deposition of gases and particles at the earth's surface (dry deposition), or the chemical conversion of trace gases to inert forms.

Aerosol particles act as nuclei for the condensation of water vapor in warm clouds and fog and for the generation of ice crystals in supercooled clouds. These nucleation mechanisms are the basis of most cloud-forming processes. As these drops grow in size, they serve as sites for the conversion of SO_2 to sulfate and adsorb soluble gases such as HNO_3, hydrogen chloride (HCl), hydrogen peroxide (H_2O_2), and formaldehyde (CH_2O). A small percentage of the droplets or ice particles will grow sufficiently large that they fall toward the ground as snow and rain. While falling, they collect other particles and soluble gases. Drops that evaporate before reaching the ground release gases to the atmosphere and the residue forms a particle; in effect, such droplets transport gases and particles to lower levels in the troposphere. This also occurs for nonprecipitating clouds, although in this case the transport may be upward. Those drops that reach the ground are an intermittent but highly efficient means of converting and removing soluble trace gases and particles from the troposphere. The key issue here is their chemical composition and deposition rate since these provide essential information on the sources and sinks of various species. It is difficult to develop estimates of global removal rates from local data sets because of the great spatial variability of precipitation events and of the concentration of particles and soluble trace gases. In addition, it is difficult to measure accurately species that are often present at trace levels. At this time there are no reliable data on global precipitation removal rates for any of the chemical cycles.

The dry deposition of particles larger than about 20-μm diameter is largely controlled by gravity. Submicrometer particles behave more like a gas; their deposition is controlled by factors such as their diffusivity and their rate of turbulent transport through the lowest layers of the atmosphere. After transport to the immediate vicinity of the surface by turbulence, trace gases and small particles are deposited on the surfaces of vegetation, soils, the ocean, and so on. For species like SO_2 and O_3, the flux to vegetation is frequently governed by biological factors, such as stomatal resistance. Reactive gases, such as HNO_3 and O_3, are removed rapidly by most surfaces, although O_3 is removed quite slowly from air over water. Dry deposition is a much slower but more continuous process than wet deposition. For gases with surface sources such as NO_2 and SO_2, dry removal is, however, even more difficult to evaluate globally because local concentration measurements are extremely dependent on the distance of the sampling site from the sources. Deposition velocities, i.e., the ratio of the flux of a substance to its mean air concentration at some reference height near the surface, have been measured for O_3 over a range of surfaces, and there are some similar data for HNO_3, NO_2, and SO_2. At the moment there exist only highly uncertain estimates of global dry removal rates for some of these trace gases and particles. A possible exception is O_3.

A transformation reaction that converts a trace gas into a form that no longer interacts in its elemental cycle may be classed as an in situ removal reaction. Excellent examples of such conversion reactions are the radical-radical reaction between the OH and HO_2 radicals leading to molecular oxygen and water, and the oxidation of CH_4 to CO_2 and water. Aqueous and heterogeneous reactions that convert gaseous species to dissolved salts—e.g., NH_3, NO_2, and SO_2 to ammonium (NH_4^+), nitrate (NO_3^-), and sulfate ($SO_4^=$)—can also be considered in situ removal reactions. Some species such as unneutralized HNO_3 would return to the gas phase if the droplet evaporated.

For long-lived compounds, such as N_2O and certain chlorofluoromethanes (e.g., CF_2Cl_2 and $CFCl_3$), the principal sinks are in the stratosphere, where high-energy ultraviolet photons, excited atomic species, and radicals are available to dissociate them. The products of dissociation, such as NO and the chlorine atom, undergo transformations and are eventually transported back into the troposphere, where they continue to react in elemental cycles until deposited at the surface. For trace gases from surface sources, the longer the tropospheric lifetime the greater is the fraction that will be destroyed in the stratosphere.

PHYSICAL EFFECTS OF TRACE SUBSTANCES IN THE TROPOSPHERE

Returning to the broader picture of the integrated chemical systems of the troposphere, we see that another rationale exists for the study of tropospheric composition. Throughout the cycle of source/transport/transformation/removal, trace substances in the troposphere have effects on important physical processes. Some species (e.g., O_3 and SO_2) absorb incoming solar ultraviolet radiation; some (e.g., elemental carbon and NO_2) absorb visible light; some (e.g., CO_2, N_2O, certain chlorofluoromethanes, and many others) absorb and emit infrared radiation; some substances in particles act as condensation or freezing nuclei in clouds and, in doing so, may alter the cloud radiative properties. The source/transformation/removal cycles of trace atmospheric materials are depicted in Figure 2.2; the primary meteorological effects in the troposphere are related to the respective categories of trace substances: gases, aerosol particles, hydrated aerosol particles, and clouds.

The absorption and emission of radiation by gases is probably the best understood of the mechanisms by which chemical species affect physical processes in the atmosphere; an assessment of these effects requires mainly the measurement of the concentrations of the relevant gases in the atmosphere. The fundamental aspects of the scattering and absorption of radiation by aerosol particles are reasonably well understood, but direct measurements must be made to determine the magnitude of the relevant parameters and their dependence on chemical processes and composition.

For example, aerosol particles can produce either a heating or a cooling of the earth-atmosphere system, the end result depending on the relative magnitude of the scattering and absorption coefficients for visible light and on the total optical extinction. Scattering is often controlled by submicrometer sulfate aerosol particles, whereas absorption is usually due to mineral components and elemental carbon, especially the latter.

The fundamental nature of the nucleation process and of the freezing of cloud droplets is also understood, but a good understanding is lacking of the dependence of these processes on particle composition and the consequent impact on the removal of materials from the atmosphere. It is known that aerosol particles play a major role in cloud processes. Some aerosol substances act as condensation nuclei for clouds, whereas others serve as freezing nuclei; these nucleation characteristics are strongly influenced by the chemical composition of the particles. In turn, as previously discussed, clouds are a major factor in controlling atmospheric composition, because precipitation is the dominant mechanism for the removal of gases and particles from the atmosphere. Again, direct measurements are needed both to estab-

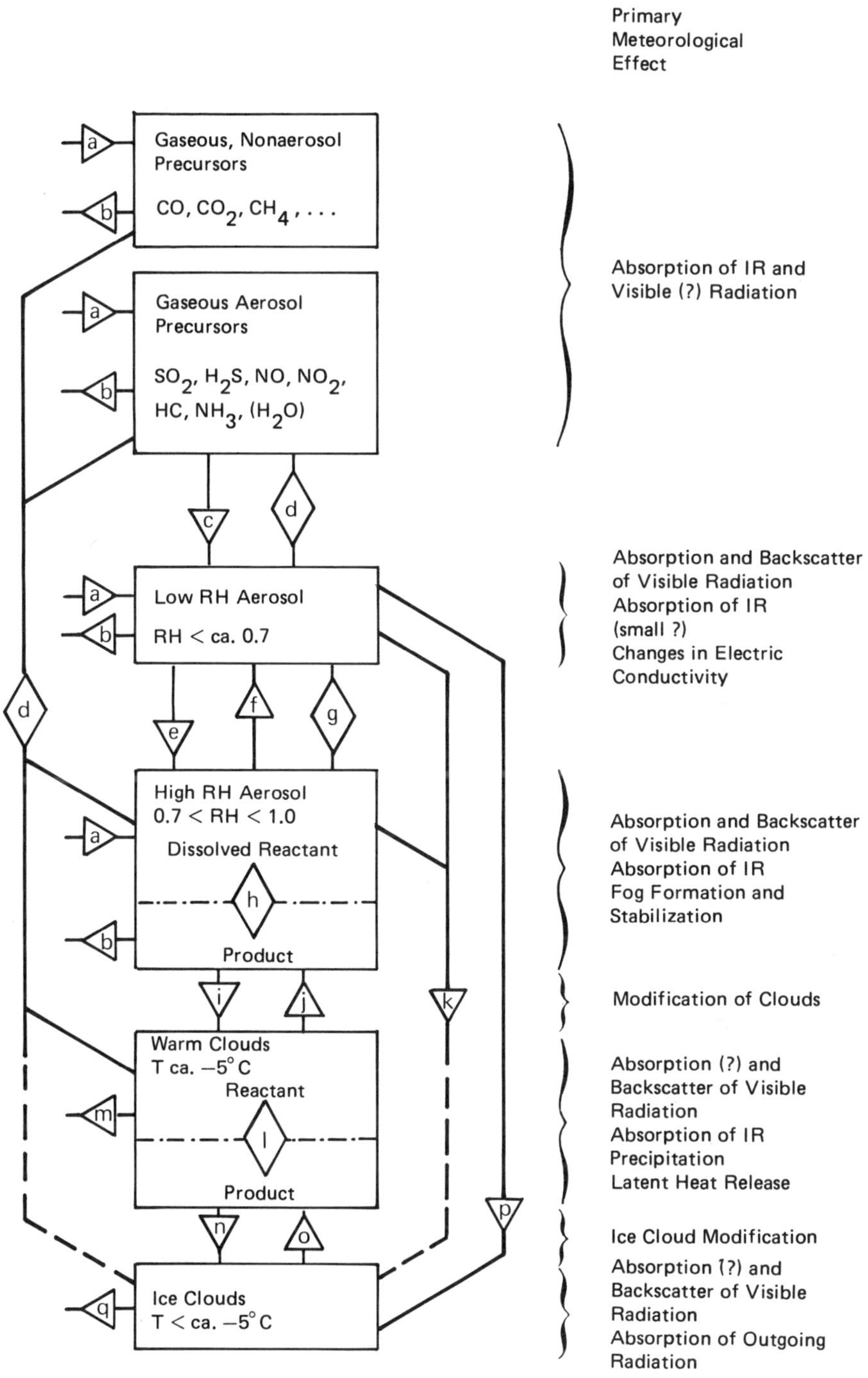

FIGURE 2.2 Phase transitions within the atmospheric chemical systems and their consequences. Rectangles denote chemical entities or physical environments in the atmosphere. The right side of the figure lists atmospheric processes affected by the species or environment identified in the rectangles. Triangles denote processes where material flows primarily in one direction; diamonds represent reversible processes: a, sources; b, sinks; c, gas-to-particle conversion; d, sorption; e, deliquescence; f, efflorescence; g, Raoult's equilibrium; h, reaction in concentrated solution droplet; i, nucleation and condensation of water; j, evaporation; k, capture of aerosol by cloud drops; l, reaction in dilute solution; m, rain; n, freezing of supercooled drop by ice nucleus; o, melting; p, direct sublimation of ice on ice nucleus; q, precipitation. (From *Atmospheric Chemistry: Problems and Scope*, National Academy of Sciences, Washington, D.C., 1975).

lish concentrations and fluxes of the relevant species in the hydrologic cycle and to assess the influence of trace substances on the physical processes themselves.

In addition, it is known that aerosol particles can alter the optical properties of clouds primarily by altering the cloud droplet size distribution (which in turn affects light scattering and absorption) and by modifying light absorption within the cloud through the direct action of the particles. These processes could change the albedo of the clouds; because the albedo of the earth-atmosphere system is strongly affected by the presence of clouds, the effects of trace substances on clouds may be of climatic importance. Further, there are likely feedbacks in which the physical effect of a substance influences its atmospheric behavior and lifetime.

SUMMARY

As discussed above, investigation of the atmospheric chemical processes related to biogeochemical cycles in the troposphere provides a unifying conceptual framework for the study of global tropospheric chemistry. These processes include the sources, chemical reactions and transformations, transport, and removal of the various species within any chemical cycle in the troposphere. A quantitative understanding of these fundamental processes will enable predictive models of the tropospheric chemical system to be developed and the physical effects of trace substances in the troposphere to be determined. Predictive models will allow the effects of future perturbations of the global troposphere to be evaluated. In the chapter that follows, the specific programs proposed in the areas of sources, transformations, transport, and removal of chemical species in the troposphere are presented in detail, as is the need for the development of global tropospheric chemistry systems models.

3 A Proposed Program

LONG-TERM GOALS AND OBJECTIVES

Over the past decade, tropospheric chemistry research has shown that the various chemical cycles in the global troposphere are interactive, complex, and of fundamental importance to the future well-being of humanity. Many essential biochemical and geochemical cycles are critically susceptible to perturbations to the global troposphere. In recognition of this, **we recommend that the United States assume a major role in initiating a comprehensive investigation of the chemistry of the global troposphere.**

The long-term goals of this *Global Tropospheric Chemistry Program* should be as follows:

1. **To understand the basic chemical cycles in the troposphere through field investigations, theory aided by numerical modeling, and laboratory studies.**
2. **To predict the tropospheric responses to perturbations, both natural and human-induced, to these cycles.**
3. **To provide the information required for the maintenance and effective future management of the atmospheric component of the global life support system.**

Attainment of these goals will require carefully designed and complementary research programs, the development of which will involve close cooperation and interaction among investigators making measurements in the field, those investigating reaction rates and mechanisms in the laboratory, and those attempting to model both the chemical systems and the meteorological processes affecting the chemical distributions. Many other laboratory investigations are essential, e.g., studies of uptake of gases by plant leaves, emission of chemicals by living systems, and cloud-scavenging simulations. It is important that experiments be designed jointly by field and laboratory scientists in conjunction with modelers and that there be a continuing exchange between the developing theoretical aspects of tropospheric chemistry and the evolving laboratory and field investigations, as illustrated in Figure 3.1.

It is known that such meteorological processes as transport and cloud and precipitation formation are intimately related to the chemical cycles in the troposphere, and that the instrumentation and platforms required to investigate tropospheric chemistry on the global scale will be expensive. Global-scale tropospheric chemical research cannot be conducted solely by individual investigators or even by small groups of investigators, although the contribution of the individual investigator has been critical in the development of tropospheric chemistry and will continue to be so in the future. Because of the complexity and diversity of the

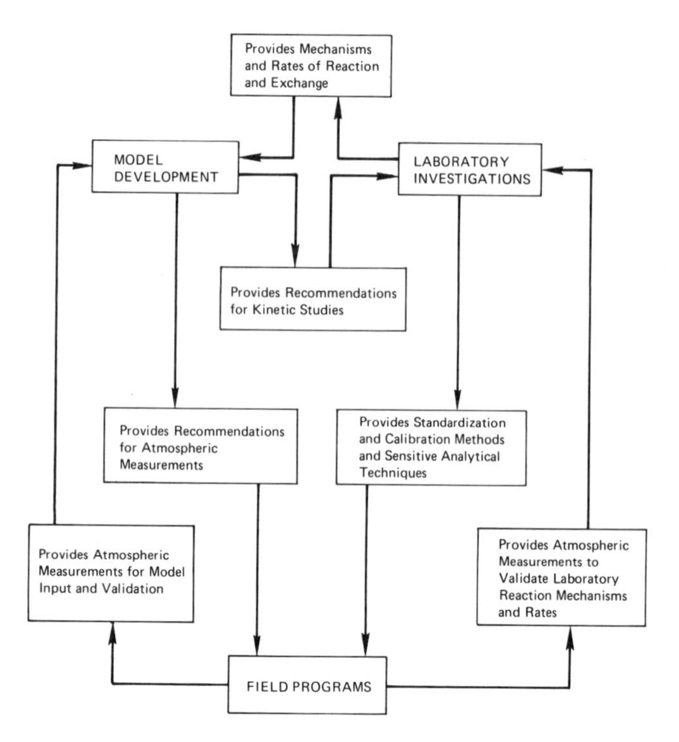

FIGURE 3.1 Schematic diagram of the essential yet interdependent functions served by field programs, laboratory measurements, and mathematical models in atmospheric chemistry research.

chemical systems and because of the myriad sampling, analytical, and modeling tools required to study them, these studies will require the joint efforts of a broad spectrum of scientists: atmospheric chemists and physicists, marine chemists, meteorologists, ecologists, plant and soil biochemists, microbiologists, plant physiologists, laboratory chemists, geochemists, engineers, and others. An effort of this magnitude requires the cooperation and participation of universities, industry, and government agencies, both in the United States and in other countries.

The four basic processes that control chemical cycles and their interactions in the troposphere—production, transport and distribution, chemical transformation, and removal—provide a unifying framework for the development of the *Global Tropospheric Chemistry Program*. Attainment of the program goals will also require the development of three-dimensional models of tropospheric chemical processes linked to meteorological and climatic processes, i.e., tropospheric chemistry systems models (TCSMs); these models will provide an overall synthesis of data obtained in the program, and they will provide theoretical guidance for the program's continued development. For these reasons, **we recommend that the *Global Tropospheric Chemistry Program* be undertaken with the following specific scientific objectives**:

1. **To evaluate biological sources of chemical substances in the troposphere.**
2. **To determine the global distribution of tropospheric trace gases and aerosol particles and to assess relevant physical properties.**
3. **To test photochemical theory through field and laboratory investigations of photochemically driven transformation processes.**
4. **To investigate wet and dry removal processes for trace gases and aerosol particles.**
5. **To develop global tropospheric chemistry systems models (TCSMs) and the critical submodels required for the successful application of the TCSMs.**

The research efforts in biological sources, photochemical transformations, and removal processes will require the development of individual submodels. The submodels would serve three functions: (1) to understand better the individual processes being investigated, (2) to extrapolate from individual observational sites to the regional and global scales, and (3) to provide components that can be used in a comprehensive three-dimensional meteorological model that is coupled to global tropospheric chemistry—a TCSM. By contrast, the research effort in global distributions and long-range transport would serve to help validate the overall performance of a comprehensive TCSM. A tropospheric chemistry systems model would focus on meteorological transport processes that are best described by atmospheric general circulation models (GCMs). These GCMs would be especially designed not only to provide large-scale tracer transport in the free atmosphere, but also to parameterize transport through the planetary boundary layer and by cloud processes. A TCSM would also require physically based cloud submodels, a good description of land surfaces, and adequate treatment of the solar radiation that drives tropospheric photochemistry.

Based on the five scientific objectives above, details of the specific scientific investigations proposed for the *Global Tropospheric Chemistry Program* are developed in the remainder of this chapter. These investigations include the *Biological Sources of Atmospheric Chemicals Study*; the *Global Distributions and Long-Range Transport Study*; the *Photochemical Transformations Study*; the *Conversion and Removal Study*; and a program for the development of global *Tropospheric Chemistry Systems Models* (TCSMs). The discussions of these specific investigations are followed by an evaluation of instrument and platform requirements for their successful completion and a brief review of the requirement for strong international participation and cooperation in the *Global Tropospheric Chemistry Program*.

BIOLOGICAL SOURCES OF ATMOSPHERIC CHEMICALS

Before an understanding of the natural or perturbed troposphere can be claimed, the flow of chemicals through it must be traced. This flow begins with entry of the chemical species into the troposphere. Although the earth's atmosphere is certainly an oxidizing environment, the actual processes of chemical and physical transformation, transport, and eventual removal depend on the chemical form and other intrinsic properties of the substance in question. Accordingly, the initial physical state and chemical properties of the substance at its source will affect its subsequent tropospheric fate.

There are many research questions, basic and applied, that require knowledge of the intensity, size, and variability of sources of tropospheric chemical species; e.g., what factors control the ambient concentration of a certain chemical, or why does the concentration increase or decrease in time, or how will human activity alter the source in question? As specific examples, one may ask why the global concentration of CH_4 is increasing secularly and why it varies with season, why the observed distribution of CO varies with latitude and season, and why there is any gaseous HCl at all in the troposphere?

For these reasons, we recommend that a major research effort be undertaken to evaluate biological sources of chemical substances in the global troposphere. The objectives of this *Biological Sources of Atmospheric Chemicals Study* would be (1) to evaluate the chemical fluxes to the troposphere from critical biological environments (biomes) and (2) to determine the factors that control these fluxes.

The experimental study of the sources of atmospheric chemicals is relatively new because the nature of many of these sources has been recognized only recently and appropriate analytical instruments are just now being developed. This is especially true for biological sources. The dominant role of biological systems as sources of tropospheric trace substances is becoming widely appreciated; a more detailed discussion of the evidence for this appears in Part II, Chapters 5 through 7.

In identifying topics and regions for research in the *Biological Sources of Atmospheric Chemicals Study*, we have employed several criteria; these are reviewed in Part II, Chapter 5, in the section by Cicerone et al. We began by focusing on key chemicals known to have important roles in atmospheric chemistry. For some of these, existing data have already shown the global importance of certain biomes as sources. In other cases, general principles from chemistry and biology suggest that certain biomes should be important. Further, we reviewed characteristics of various biomes and arrived at criteria to estimate their hemispheric or global importance as sources. These considerations led to the field and laboratory investigations of biological sources of tropospheric chemicals that are proposed here.

Much of the early field research in the *Biological Sources of Atmospheric Chemicals Study* must be exploratory in nature. We suggest a research strategy of preliminary investigations of the various sources using relatively small research groups. When the nature and importance of specific sources are better defined, more detailed and refined field and laboratory investigations can follow. This strategy will lead to an understanding of the underlying physical, chemical, and biological factors regulating production of the compounds of interest. In situations where sources are evident but no direct mechanisms are apparent, investigation of indirect paths must be undertaken. Such investigations would require plausible mechanisms for the relevant kinetics and for application of the kinetics under field conditions, and coordinated measurements to verify and test proposed mechanisms. One might expect this process to involve successive iterations among field measurements, laboratory measurements and theory, and ultimately to require studies designed to identify processes important for biochemical and microbiological production of primary chemical species—isoprene, for example.

Certain important abiological sources (e.g., industrial emissions, volcanoes, and lightning) are not discussed here. Also, some field studies of CO_2 exchange are proposed. Even though CO_2 is not important in tropospheric chemistry through its reactions, CO_2 exchange rates can provide important information on the mechanisms and rates of exchange of other gases.

Investigations of Specific Sources

Tundra, Taiga, and Freshwater Marshes

The size of the areas covered by tundra, taiga, and freshwater marshes and some of their unique properties suggest that these regions may be important contributors of CH_4, N_2O, NO, and other volatile species to the troposphere. On the ocean border, volatile organic halides and reduced sulfur species are also of interest.

Site selection and analytical techniques for these investigations are influenced by the remoteness of most representative locations. For these reasons and because so few data are available now, the initial phase of a research program probably is best accomplished by small ad hoc expeditions to acquire the basic information and exploratory data necessary to design more systematic and extensive studies.

TABLE 3.1 Measurement Needs in Various Source Regions

	N_2O	NO_x	CH_4	H_2S DMS/RSH	NMHC	RX
Seaboard tundra	X	X	X	X	X	X
Taiga	X	X	X		X	
Marshland	X	X	X		X	

NOTE: RX represents organic halogen species, RSH represents mercaptans, and NMHC represents nonmethane hydrocarbons. Determination of emission rates of CO and CO_2 is also needed; CO_2 fluxes could provide valuable mechanistic information.

Sites to be investigated should be logistically favorable, uncontaminated, and ecologically representative.

The present understanding of these source regions suggests emphasis on the measurements indicated by an "X" in Table 3.1. Climate characteristics and local conditions will probably constrain the sampling periods narrowly, as could the availability of nearby laboratories and analytical capabilities, but data are needed from various seasons, particularly during transition times.

Tropical Forests

Tropical forests represent a biome of considerable importance for tropospheric chemistry. They include a significant fraction of the total carbon content of the global terrestrial biosphere. They may be important sources for CH_4 and N_2O, gases whose global concentrations have increased significantly over the past several decades, and tropical forests may play a significant role in the global budgets of CO, nonmethane hydrocarbons, NO_x, and various forms of volatile sulfur. Studies of tropical forests will also lead to an increased understanding of the role of microbially mediated reactions, which should be markedly faster in the high-temperature, high-humidity tropical regions.

An orderly strategy for the study of tropical forest biomes might begin with careful measurements of the composition of the local troposphere to identify species whose concentrations are elevated with respect to typical ambient tropospheric background levels. An investigation of a region such as the Amazon Basin might be particularly instructive. The prevailing winds are typically from the east, and the concentration of key species might be expected to increase with time as air masses penetrate deeper into the basin. Measurements determining the spatial and temporal change in gas concentrations, in combination with careful meteorological analysis, could be used to derive an empirical estimate for the total net flux of selected species emanating from the local biosphere. These data, in turn, could provide an estimate of the significance of specific individual sources within the basin.

The extensive measurement program should be complemented by intensive studies of selected microenvironments within the larger biome. For example, if the extensive program should establish an important distributed source for CH_4, the nature of the source should be clarified through more intensive investigations. Exploratory studies of rivers, flood plains, and other environments should aid in the identification of important source regions.

Studies of tropospheric chemistry in tropical forests should improve the understanding of the tropical biome as an integrated physical, chemical, and biological system. To this end, data relating to the transformation and redistribution of essential nutrients, such as nitrogen and sulfur, are particularly relevant and should be available as a by-product of the proposed research.

Biomass Burning

Some of the most easily recognized sources of tropospheric chemicals can be described as point sources. Dramatic in appearance and possibly in their actual impact, these are exemplified by biomass burning, lightning, volcanoes, animal feedlots, and industrial or urban combustion and waste plumes. Both the nature and the magnitude of these sources make them important for global or hemispheric tropospheric chemistry. High-temperature processes that synthesize otherwise unnatural substances and other activities that process large amounts of raw materials, e.g., biomass burning and the refining of metals and petroleum, are particularly potent. We emphasize here the need for coordinated field investigations of biomass burning as a source of atmospheric chemicals, although much research is also needed to quantify other point sources.

Biomass burning produces a variety of gases and particles. The less reactive gases and the smaller particles can affect the global troposphere. Regional effects are likely from the more reactive gases and the larger particles.

We recommend a two-phased field investigation of biomass burning. The objectives for the early phase would be (1) to extend available methods for making quantitative measurements of the emissions of CO_2,

CO, N_2O, CH_4, NO_x, nonmethane hydrocarbons, trace elements, and particles of various sizes from biomass burning, and (2) to obtain exploratory data (as quantitative as possible) for emission rates of COS, various cyanide compounds (RCN—mostly hydrogen cyanide (HCN) and methyl cyanide (CH_3CN)), and CH_3Cl. Also, as in the later phase, the measurements must be made with an awareness of soil and vegetation types and conditions, and the amounts of charcoal produced and areas burned must be measured because of the possible importance of charcoal in the global carbon cycle. Methods to determine charcoal production and the sizes of the affected areas need refinement to permit global assessments. An initial phase might focus on the North American continent to minimize logistical problems and to take advantage of related expertise available through forest meteorologists and forest research programs. There are, for example, prescribed, controlled forest fires set for research purposes in the southeastern and northwestern United States. Forest fire research laboratories could be of considerable use in exploratory programs. The goals of the early phase could be accomplished more easily at such locations, or in Alaska, than in the tropics (where much biomass burning occurs). Some sites should be amenable to the use of instrumented towers and to long-path spectroscopic absorption methods.

In the later phase, field measurements would concentrate on tropical fires. These are extensive and frequent, and they would present greater logistical difficulties. On the basis of the results and experience already gained, it should be possible to quantify emissions of CO_2, N_2O, CH_4, CO, and the more stable nonmethane hydrocarbons from these large fires, and to obtain fairly reliable estimates of emissions of COS, RCN, CH_3Cl, trace metals, particles, and nitrogen oxides.

In all phases, ground sampling of gaseous and particulate species should be undertaken. Aircraft should be used extensively, however, because there is rapid upward motion of the emissions, large areas must be covered, and sampling at representative ground locations could be difficult.

Coastal Wetland and Estuarine Environments

Coastal wetland and estuarine environments are hot spots for the production and emission to the troposphere of many chemically reduced gaseous species of carbon, nitrogen, and sulfur. Most natural anoxic sediments that are in close contact with the troposphere are found in these environments. For gases such as N_2O, CH_4, CO, COS, CS_2, $(CH_3)_2S$, H_2S, and perhaps a few others, it is reasonable to propose initiating immediately a field research program emphasizing basic processes of gas production and exchange with the tropospheric boundary layer. For other species such as NO, NH_3, and volatile metals, existing technology is inadequate for quantitative biosphere-atmosphere exchange studies, even in relatively intense source areas. For almost all reduced gas species, measurement technology is currently inadequate for flux studies of weak sources and sinks.

Studies of gas fluxes in coastal wetland and estuarine environments might focus initially on subtropical and tropical salt marsh and mangrove habitats, because these typically contain extensive anoxic sediments and are characterized by high organic inputs, warm temperatures, and high levels of microbial activity—factors that enhance the production of reduced gas species. A number of characteristics (e.g., vegetation, exposed sediment surfaces, and overlying water) deserve separate study. On the basis of existing data, one can expect gas fluxes from these environments to be highly variable in both time and space on scales of minutes to hours and meters to kilometers, respectively. Critical forcing variables with regular periodicity (e.g., tides, sunlight, plant physiological status, and sediment temperature) interact with episodic events such as severe weather to influence both production and exchange rates.

Both exploratory and intensive field studies of trace gas fluxes are needed in coastal wetland and estuarine environments. Exploratory measurements of fluxes from tropical mangrove and estuarine environments in South America, Africa, and Asia are needed to assess the relative magnitude of emissions from these sources compared to those from more accessible sites in the southeastern United States. Intensive, process-oriented studies of carbon, nitrogen, and sulfur emissions from salt marsh and mangrove habitats could be initiated in areas such as the southeastern U.S. Atlantic coastal region, Mississippi River delta, and Florida Everglades. Together, these studies would improve the understanding of geographical variability in biogenic gas emissions to the troposphere and the processes that control these fluxes.

A program of exploratory gas flux determinations requires research teams focused specifically on such studies, and every effort should be made to determine gas fluxes at locations where complementary meteorological and ecological or biogeochemical characterizations are available. Each gas flux determination should be accompanied by measurements of sediment and/or water column parameters such as water content, temperature, organic content, pH, and Eh (the oxidation-reduction potential). Surveys should include measurements at each site over vegetation, exposed sediment surfaces, and water.

Intensive studies would require more continuous efforts of research teams at sites where fluxes are determined in conjunction with biological and geochemical studies of sedimentary biogeochemical processes.

Agricultural Biomes

Agricultural areas are enormous potential sources of tropospheric chemicals. For example, rice growing, because of its large global extent and waterlogged anoxic soils, appears to be a major source of atmospheric CH_4. Similarly, nitrogen-intensive grain growing (corn, cotton, wheat, legumes, and some rice) can emit enough volatile nitrogen compounds to influence large regions (if NH_3 or NO_x) and the globe (if N_2O). Because of the large areas under cultivation, the growing usage of chemicals, and the inherently large turnover rates of nutrient elements with volatile compounds, we propose a program of extensive field measurements. The focus of these efforts would be on rice paddies and on heavily fertilized crops such as those in the U.S. corn-soybean belts.

In Part II, Chapter 5, in the section by Cicerone et al., we discuss the potential of emissions from rice paddies to exert a strong influence on regional and global tropospheric chemistry. Particular attention should be paid to volatile species containing nutrient elements in reduced valence states, e.g., CH_4, N_2O, NH_3, NO, methylated metals, isoprene, and possibly CO.

For CH_4, detailed and systematic field measurements are required because of the apparent importance of rice paddies and because CH_4 emission rates from rice paddies depend on several factors. To determine the necessary CH_4 fluxes, at least two parallel efforts would be required. The first would simply extend the current data base by using available techniques. For the second, it would be necessary to complete the development and employ a state-of-the-art meteorologically based flux-measurement technique using turbulence-correlation or gradient measurements. Methane fluxes must be determined over complete growing seasons, as functions of nitrogen fertilization rate and in organic-rich rice paddies. Soil temperature, pH, and Eh should also be measured.

Most of the data for other trace gases could be obtained while the CH_4 flux investigations were being made, largely because the initially required investigations would be more exploratory than systematic. If the results indicated significant emissions, then more extensive and systematic studies would ensue, similar to those outlined above for CH_4. For NH_3 and NO a similar sequence of investigations would be required. For NH_3, adequate chemical trapping methods are available, but NO is more problematic; a field-compatible technique is needed.

Potentially large emissions from various heavily fertilized agricultural biomes require investigation. Modern agriculture often relies on intensive management of resources, both physical and financial. Bulk quantities of nitrogen, phosphorus, and sulfur are applied commonly, as are large quantities of trace elements. Specialized chemicals and biochemicals, including enzymes and brominated organics, are also used to regulate processes and to control insects and pests. To assess the role of these intensively managed agricultural biomes as sources of atmospheric trace chemicals, some relevant research is under way in the agriculture research community. We propose complementary research on (1) measurements of the emissions of species such as N_2O, CH_4, and CO that could have global and hemispheric tropospheric effects, and (2) field measurements of shorter-lived gases that have been impractical up to now. Topic (2) could include systematic measurements of NH_3 emissions from nitrogen-fertilized plots and exploratory studies to detect emission of nitrogen oxides, volatile phosphorus, and metals. Gaseous emissions of NH_3, along with those from animal feedlots and natural sources, could lead to NH_4^+, $(NH_4)_2SO_4$, $(NH_4)HSO_4$, and NH_4NO_3 in aerosol particles in remote areas of the globe. Further, NH_3 lost from fertilized fields could also be a key buffer of precipitation acidity. Measurements similar to those proposed for rice paddies would address these questions.

Open Oceans

The oceans are a large-area, low-intensity source of reduced sulfur compounds to the lower troposphere. Preliminary studies have identified a number of sulfur compounds in seawater that are presumed to be biogenic, including H_2S, COS, CS_2, dimethyl sulfide (DMS), dimethyl disulfide (DMDS), and dimethyl sulfoxide (DMSO). Dimethyl sulfide appears to be the most abundant species, contributing an estimated flux of approximately 40 Tg S/yr to the marine boundary layer. Once in the marine boundary layer, DMS is probably oxidized by photochemical processes to produce SO_2, with intermediates such as DMSO and methane sulfonic acid (CH_3SO_3H). Qualitatively, the concentration of reduced sulfur compounds in surface seawater is correlated with indicators of algal biomass.

A research effort to investigate the sulfur cycle in productive areas of the world's ocean should be initiated to elucidate sources of reduced sulfur in the ocean and their role in the global sulfur cycle and budget. Particularly important would be studies of in situ biogenic production of sulfur species in the water column, fluxes across the sea-air interface, and chemical processes in the marine boundary layer determining transport and fate. Most of the research could be conducted from a major research vessel with periodic aircraft overflights to measure the vertical distribution of sulfur species in the troposphere and estimate rates of exchange between the boundary layer and free troposphere. Sites for such studies might include continental shelf waters and a major upwelling area.

The nitrogen cycle in the sea is strongly modulated by biological processes. Evaluation of productive oceans as a source of atmospheric N_2O is particularly relevant to an improved understanding of global carbon, nitrogen, and phosphorus cycling. Capabilities should be developed to explore the possibility of significant NO emissions and to measure sea-air N_2O fluxes more directly. A program of continued oceanographic studies of N_2O production and distribution should be pursued while technology is being developed for measuring low-level N_2O and NO fluxes from surface and aircraft platforms.

Many atmospheric halogen compounds appear to have significant oceanic sources (see Part II, Chapter 7). The most pressing requirement is to estimate the oceanic source strengths of CH_3Cl, CH_3I, CH_3Br, and possibly other organo-bromide compounds. Estimates of the magnitude and distribution of these sources are needed before even the most rudimentary understanding of tropospheric halogen budgets and the role of the oceanic sources in stratospheric chemistry can be claimed. The only available experimental data on oceanic fluxes of methyl halides (RX) are based on measured supersaturation of surface waters. Much more extensive measurements such as these are strongly suggested.

The productive oceans may be significant, albeit secondary, sources of CO and CH_4 to the global troposphere. Again, as in the examples above, the production of these species is related to biological processes, is patchy in distribution, and often the sea-air fluxes are below the detection limits of existing measurement systems.

As we prepare this document, oceanographers are revising estimates of open-ocean biological productivity, upward by perhaps an order of magnitude. As major oceanographic research programs of oceanic productivity are developed, every effort should be made to include simultaneous studies of trace gas (e.g., DMS, N_2O, CO, CH_4, and RX) production and emissions to the troposphere.

Temperate Forests

Temperate forests occupy a substantial fraction of the global land area. Their annual production of dry matter represents approximately 10 percent of the total global production by the biosphere. The forests vary greatly in species composition from almost pure stands of conifers to the mixed hardwoods of temperate North America and Europe, and they also vary in soil type, moisture, and climate. For these reasons, they have a broad range of properties of interest to the tropospheric chemist.

Volatile emissions from forests certainly occur, but there is little quantitative information on these emissions. Various high-molecular-weight aromatic and aliphatic hydrocarbons and CH_4 are potentially of interest as forest emissions. The direct emission of significant quantities of aerosol particles by trees has been suggested, but little quantitative information is available.

To obtain the minimum necessary information on these emissions, sampling should be conducted at representative sites for the major forest types. It is desirable to design a program that is flexible both in time and space, starting with exploratory survey analyses in selected representative locations and expanding the number of sites and intensity of analysis as results dictate.

Wherever possible, sites should be selected where other data are available from ongoing ecological, physiological, and other studies on principal forest types (conifers, deciduous hardwood, etc.).

The current state of knowledge of volatile emissions and reabsorption by forest ecosystems is so limited that many exploratory studies are required to guide the direction of more intensive research. During early studies, analysis at several sites, times of day, and seasons for CH_4, CO, nonmethane hydrocarbons, and NH_3 should be undertaken. Measurement methods to provide gradient information will be needed.

Savannas and Temperate Grasslands

Significant emissions of biogenic gases may occur from temperate grasslands and from savannas (tropical and subtropical). Both direct and indirect indicators point to the need to obtain estimates of emission rates for several gases. Savannas cover about 4 to 5 percent of the earth's surface, and they display high cycling rates for nutrients, i.e., high rates of gross and net primary productivity. Although they store less material in their shrubs and grasses than is found in the wood of tropical forests, the biological material of savannas has a shorter lifetime and higher nitrogen/carbon and sulfur/carbon ratios than hardwood. From the rapid turnover rates, the chemical composition of the material, and current data that suggest a large role for termites and herbivorous insects, one may conclude that significant volatile emissions are quite likely from certain dry tropical areas. In particular, there is potential for large emissions of CH_4, CO, CO_2, many nonmethane hydrocarbons, N_2O, and possibly methylated metals. Initial measurements should focus on sites and processes that concentrate nutrients (termite colonies, for example), but the large land areas involved could release significant emissions from lower intensity sources distributed over large areas.

Although they cover only about 60 percent as much area as savannas, and their net primary productivities are only about 50 percent those of savannas, temperate grasslands may be an important source region on the global scale. As with savannas, we focus attention on CH_4, CO, CO_2, nonmethane hydrocarbons, N_2O, and possibly volatile metals. Initial measurements to quantify these fluxes should be similar to those for savannas, although site selection might be made so as to coincide with grassland sites that are already well-characterized ecologically. Further, consideration should be given to simultaneous measurements of downward deposition rates of O_3, NO_x, HNO_3, SO_2, and CO_2 if techniques permit.

Needs for Instruments and Verification of Methods

In the discussion of the *Biological Sources of Atmospheric Chemicals Study*, we have emphasized selected biomes and processes that are known or suspected to be important sources of key tropospheric chemicals. Although the proposed program is broad in scope, it is not complete nor are all means of investigation and an exact sequence of projects spelled out. For example, we have not emphasized research on industrial or volcanic emissions, or lightning as an in situ source, nor have we emphasized sources of primary particles, although it is clear that these topics require research. In Part II, Chapters 5 through 8, we discuss practical difficulties that affect a researcher's ability to obtain meaningful global or regional emissions inventories, e.g., for biogenic sources, and the need for seasonal and annual information.

For the early phases of the field measurements of surface sources, much of the necessary instrumentation exists. This is especially true for those studies that focus on (1) identifying hot spots (intense sources), (2) obtaining relative data, for example, day-night or summer-winter differences, and (3) fluxes of relatively stable species that are amenable to storing of samples or species amenable to chemical derivatization methods. By contrast, field investigations that need further development of research instrumentation and/or methodology include those focused on (1) assessments of the significance of less intense sources that cover a large geographical area, (2) obtaining fluxes to high absolute accuracy, and (3) fluxes of species that are relatively reactive and are difficult to store.

There must be intercomparisons of methods to establish the capabilities for absolute flux determinations; the effectiveness of chamber enclosures, vertical gradients, and fast-time response meteorological eddy-correlation methods will need evaluation. It is not possible or desirable that all methods be evaluated generally for every species and application. Very specialized evaluations are needed instead. For example, the use of vertical gradient methods is limited to those species that exhibit measurable differences of concentration between levels in the surface layer—relatively inert gases (e.g., CH_4) usually are mixed to within a percent or two in the surface boundary layer (if not the entire troposphere) and reactive gases (e.g., NH_3) can disappear almost entirely in the lowest 20 m. Sound general practice will demand that more than one independent technique be used simultaneously and that all relevant environmental parameters be measured during each field measurement. Fast time-response sensors must also be developed for key species—as of 1984 only O_3, NO, CO_2, and CO can be measured with time resolutions of 0.1 to 0.2 s (needed for successful application of meteorological-correlation techniques). Fast-response sensors for CH_4, SO_2, N_2O, and several other species seem attainable. The general principles and instrument characteristics for fast-response sensors in flux determinations are outlined in Part II, Chapter 8. Methods for determining fluxes within and above forest canopies and forest fires require even more specialized techniques based on aircraft or towers and balloons.

Field-compatible and rugged instruments are needed for advanced determinations of fluxes from surface sources or sinks (downward depositing fluxes). They must be stable and sensitive enough to measure vertical differences in the lowest 20 m and/or to measure fluctuations in chemical concentrations with time resolutions of 0.1 to 0.2 s. Further, both independent and interdependent tests of flux-determination methods must be performed.

GLOBAL DISTRIBUTIONS AND LONG-RANGE TRANSPORT

Major developments have occurred in global meteorological modeling during the past three decades. Much of this progress was possible through advances in meteorological theory, numerical methods, and computer power. However, the attendant development of *Tropospheric Chemistry Systems Models* (TCSMs) has been much slower (see Part II, Chapter 5). A major problem has been the dearth of chemical measurements throughout most of the global troposphere, particularly above the planetary boundary layer. Such data are essential for the validation of models, and they serve as the basis of model development. Thus a coordinated research program of tropospheric chemical measurements and transport/chemistry model development is required. Consequently, **we recommend a *Global Distributions and Long-Range Transport Study* through the establishment of a *Global Tropospheric Chemistry Sampling Network*. The primary objective of this network would be to obtain tropospheric chemical data that can be used to identify important meteorological transport processes and to validate and improve the ability of models to simulate the long-range transport, global distribution, and variability of selected chemical species. Ancillary objectives are (1) to determine the distribution of those chemical species that play important roles in the major chemical cycles of the troposphere, and (2) to detect, quantify, and explain long-term trends in environmentally sensitive trace gases and aerosol particles, especially those that are radiatively important.**

It is not possible at this time to define precisely the structure of this network and its sampling protocol because there is insufficient knowledge of the global concentration fields for the chemical species of interest and of the large-scale transport processes that affect these distributions. Nonetheless, one can anticipate the general features of such a network and its sampling protocol. We stress that the design and implementation of this network will be, of necessity, an evolutionary process. The network and its protocol will be periodically redefined in response to the growing knowledge of the chemical and physical properties of the troposphere.

We also anticipate that it will not be necessary for all stations to follow the same protocol. Requirements such as the spatial distribution of sampling sites and the frequency of sampling will vary depending on the distribution of sources and sinks for a particular species, the atmospheric lifetime of that species, and the intended use of the data. Consequently, **we suggest that within the framework of the overall network there exist three subsidiary networks: a *Global Distributions Network,* a *Long-Term Trends Network,* and a *Surface Source/Receptor Network.***

To meet the overall objectives, the networks must be well-organized, the protocols must be carefully formulated, and activities must be tightly coordinated. The objectives can be met by networks composited from sampling programs that are under the control of individual investigators or national and international government agencies. Many current sampling activities would play a vital role in the proposed program, although in some cases the present sampling protocols might have to be modified or expanded. However, it is essential that the people involved in these programs be scientists who are active in research and who are professionally motivated to obtain good data, to maintain a high degree of quality control, to interpret the results, and to publish them promptly. Consequently, **we recommend that the *Global Tropospheric Chemistry Network* be coordinated through an international committee of prominent atmospheric scientists who are active in relevant areas of network studies.**

The *Long-Term Trends Network* would consist of a few stations located in remote environments that are relatively unaffected by significant local sources, either natural or anthropogenic. These stations would be dedicated to the long-term monitoring of environmentally important trace gases that have relatively long tropospheric residence times. The accurate measurements necessary for this program require skilled technical support personnel and sophisticated measurement devices. The *Long-Term Trends Network* could be based on existing efforts such as the *NOAA Global Monitoring for Climatic Change (GMCC) Network* and the *Chemical Manufacturers Association (CMA) Atmospheric Lifetime Experiment (ALE) Network*, with the careful intercalibration of current measurement techniques and the addition of a number of other measurement programs.

The *Global Distributions Network* would consist of a greater number of stations operating over a relatively limited period of time (3 to 5 years); these would provide greater spatial resolution for trace constituents with tropospheric lifetimes of a few months to a few years that are important both to transport studies and to chemical cycles.

The *Surface Source/Receptor Network* would incorporate a larger number of stations that employ relatively inexpensive and low technology techniques. They would measure the surface air concentration of selected species in airborne particles and the concentration and rate of deposition of these species in precipitation.

Chemical Species and Measurement Techniques

It is not now technically or logistically possible to measure the global distributions of all important chemical species in the global troposphere. Furthermore, because of the inadequate knowledge of the chemical processes affecting these species and of their meteorological transport, it would not be possible to properly design such an experiment. Rather, **we recommend that species be selected on the basis of the three objectives of the *Global Tropospheric Chemistry Sampling Network* and with the additional requirement that the species be readily measurable with existing technology that can be implemented at relatively remote sites.**

Initially, the focus would be on gases having simple chemical destruction pathways, but as model development progresses, measurements would be needed of medium- and short-lived species that have a more complex transformation chemistry. The data on short-lived species would be used to test not only the simulation of long-range transport but also the parameterization of important subgrid scale transport processes such as convection and boundary layer turbulence.

The species selected on the basis of these criteria are listed in Table 3.2. A number of very important species (e.g., NO, NO_2, HNO_3, and SO_2) are not listed at this time because the appropriate, easily transferable measurement technology does not now exist. When suitable techniques are developed, these species would be added to the network protocol. Each substance in Table 3.2 is classified as to its major source (anthropogenic or natural), the primary application of the data (the development of transport models, importance to climatic processes, or elucidation of chemical cycles), its relative lifetime in the atmosphere (short—months or less, medium—a few months to a few years, or long—more than a decade), and the network in which it would be measured (*Global Distributions*, *Surface Source/Receptor*, or *Long-Term Trends*).

The chlorofluoromethanes, e.g., CCl_3F and CCl_2F_2, and other halocarbons can provide a quantitative test of long-range transport models because the distribution and strength of their sources are relatively well known. The species CO_2, CH_4, CO, and the alkanes ($<C_5$) play a role in the carbon cycle; once we have an improved understanding of chemical transformations in this cycle, these data could be used to validate more complex chemical transport models. CH_4 is also radiatively active and its concentration is believed to be increasing in the troposphere. Natural sources are dominant for these species; however, there are significant anthropogenic sources for CO in some regions, especially in the northern hemisphere. Carbon dioxide measurements are included to provide data on the anthropogenic emissions of this substance, and information about meteorological and oceanographic processes and interactions with the biosphere. Nitrous oxide is also

TABLE 3.2 Possible Species to be Measured in the Global Tropospheric Chemistry Sampling Network

Medium	Species	Source[a]	Use[b]	Lifetime[c]	Network[d]	Notes
Gas	CCl_3F	A	T/R	L	GD, LTT	Fluorocarbon 11
	CCl_2F_2	A	T/R	L	GD, LTT	Fluorocarbon 12
	$CHCl_2F$	A	T/R	M	GD	Fluorocarbon 21
	$CHClF_2$	A	T/R	L	GD	Fluorocarbon 22
	CH_3CCl_3	A	T/R	M	GD, LTT	
	C_2Cl_4	A	T/R	S	GD	
	C_2HCl_3	A	T	S	GD	
	CH_4	N	T/C/R	M	GD, LTT	
	Alkanes $< C_5$	A/N	C	S	GD	
	CO_2	A/N	T/C/R	L	LTT	
	CO	A/N	C	S	GD	
	COS	A/N	C	L	GD	
	CS_2	A/N	C	S	GD	
	N_2O	A/N	R	L	LTT	
	O_3	N	T/C/R	S	GD, LTT	
	H_2O	N	T/C/R	S	GD	
Particles	Na^+,Cl^-	N	C	S	SSR	Sea salt
	$SO_4^=$	A/N	C	S	SSR	
	NO_3^-	A/N	C	S	SSR	
	NH_4^+	A/N	C	S	SSR	
	Soil	N	T/C	S	SSR	
	Excess V	A	T	S	SSR	
	^{210}Pb	N	T	S	SSR	
	Total C	A/N	C/R	S	SSR	
	Total Organic C	A/N	C	S	SSR	
Precipitation	Same species as particles plus,					
	HCOOH	N	C	S	SSR, LTT	
	CH_3COOH	N	C	S	SSR, LTT	
	K^+, Mg^{++}, Ca^{++}	N	C	S	SSR, LTT	

[a] A, anthropogenic; N, natural.
[b] T, transport; C, cycles; R, radiatively active.
[c] S, short (months or less); M, medium (from a few months to a decade); L, long (greater than a decade).
[d] GD, global distributions; SSR, surface source/receptor; LTT, long-term trends.

radiatively active, and its concentration is increasing in the troposphere.

Carbonyl sulfide and carbon disulfide play a significant role in the sulfur cycle, especially in the stratosphere. There is considerable uncertainty regarding sources for these species, but both industrial and natural sources are believed to be significant for CS_2.

Most of the species above can be readily measured by collecting on-site samples with flasks or trapping techniques and returning the samples to a central laboratory for analysis. Many species can also be measured on-site by using automated gas chromatography.

Ozone plays a critical role in the chemistry and radiation balance of both the troposphere and stratosphere. Because it is involved in many transformation processes in the troposphere, knowledge of the distribution of O_3 is essential for the development of a TCSM. Tropospheric O_3 can be used as a tracer for investigating the exchange between the troposphere and the stratosphere and for studying long-range transport above the boundary layer. However, it will be necessary to make frequent measurements of the vertical profile of O_3 into the stratosphere. Ozonesondes have been used routinely with some success for many years, and a number of stations are currently in operation, most of them in the midlatitudes in the northern hemisphere. However, improvements in measurement accuracy are required. New O_3 sounding methods are now being developed, and when operational their implementation should be encouraged.

Water vapor is a critical species for most tropospheric chemistry processes. The global data set for water is more comprehensive than for any other trace gas in the lower troposphere. Data obtained in the standard atmospheric sounding programs are generally quite poor in the middle and upper troposphere where water concentrations are low. Although there have been some excellent research sounding programs measuring water vapor, their spatial coverage has been sparse. Improved instruments for measuring the low-water-vapor concentrations characteristic of the upper troposphere should be developed, and this instrumentation should be used in the O_3 sounding program.

The concentration and composition of aerosol particles could be measured by using samples collected on filters. The species $SO_4^=$, NO_3^-, and NH_4^+ are of interest because of their role in the sulfur and nitrogen cycles. The measurements of soil and sea-salt aerosol particles will provide information on the budgets of the two most massive components in the global aerosol flux. Soil aerosol particles and ^{210}Pb also serve as natural tracers for the transport of air from continents to the oceans. Vanadium (V) serves as a tracer for emissions from many oil-fired anthropogenic processes, particularly those using heavy residual fuel oils. The measurement of total carbon and total organic carbon will provide data necessary for the development of global carbon budgets. Finally, the measurement of light absorption on the filters will yield data on the corresponding properties of aerosol particles (primarily that of elemental carbon or soot); these data will be vital for assessing the impact of aerosol particles on global climate. Other optical properties, e.g., the scattering component of extinction, are already being measured at GMCC sites and may be included when possible.

The precipitation studies will focus on the same suite of species measured in the aerosol particle studies. In addition, analyses will be made for weak organic acids, such as formic, acetic, and oxalic; these acids are significant components in precipitation in some regions. Finally, the samples will be analyzed for K^+, Mg^{++}, and Ca^{++}. By using the concentrations of these and other dissolved species, it is possible to compute the pH of precipitation and check the directly measured value by considerations of ionic balance.

Experimental Design

Measurement Validation

A validation program should begin with an exchange of calibration standards followed by an exchange of blind standards. Past experience with N_2O calibrations has demonstrated that this procedure alone will not be sufficient to establish the accuracy and precision of the techniques. A critical second step will be to carry out on-site comparisons among the candidate systems and to compare these systems with state-of-the-art instruments that are not yet ready for routine use at remote sites. Recent intercomparisons have proven to be extremely useful in a number of current programs. The validation of the absolute accuracy and precision of the measurements is critical for long-term trend measurements and for measurements of long-lived gases. In both cases, the detection and quantification of small concentration variations are being sought. Those techniques that survive these intercomparisons should then be compared with all other state-of-the-art measurements under a variety of environmental conditions that are representative of those expected at the planned network sites. Only after successful techniques have been identified should network operations begin.

Observational Protocol

Concurrently with the measurement validation program, a measurement protocol should be developed. The sampling frequency and duration required to yield a true mean value of the trace substance concentration and to capture most of the variance must be determined. For vertical profiles, the vertical resolution needed to provide a representative tropospheric profile for a particular species must be defined. To establish these parameters, it will be necessary to make frequent measurements with a fine temporal and spatial resolution. A possible strategy for making vertical profile measurements in the networks is discussed below. For some substances, useful data may already exist; if so, these data should be analyzed for loss of variance and distortion of mean value as a more coarse time and height grid is used. It is generally assumed that the long-lived gases are well-mixed in the vertical and that they have a very low variance over a seasonal time scale, but this assumption should be confirmed. An analysis of the Atmospheric Lifetime Experiment (ALE) and GMCC time series should be helpful, but some vertical profile data are also needed. The distribution variance of the medium-lived industrial gases and natural carbon compounds is generally unknown; for example, there are time series data for CH_4 and CH_3CCl_3, but there are no detailed profile data except for CH_4 over southeastern Australia. Vertical profile data already exist for H_2O and O_3 but not in time series. Vertical time series data are particularly important for a highly variable gas such as O_3.

Any measurement strategy will require compromises in establishing a protocol to measure the vertical, horizontal, and temporal distribution and variability of a species. The question of instantaneous versus time-averaged samples must also be examined. Studies have been made in which grab samples of CO_2, CH_4, and CH_3CCl_3 are compared with continuous measurements, and these data should be useful. Similar data will be needed for other gases and for particles. Since many of these data will be used either to identify a long-term trend or to validate model-generated transport, it is very important that the measured values truly represent atmospheric mean values and that one can account for all major sources of natural variance on time scales shorter than the observed trend.

Principles for Network Design

Long-Term Trends Network The four existing GMCC stations (Barrow, Alaska; Mauna Loa, Hawaii; American Samoa; and the South Pole) and the five ALE stations (Adrigole, Ireland; Cape Meares, Oregon; Ragged Point, Barbados; Cape Grim, Tasmania; and American Samoa) serve as an excellent foundation for a small *Long-Term Trends Network* for the accurate measurement of radiatively active trace species in the troposphere. An analysis of the existing GMCC and ALE time series for CO_2, N_2O, CH_3CCl_3, CCl_3F, and CCl_2F_2 would help to determine the minimum number of stations required. On the basis of current meteorological knowledge, we would select at least five stations—one in the tropics and one in the midlatitudes and high latitudes of each hemisphere.

It would be very useful if at least one station were available for instrument development, testing, and validation, and for observational studies. Therefore, the establishment of a station in the continental United States would be desirable. It would be necessary to carefully intercalibrate the instruments with the measurement devices previously used in the ALE and GMCC networks so that information from past time series is closely tied to future work. Subsequently, it would be necessary to establish a common measurement system and observational protocol at all stations. The *Long-Term Trends Network* would measure a number of gases (e.g., O_3 and CH_4) not measured in the current GMCC and ALE protocols.

Global Distributions Network The *Global Distributions Network* would provide greater spatial resolution than that obtained from the *Long-Term Trends Network*. This network would be needed for the relatively short-lived species, such as O_3, H_2O, CO, and for the reactive nitrogen and sulfur species when techniques are available for their measurement. The network may also be needed to establish global fields for medium-lived species, such as CH_4 and CH_3CCl_3. Vertical profiles would be required for the short-lived species and possibly medium-lived species.

The analysis of the existing data for CH_3CCl_3 would help to determine whether the present stations provide sufficient spatial resolution for medium-lived gases. Periodic flights between the long-term stations with an aircraft capable of measuring most species in Table 3.2, as well as others requiring more advanced technology, would provide information about gradients in latitude, if any; such gradients might not be discernible from network measurements alone. However, such flights can be expensive, and they must be carefully justified. The *Global Distributions Network* would not resolve the spatial variability for short-lived gases such as O_3, CO, and H_2O. Aircraft measurements between the sites could provide higher spatial resolution for these species, but at some extra cost.

A time series of accurate vertical profiles, at least three a week for O_3, would provide considerable three-dimensional information about atmospheric transport. Such time series would also rigorously test a model's ability to simulate correctly transport on time scales ranging from synoptic to seasonal. In particular, a coarse network of reliable O_3 sondes would yield a wealth of information about the atmospheric transport of O_3 and its climatology.

Surface Source/Receptor Network This network would focus on the sampling of suspended particles and precipitation in representative source regions and receptor areas. Relatively simple sampling procedures would be used. A number of investigators have used such a strategy, with volunteer personnel, in remote marine and continental field stations for several years with success. Such operations are relatively inexpensive. Few existing networks have an adequate sampling protocol for aerosol particles. One important goal of the *Surface Source/Receptor Network* would be to establish a common sampling protocol and to standardize measurement techniques for particulate matter.

The surface-level measurements provide information on the transport of dust and anthropogenic particles from major source regions. Measurements of the concentration and rates of deposition of mineral aerosol particles from major desert regions as a function of latitude, longitude, and season would provide significant tests for transport models and considerable information about long-range transport mechanisms. These data would be especially important for developing event models as would the information about the surface air and precipitation concentrations of NO_3^- and $SO_4^=$. These are important elements of the global and regional nitrogen and sulfur budgets.

A number of large regional surface networks have been or are now being established (see Part II, Chapter 5, section by Prospero et al.). Analysis of these data sets should be helpful in designing the *Surface Source/Receptor Network*. Of particular importance is the major effort managed by the World Meteorological Organization (WMO), the Background Air Pollution Monitoring Network (BAPMoN) program. Some of these programs could be especially useful in characterizing sources.

Because the primary sources of most species to be measured are located on the continents, ocean stations act as receptor sites and can provide excellent background data. In some areas, the ocean serves as the dominant source for some species—for example, reduced sulfur that is ultimately oxidized to $SO_4^=$. In such cases, these stations provide valuable data on sources. Stations in the southern hemisphere would be especially important because of the severe dearth of data from this region. We suggest a preliminary ocean network that would consist of one station in each major wind regime in each ocean and one in each region of large-scale subsidence. There are 12 to 16 islands whose geographic locations appear to satisfy these *Surface Source/Receptor Network* requirements for ocean sites. Some are currently active as network sites.

Strategy for Obtaining Vertical Distribution Data

Vertical distribution data are vitally important, especially for medium- and short-lived chemical species. However, it is unrealistic to have highly instrumented aircraft deployed at network stations except for short periods of time in conjunction with intensive field experiments. There are so few data on vertical distributions that a valid sampling protocol for such aircraft could not be specified at this time.

We propose a stepped approach for obtaining vertical distribution data. First, vertical distributions must be measured in a limited number of source and sink environments where concentration profiles and their temporal and spatial variances can be measured and related to sources (or sinks) and to the local meteorology. The data obtained from these studies would enable an appropriate protocol to be specified for obtaining useful vertical profile data on a global scale.

Extended time series data taken above and below the boundary layer are also needed. Initially, these data could be obtained by making a concurrent series of measurements at a station on a mountaintop above the boundary layer and at another at ground (or sea) level. Although sampling on mountaintops is difficult, but not unprecedented, we recommend that the present program of mountaintop/surface level sampling at Mauna Loa and a coastal site in Hawaii be intensified. The protocol should include all of the species listed in Table 3.2. Within BAPMoN, there are a number of other mountain sites either in existence or planned. Some of these may be suitable for concurrent sampling of the boundary layer and the free troposphere.

Extensive vertical profile data, however, can only be obtained with aircraft (except for O_3 and water vapor, for which sondes can be used). Logistical and financial considerations dictate that long-term studies must be carried out with locally available light aircraft. The sampling protocol would likely be limited to those species that could be sampled over a relatively short time period by using "bolt-on" instrumentation. An effort must be made to develop light sampling equipment and data logging packages for use aboard small aircraft. A special effort should be made to improve aerosol particle sampling techniques; present techniques require excessive amounts of power and flying time. Early in the program, field experiments should be made to compare vertical profiles obtained by aircraft making high-time-resolution measurements with those made at paired mountaintop/surface level sites. It seems logical to make these first measurements in the Hawaii area so as to capitalize on the existing facilities on Mauna Loa.

The Program

The *Global Distributions and Long-Range Transport Study* would evolve in a dynamic way. Here we outline the logical progression that a developing global measurement program might follow.

Phase I

In *Stage One* of Phase I, the objectives should be (1) to develop and test collection systems and (2) to determine the accuracy and precision of the measurements by intercomparisons with state-of-the-art techniques. When possible, network design and sampling protocols should be determined empirically from existing data or exploratory measurements. A special effort should be made to develop relatively simple and reliable sampling techniques for such species as NO, NO_2, HNO_3, and SO_2.

We anticipate that ultimately the subnetworks in the *Global Tropospheric Chemistry Sampling Network* will be operated by groups of scientists. Specific sampling and analysis tasks will be carried out by scientists actively engaged in measurements and interpretation. Such active participation ensures that the program will remain responsive to the needs of the scientific community. We stress that, although the sampling protocol should be strictly defined, it should always be subject to change as new and better techniques are developed. New techniques should be implemented cautiously. They should be tried in the field with the technique that is to be replaced, and this trial should be carried out over an annual cycle.

In *Stage Two*, the objectives should be (1) to test equipment under field conditions; (2) to develop procedures for logistics, sample analysis, and data handling; and (3) to provide data that can be used for planning full-scale network operations. Three or four prototype stations should be established first in representative marine and continental environments. Studies should be made to determine the required vertical sampling resolution and frequency of sampling. After these tests are completed, the three networks should be developed and deployed. In this stage there should be extensive interaction between the network studies and the model development program.

In *Stage Three*, all three networks—*Surface Source/Receptor*, *Long-Term Trends*, and *Global Distributions*—should be fully operational. The modeling and measurement efforts should be sufficiently advanced that planning can begin for large-scale regional experiments that require a greater density of network stations and a higher sampling frequency. These experiments would coincide with one or more of the other major studies proposed as part of the *Global Tropospheric Chemistry Program*.

Independent investigators could carry out specialized measurements or study short-term processes at a network site, thus benefiting from a long-term data record for possibly related species. The results of a limited set of ancillary measurements could then be extended to a larger time and space scale through modeling efforts. Consequently, the network stations should be designed to accommodate expansion without compromising the protocol experiments.

Phase II

The activities in Phase II can be discussed only in general terms. The *Long-Term Trends Network* would continue, and a major effort should focus on model development. Phase II would evolve into independent research programs having a number of objectives: the generation of global fields from the refined station data; the testing, validation, and analysis of dynamical/transport models; the elucidation of the mechanisms that control transport processes; and, finally, the development of comprehensive *Tropospheric Chemistry Systems Models* (TCSMs). The *Global Distributions* and *Surface Source/Receptor Networks* could be further refined through model development and interactions with the modeling research program.

We anticipate that satellites eventually will play a major role in measuring the global distribution of some species. The remote sensing community is currently examining the feasibility of remote sensing for a number of important species. If such satellite packages are deployed, it will be necessary to verify instrument performance against ground truth data. We would expect that the *Global Tropospheric Chemistry Sampling Network* stations and a program of research aircraft flights would play a significant role by making measurements of these selected species. To facilitate such cooperation, close contacts should be maintained between the *Global Tropospheric Chemistry Program* and the remote sensing community.

PHOTOCHEMICAL TRANSFORMATIONS

Current understanding of tropospheric photochemistry (see Part II, Chapters 5 through 7) is based largely on laboratory measurements of reaction-rate coefficients and on modeling exercises. Modeling involves mathematical synthesis of the ambient atmosphere using elementary gas-phase reactions as basic building blocks. Certain aspects of atmospheric photochemical theories have been tested by comparing results from photochemical models with data from isolated field observations. In no case have the postulated basic mechanisms been validated by comprehensively measuring all relevant photochemical species in a well-defined atmospheric setting. Thus **we recommend the initiation of a major research effort on photochemical transformations in the troposphere. The primary objective of this *Photochemical Transformations Study* would be rigorous testing of photochemical theory through field and laboratory investigations of photochemically driven transformation processes.** This would be accomplished through a series of *Theory Validation Experiments*.

A second objective of the *Photochemical Transformations Study* would be to establish a comprehensive data base on tropospheric concentrations of the primary species involved in photochemical transformations. This would be accomplished through a *Concentration Distribution Experiment* coordinated with the *Global Distributions and Long-Range Transport Study* discussed in the previous section.

Theory Validation Experiments

Several important facets of fast photochemistry can be addressed in a field measurements program. These include the following:

1. Studies of the H_xO_y photochemical cycle with special emphasis on the central species OH;
2. Studies of the N_xO_y photochemical cycle;
3. Studies of the photochemical sources and sinks of the centrally important species O_3;
4. Diagnostic studies to assess the possible role of photochemically induced processes within clouds. Research in this area is addressed in somewhat greater detail in the *Conversion and Removal Study* presented in the section below entitled "Conversion, Redistribution, and Removal."

Figure 3.2 presents a simplified reaction scheme for the photochemical cycles of H_xO_y, N_xO_y, and O_3. The major reactions are shown for each individual species to facilitate the following discussion. These reactions are discussed in more detail in Part II, Chapter 5, in the section by Davis et al.

For each of the first three photochemical research areas, we have examined possible field sampling scenarios in terms of (1) the identification of critical gas-phase species requiring measurements, (2) sampling strategies, and (3) current and future instrument readiness. The purpose of this exercise has been to present an overview of the many scientific opportunities available, without endorsing any final approach or strategy.

Identification of Critical Measurements

To assess the critical measurements needed for a given photochemical field experiment, each experiment has been considered at two levels. At Level 1, a critical list has been proposed that defines (1) measurements that are considered scientifically essential, and (2) measurements for which there appear to be either existing instruments or for which the technology is imminent. Our proposed guideline would be that if an experiment cannot be carried out at Level 1, it probably should not be undertaken in other than an exploratory form.

At Level 2, the proposed critical measurements list attempts to define all those variables needed to carry out a comprehensive study of a given fast-photochemical system. The assumption is that all required instruments will be available. Experiments at this level should address most, if not all, of the major scientific questions related to the photochemical system under investigation.

Although the photochemical processes that we pro-

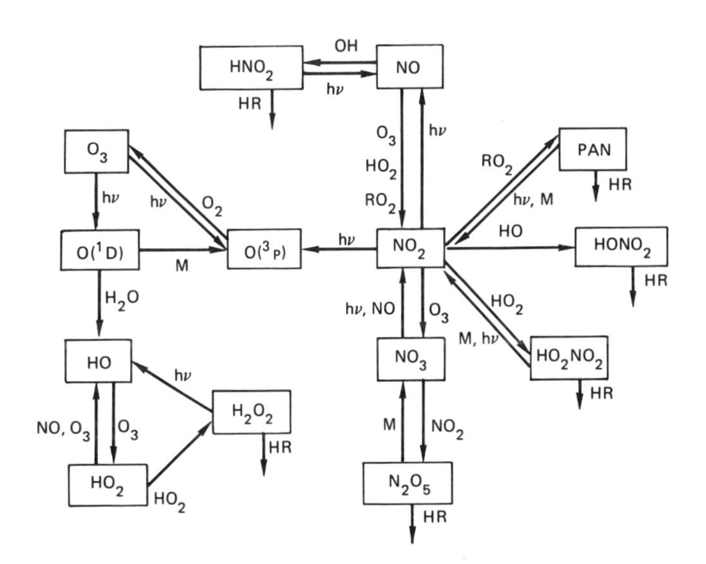

FIGURE 3.2 Major atmospheric reactions of H_xO_y, N_xO_y, and O_3, where M denotes N_2 and O_2, HR refers to heterogeneous removal, and hν indicates radiation required.

pose to study are relatively fast and not significantly affected by transport processes, simultaneous measurement of meteorological parameters such as wind velocity, pressure, temperature, and humidity is essential. Temperature and humidity usually affect photochemical processes directly. In addition, knowledge of the history and trajectory of the air mass under study will provide information on the distributions of long-lived species that will be important for data interpretation.

Table 3.3 summarizes the Level 1 and Level 2 critical measurement requirements for three of the elements in the proposed *Theory Validation Experiments*. These include H_xO_y experiments, N_xO_y daytime experiments, and O_3 experiments.

H_xO_y Experiment At Level 1, the measurements list reflects the ongoing hypothesis that the OH radical plays a central role in photochemical theory. The list therefore encompasses all those species that normally would be used to define the simplest production and loss reaction scheme for OH (see Part II, Chapter 5, section by Davis et al.). The latter scheme includes the primary production of OH from the photolysis of O_3 as well as two OH loss pathways involving reaction with CO and CH_4. To test this simple five-step mechanism requires the measurement of seven variables: OH, H_2O, O_3, CO, CH_4, ultraviolet (UV) spectral radiant flux density (and/or relevant photodissociation rates), and total pressure. In addition, we have added an eighth variable to Level 1, NO, a species that provides the "simple" mechanism with a feedback step. In this case, the NO species regenerates OH via its reaction with HO_2.

A Level 2 investigation of this same chemical system would require measurements of several other species, all of which involve reactions that either form secondary sources of OH or additional loss pathways for it. New measurement requirements therefore include H_2O_2, HO_2, CH_3OOH, and CH_2O. If the sampling environment contained high concentrations of hydrocarbons, additional species would probably need to be measured. The total number of new variables is difficult to assess at this time (i.e., with the current state of knowledge of hydrocarbon degradation mechanisms), but it could run as high as 5 to 10.

N_xO_y Experiments The Level 1 measurements list for the N_xO_y daytime experiment includes six variables: NO, NO_2, O_3, HO_2, CH_3O_2, and the UV flux. Measurements of these variables would permit a first assessment of the basic photochemical equilibrium expression:

$$\frac{(NO)}{(NO_2)} = \frac{J}{[k_1\,(O_3) + k_2\,(HO_2) + k_3\,(CH_3O_2)]}.$$

(See equation (5.24) in Chapter 5 for an explanation of symbols.) This expression has never been rigorously tested under clean tropospheric conditions. A Level 2 N_xO_y daytime experiment would require two additional N_xO_y species, NO_3 and HNO_2. Both of the latter species are believed to be present in photostationary state concentration levels. Finally, for completeness, serious consideration must be given to the measurement of OH. If the sampled environment contained high levels of hydrocarbons, as discussed previously, measurement of several RO_2 (R = organic group) radical species may also be required.

Two additional N_xO_y experiments, a nighttime experiment and a budget experiment, can provide a more complete understanding of many facets of nitrogen ox-

TABLE 3.3 A Comparison of Critical Measurement Requirements for H_xO_y, N_xO_y Daytime, and O_3 Experiments

Experiment Type	Measurement Type																
	OH	H_2O	H_2O_2	HO_2	O_3	CH_4	CH_3O_2	CH_3O_2H	CO	CH_2O	NO	NO_2	NO_3	HNO_2	UV Flux	Visible Flux	Meteorological Parameters
Level 1																	
H_xO_y	X	X			X	X			X		X				X		X
O_3	X	X			X				X		X	X			X		X
N_xO_y daytime				X	X		X				X	X			X		
Common measurements					X						X				X		
Level 2																	
H_xO_y	X	X	X	X	X	X		X	X	X	X				X		X
O_3		X		X	X		X		X		X	X			X		X
N_xO_y daytime				X	X		X				X	X	X	X	X	X	
Common measurements				X	X						X				X		

ide chemistry. One of the key questions that might be raised in the N_xO_y nighttime experiment is whether the NO_3 radical converts the NO_2 species to N_2O_5 under nighttime conditions. If it does, other reaction channels that might readily consume or divert N_2O_5 and/or NO_3 to other nitrogen oxide forms such as HNO_3 or particulate NO_3^- should be examined. At Level 1, we propose that the critical species include NO_2, NO_3, N_2O_5, and humidity. At Level 2, we recommend that the nighttime experiment be expanded to include three additional variables: HNO_2, HO_2NO_2, and aerosol NO_3^-.

Concerning the proposed N_xO_y budget experiment, it is logical that this experiment be performed both during the day and at night. At Level 1, we propose that a minimum of five variables be considered critical: NO, NO_2, HNO_3, peroxyacetyl nitrate (PAN), and NO_3^- (aerosol). For a comprehensive N_xO_y budget experiment, virtually all active nitrogen species need to be included. The Level 2 critical measurements list includes NO, NO_2, NO_3, HNO_2, N_2O_5, HNO_3, HO_2NO_2, PAN, and NO_3^-. For sampling areas having high hydrocarbon levels, it seems prudent to add the measurement of other organic nitrogen compounds to the critical list.

O_3 Experiments At Level 1, the proposed O_3 experiments would require measurement of nine variables: CO, NO, NO_2, O_3, H_2O, OH, UV spectral radiant flux (and/or relevant photodissociation rates), O_3 flux, and total pressure. With these measurements, two experiments could provide further information on the photochemical sources and sinks of tropospheric O_3. In Experiment 1, simultaneous measurements of the vertical profiles of CO, H_2O, O_3, NO, and NO_2 are proposed. If performed on an aircraft platform, the measurements would be taken at a sampling rate of approximately one every 5 s. Under these conditions, calculations of the appropriate correlation coefficients could provide a qualitative to semiquantitative evaluation of the influence of photochemical and transport processes on the O_3 distribution.

In Experiment 2, we propose that a direct assessment of the difference between photochemical production and loss be attempted. When horizontal O_3 advection is negligible, this experiment can be summarized in mathematical form as follows:

$$\frac{\partial\overline{(O_3)}}{\partial t} + \frac{\partial\overline{(w'c')}}{\partial z} = \overline{P}(O_3) - \overline{L}(O_3).$$

Here w' and c' are the deviations in the mean value of the vertical wind and the concentration of O_3, respectively, and the overbar is an average over a time or distance long enough to obtain a stable estimate. The time rate of change of O_3 and the average flux divergence are equated to the difference between photochemical production (P) and loss (L) of O_3. Chemical measurement requirements consist of O_3 (as a function of time) and the O_3 vertical flux divergence.

At Level 2, the measurements list for the O_3 source/sink experiment should include two additional variables, HO_2 and CH_3O_2, both of which are needed for a detailed mechanistic understanding of the photochemical sources of O_3. Both HO_2 and CH_3O_2 radicals react with NO to generate NO_2. The photolysis of the latter regenerates NO and also produces atomic oxygen, a species that is quickly titrated by O_2 to form O_3. Under clean atmospheric conditions, the reactions of HO_2 and CH_3O_2 with NO are believed to constitute the dominant photochemical source of O_3. For environments containing high concentrations of hydrocarbons, there may be significant levels of RO_2 (other than HO_2). These radicals can also provide a photochemical source of O_3. In these environments, the Level 2 critical measurements list should be expanded to include as many as 4 to 10 new species.

Sampling Strategy

For the selection of global sampling sites for the *Photochemical Transformation Study* experiments, the global distribution of solar irradiance must be considered. This would argue strongly for the tropics and subtropics as an important region requiring investigation. High priority must also be given to the midlatitude regions that reflect the strong influence of chemical input from industrialized nations. Tropical rain forests with their rich chemical environment are also regions of considerable interest. From the point of view of "chemical simplicity," available evidence points to the marine environment or a clean continental site as preferred locations to start the experiments. The latter consideration could prove to be important near the beginning of any new program when the number of instruments available for critical measurements may still be limited.

For the selection of appropriate sampling platforms, it is obvious that ground-based or ship platforms can provide long-term, continuous sampling and relatively simple logistics compared to aircraft platforms. However, aircraft platforms are essential for sampling at various altitudes and for sampling over wide areas and remote parts of the world. Both types of platforms will be required.

Instrument Readiness

The objective of the proposed fast-photochemical *Theory Validation Experiments* is to take a chemical "snapshot" of the troposphere. Ideally, many snapshots of as many different chemical environments as possible should be taken. The interpretative chemical analysis that follows from this type of experiment may best be described as zero-dimensional, or in special cases, perhaps one-dimensional. Therefore there is little need for an elaborate meteorological data base. However, this type of experiment does impose significant time resolution requirements on the instruments. If a chemical snapshot is to be taken, the "shutter speed" cannot be too long, or dynamic processes will become a major factor controlling concentration levels. For most species that are an active part of fast-photochemical cycles, the time to reach a photostationary state varies from less than a second to perhaps as much as a few minutes. Thus the time response for all instruments used in these fast-photochemical experiments ideally should be equal to or better than 1 min.

The relatively fast time response required of most instruments involved in this type of experiment raises an important question: What is the readiness of current or near-term technology to handle a Level 1 or Level 2 study? This is reviewed below.

H_xO_y Species Currently, adequate instrumentation exists for the measurement of at least two H_xO_y species, viz., H_2O and O_3. The instrumentation is suitable for both ground and airborne sampling. Instrumentation needed for the measurement of OH is not yet fully developed, but several major efforts are under way, and instruments may be available within 1 to 2 years. The major problem appears to be the development of adequate in situ sensors for detecting the peroxy species: HO_2, H_2O_2, and CH_3OOH. For each of these species, at least one method is now being tested.

N_xO_y Species For both NO and NO_2, appropriate instrumentation either is on hand (e.g., NO), or is likely to be available within a year or two (e.g., NO_2). For NO_3, one in situ method is currently being developed. There is also an established long-path absorption NO_3 measuring method, and its sensitivity is currently being improved. Nevertheless, some additional development effort may be needed for detecting NO_3. One method is now being field tested for PAN, with some field measurements already reported, but whether this method can be extended to other peroxynitrates is not clear. Further development of new instrumentation is required. N_xO_y species for which no in situ instrumental technique now exists include N_2O_5, HNO_2, and HO_2NO_2 (in the case of HNO_2, however, a long-path absorption method does exist). Development of instrumentation to measure N_2O_5 and HNO_2 should be emphasized in future planning.

Hydrocarbons Although a grab-sampling approach may be used to determine many hydrocarbons (in conjunction with laboratory-based gas chromatography/mass spectrometry systems), major problems remain for those species that are reactive. At this time with the limited knowledge of the hydrocarbons present in remote areas, it is not feasible to present a comprehensive list of species that should be measured. The uncertainty is somewhat less for midlatitude industrialized regions where more extensive measurements have been made as part of air pollution investigations. For some key photochemical species, e.g., CH_2O, concentrations have been measured even in remote areas. These measurements have primarily involved grab samples, and there remains a significant need for new measurement instrumentation for CH_2O. A high priority should be given to instrument development for the RO_2 free radical species, particularly CH_3OO.

Summary It appears that within the next 2 to 3 years instruments will be available to carry out most of the proposed Level 1 photochemical experiments. On the other hand, very few Level 2 experiments will be possible in this same time period. Projections of the time required to develop instrumentation for Level 2 experiments are difficult to make, but our best guess would be 5 to 10 years.

Concentration Distribution Experiment

Sampling Strategy

Validation of the various facets of photochemical theory should be an intermediate goal, not an end in itself. A major improvement in understanding photochemically driven transformations can be achieved only when the theory can be applied to a data base. Unfortunately, formidable problems can arise when designing experiments to collect a global data base.

As discussed in Part II, Chapter 5, in the section by Prospero et al., the shorter the lifetime/residence time of a chemical species, the larger the spatial and temporal variability in the mixing ratio of that species. This variability has local and zonal components. If the time mean values do not vary much in a particular latitude zone, then one time series is sufficient. In the latter case, the *Global Distributions Network* should provide sufficient latitude resolution. On the other hand, if the time mean values vary significantly as a function of longitude, a regional network will be needed to develop an accurate picture of the species distribution. One such representative region is the tropical Pacific Ocean. However, others should also be considered.

Design of a regional network requires considerable in-depth planning. The *Theory Validation Experiments*, which would precede any regional *Concentration Distribution Experiment* by several years, should provide some insight into this problem. Furthermore, the *Global Distributions Network* should provide a very coarse resolution picture of the time mean fields for CO, O_3, H_2O, and possibly total reactive nitrogen. It is possible that higher resolution model fields can be generated with the use of general circulation/transport chemistry models. With these data, it should be possible to determine both the fraction of the local variability resulting from zonal variability and the magnitude of that zonal variability for the key photochemical species. Each of these different types of information must be considered when defining a regional sampling network.

Defining the species to be measured in the *Concentration Distribution Experiment* can be approached by imposing the boundary condition that the subset of species to be measured must permit (using validated theory) the calculation of virtually all other critical photochemical species. A basic core of chemical variables might consist of H_2O, O_3, CO, and NO_x. The UV spectral radiant flux and other meteorological parameters should also be included. However, this core grouping would leave major uncertainties in the evaluation of important species OH, HO_2, H_2O_2, and CH_2O. In principle, a measurement of OH or HO_2 would resolve this problem, but such a measurement suffers from several other problems, among them the fact that OH is a very short-lived species. An alternative might be the measurement of the somewhat longer-lived species, e.g., H_2O_2, CH_3OOH, and CH_2O.

Instrumentation

Unlike the *Theory Validation Experiments*, the instrumentation required to assess mean concentrations of key photochemical species will not have major time resolution restrictions. Thus both indirect and grab-sampling measurement systems could be seriously considered provided extensive instrument intercomparison tests were employed to establish the reliability of these systems. There are significant advantages to the use of such low-to-medium technology instrumentation at remote sampling sites, locations that generally are staffed by technicians only. If balloon-platform sampling were to be employed at these remote sites, reliable chemical instrumentation (e.g., small in size, weight, and electrical power consumption) would need to be developed. Within the next few years, reliable instrumentation should be available for the measurement of most of the species indicated above.

Platforms

The majority of sampling in the *Concentration Distribution Experiment* would likely occur at ground stations. However, some balloon, aircraft, and ship-sampling time should also be expected. Rough estimates for aircraft flight time range from 800 to 1500 h over a 5- to 10-year period.

Laboratory Measurement Requirements

Coordinated laboratory measurement programs are required to improve the inherent accuracy of current chemical models. The uncertainties in these models are detailed in Part II, Chapters 5 and 6. Laboratory investigations are needed in three areas: (1) mechanistic studies of family chemical systems; (2) measurements of individual gas-phase rate coefficients under tropospheric conditions; and (3) fundamental spectroscopic and photochemical studies. The last is proposed primarily as a means of further supporting the development of both new laboratory and field-measurement technology.

Mechanistic Studies of Family Chemical Systems

Hydrocarbons The largest deficiency in the understanding of tropospheric transformation processes is related to the oxidation mechanisms initiated by OH and O_3 for CH_4 and larger hydrocarbons. Characterization of the relevant species and their reaction pathways is urgently needed.

Nitrogeneous Compounds Critical gaps exist in the knowledge relative to the transformation of several nitrogen species, particularly NO_3, N_2O_5, and organic and inorganic peroxynitrates. More detailed information on reaction mechanisms is required for an accurate assessment of nighttime N_xO_y chemistry and the overall N_xO_y budget.

Sulfur Compounds Details of the oxidation mechanisms of SO_2 and reduced sulfur species are not known. Although these species are not likely to control transformation rates involving H_xO_y, N_xO_y, and hydrocarbon species, an understanding of these mechanisms is necessary to evaluate the global cycling of tropospheric sulfur.

Measurements of Individual Gas-Phase Rate Coefficients

A large number of reactions of fundamental importance in the troposphere should be investigated under typical conditions of atmospheric pressure and chemical composition. Species that might be influenced by these conditions include OH, HO_2, RO, and RO_2. The highest priority should be given to reactions involving OH and HO_2.

Fundamental Spectroscopic and Photochemical Studies

Photodissociation processes play a critical role in the tropospheric fate of many trace gases. Little information is available on the UV absorption cross sections and quantum yields of oxidation intermediates derived from large hydrocarbons, particularly carbonyl and peroxy compounds. There are even some uncertainties surrounding the quantum yields of key species like NO_2. Measurements of the spectroscopic properties and associated chemical relaxation processes are needed for the development of new detection methodology. These new measurement techniques are required for both laboratory and field studies of key atmospheric species, particularly those listed in the proposed *Theory Validation Experiments* (see Table 3.3).

Photochemical Modeling

Because of the highly nonlinear nature of atmospheric photochemical transformations, numerical and mathematical models will be an essential tool in the *Photochemical Transformations Study*. Photochemical models will be a key element leading to the development of a detailed sampling strategy and the analysis and interpretation of the data. As outlined in Part II, Chapter 6, of particular importance will be the testing of photochemical theory by comparing the measured concentrations of H_xO_y and N_xO_y species and O_3 production and destruction rates with those calculated from a photochemical model. Because the chemical time constants for the species of interest are short in comparison with typical tropospheric mixing times, normally it will not be necessary to simulate transport, dynamics, and surface processes in this modeling exercise. Relatively simple box models can be employed.

When discrepancies between model calculations and field measurements arise, refinements in the photochemical mechanisms used in the model will be considered, leading to a more accurate and complete understanding of atmospheric transformations. Once the mechanisms used in photochemical models have been confirmed by the *Theory Validation Experiments*, they can be incorporated into a global diagnostic model for photochemically reactive species. This model development should lead to a more accurate simulation of tropospheric chemical cycles.

CONVERSION, REDISTRIBUTION, AND REMOVAL

Understanding the basic chemical cycles in the troposphere requires a detailed knowledge of the processes by which gases and aerosol particles undergo chemical conversion and redistribution, both in clear air and by clouds and precipitation. It is also necessary to know how substances are eventually deposited at the earth's surface. The chemical conversion processes of primary concern in this section are the gas- and aqueous-phase reactions that increase the oxidation state of the central atomic species, e.g., SO_2 to H_2SO_4, NO_2 to HNO_3, NH_4^+ to NO_3^-. Precipitating clouds scavenge airborne gases and particles from a considerable fraction of the troposphere and deposit them as "wet deposition." Processes associated with "dry deposition" are slower and are largely confined to air near the surface. Dry deposition encompasses the turbulent and diffusive flux of trace gases and aerosol particles at the surface and the gravitational settling of larger particles. In general, wet deposition dominates at large distances from sources or in areas with copious precipitation. Dry deposition is more important near sources or in areas with little precipitation. Removal processes are discussed in more detail in Part II, Chapter 5, in the section by Hicks et al.

The *Global Tropospheric Chemistry Program* must address several basic questions on the conversion, redistribution, and removal of key substances in the troposphere. These questions include the following:

1. Do we have quantitative agreement of fluxes between deposition and sources as a measure of the overall atmospheric cycles of key chemical species?
2. How important are clouds and precipitation on a global, regional, and local scale in controlling chemical conversion and removal within key chemical cycles?
3. To what extent do clouds and precipitation influence the vertical distribution of trace substances in the atmosphere and how does this affect the global, regional, and local concentration fields of major tropospheric species?
4. What is the composition of atmospheric aerosol particles in various regions of the troposphere and how does this composition relate to the coincident and preceding gas-phase chemistry and cloud and removal processes?

These fundamental questions call for intense and well-coordinated research efforts to obtain data on the relevant processes and to develop models for embodying them in tropospheric chemistry systems models. For these reasons, **we recommend initiation of a major research effort to investigate the processes of conversion, redistribution, and removal in the troposphere. The primary objectives of this *Conversion and Removal Study* would be as follows:**

1. **To obtain chemical conversion and removal rates for selected cases of clear and cloudy air, including condensing and evaporating clouds, and to develop methods that can be applied to experiments investigating weather systems that are dominant in different regions.**
2. **To investigate the processes that control the rate of dry deposition of gases and aerosol particles to the earth's surface.**

At present, only limited capabilities exist to carry out detailed large-scale field studies. Constraints arise from a lack of crucial instruments and tested methods in this relatively new field of chemical meteorology. Thus a general research strategy is to call for near-term field experiments to attack specific questions with available technology and, as knowledge grows from these experiments, and new instruments are developed, to broaden the scope of effort accordingly, both scientifically and geographically. As progress is made toward the attainment of these objectives, a framework can be developed for major future field experiments to quantify the complex physical and chemical processes linking clouds, precipitation, and both wet and dry deposition with the major chemical cycles. Such experiments will be essential for the verification of global- and regional-scale tropospheric models.

We recommend a *Wet Removal Experiment* to assess the flow of specific chemical species into and out of selected types of clouds and subsequent redistribution of those species in the troposphere. This *Wet Removal Experiment* would involve cloud and precipitation chemistry research and a significant investment in meteorological data collection and analysis. We also recommend a *Dry Removal Experimental Program* that would concentrate initially on development of fast-response and high-precision chemical sensors and the investigation of important surface exchange processes as well as the characteristics of the air near the surface.

Experimental Constraints

The dominant tropospheric gas-phase reactions and mass transport phenomena can be described for only a few species in the troposphere. The same is true for aqueous-phase processes. The primary reason for slow progress in this area has been the lack of instruments capable of detecting species of interest with sufficient sensitivity, accuracy, or time resolution. Several potentially critical compounds have not yet been detected in the troposphere. High-relative-humidity environments (e.g., clouds) preclude operation of some current instruments. Detection capability for several important chemical species can be summarized as follows:

1. Concentrations (1 to 3 ppb) of most sulfur compounds can now be measured either in real time or with time resolution of minutes to hours. However, virtually all instruments for these measurements are laboratory prototypes that have become available for field testing only recently. An important goal is the development of instruments for the detection of SO_2 with detection limits in the higher parts per trillion range and with time resolutions of about 0.1 s.
2. Many nitrogen compounds can be measured at low concentrations, either in real time or intermittently, with time resolutions of 1 s to several tens of seconds. Most of the instruments—with the exception of those for NH_3, NO_3, and N_2O_5—have been or are being field tested.
3. Most carbon compounds, except CO and CO_2 (gas, aqueous, and aerosol phase), cannot be measured in real time, but require sampling times of several minutes to several hours. Considerable instrument development is required to improve sensitivity, speciation determination, and time resolution for the many tropospheric carbon compounds.
4. Some gaseous halogen compounds of interest can be detected with time resolution of minutes. A real-time instrument for HCl in the lower parts per trillion range is needed.
5. Field-tested instruments exist to measure O_3 with the required sensitivity and time resolution. However, considerable confusion still exists as to the validity of H_2O_2 detection techniques. Gaseous H_2O_2 cannot be measured at present, although at least one method is being developed. The aqueous-phase concentration of H_2O_2 can be measured with sufficient sensitivity, but significant discrepancies among the various methods must be resolved.

Measurement capabilities for relevant physical properties of the troposphere can be summarized as follows:

1. Visible light scattering and absorption can be measured accurately from aircraft or on the ground. Scattering can be measured with a response time of a second, while absorption requires several minutes. Infrared optical properties are as yet much more difficult to assess.
2. Cloud condensation nuclei measurements can be made on a relatively routine basis, but measurement of ice nucleating properties may require further instrument development. The relationship between cloud al-

bedo and aerosol particles should be a target of future research.

3. Cloud water collection as a function of droplet size is important for chemical studies, as is measurement of cloud liquid water content. Further instrument development may be required in these areas.

Field programs investigating conversion and removal processes in several tropospheric chemical cycles can be initiated with currently available instruments. However, many fundamental questions in the sulfur, nitrogen, halogen, carbon, and trace element cycles, and the complex relationships among the various cycles, can be answered only following further instrument development and as part of a well-integrated research strategy for field experiments. Instruments that have been developed for fast-response eddy-correlation flux determination for some sulfur and nitrogen compounds must be extensively field tested. The development and testing of fast-response instruments, i.e., with at least 1-Hz frequency response for surface-based measurements and at least 10-Hz frequency response for aircraft measurements, or of some technique for making such direct determinations of eddy fluxes by using more slowly responding instruments, is of vital importance for future field programs; see Part II, Chapter 8 for a discussion of related criteria.

Recommended Field Experiments

Wet Removal Experiment

The objectives of the *Wet Removal Experiment* are to obtain overall conversion and wet removal rates in a variety of cloud systems. While measurements should be undertaken of species in several tropospheric chemical cycles, early emphasis will likely be on studies of conversion and removal rates of species in the sulfur and nitrogen cycles because measurement capabilities are generally more advanced for these substances. In Phase I of the *Wet Removal Experiment*, diagnostic studies should be carried out to test, where possible, the postulated oxidation mechanisms for sulfur, nitrogen, and other elemental species and to determine the efficiency of in-cloud scavenging mechanisms. We use the term "diagnostic study" to refer to a relatively modest research project that is carried out to answer limited research questions. A successful test of experimental concepts and newly developed instruments would subsequently enable them to be applied in Phases II and III to more comprehensive field studies, which would include more detailed studies of carbon, halogen, and trace element chemistry and increasingly complex weather systems.

For studies in all phases of the *Wet Removal Experiment*, a thorough aerosol characterization effort should be maintained such that key chemical and physical variables can be adequately studied and understood. This effort should also be made in other large-scale measurement programs in the *Global Tropospheric Chemistry Program*.

Phase I—Diagnostic Studies Examples of recommended diagnostic studies that would provide significant insight into chemical conversion, redistribution, and wet removal are as follows:

1. The relative importance of competing oxidants for SO_2 oxidation, such as molecular oxygen (O_2), O_3, H_2O_2, or OH, as well as homogeneous or heterogeneous catalysts such as transition metals, could be evaluated from the measurement of the chemical composition of cloud droplets and interstitial air between the droplets. Applying simple chemical models would enable an estimate to be made of the relative importance of the various proposed mechanisms.

2. Strong oxidants such as H_2O_2 are frequently present in cloud water and precipitation. The origin of this H_2O_2 is not clear. Possible formation pathways involve both gas and liquid phases. Measurements of H_2O_2 and other chemical species in cloud water, rainwater, and interstitial air at different elevations within a given cloud environment should aid in defining the origin of H_2O_2.

3. Fog, cloud water, and rainwater in different regions apparently contain different amounts of formaldehyde (CH_2O). Aldehydes react with dissolved SO_2 via a nucleophilic attachment on the carbonyl carbon to form bisulfite addition complexes. With CH_2O, the complex formed is hydroxymethane sulfonic acid. Formation of these adducts would increase the concentration ratio of sulfur (IV) to sulfur (VI) in cloud water in which CH_2O is found. The role played by aldehydes in regulating the oxidation state of sulfur in cloud systems could be evaluated by analysis of cloud water.

4. A more general type of field experiment could broadly integrate several facets of in-cloud fast photochemistry. As discussed in Part II, Chapter 5, in the section by Davis et al., the penetration of ultraviolet radiation into a cloud may produce significant quantities of OH and HO_2. These reactive species could lead to gas-phase conversion reactions in the interstitial air, and/or the reactive species could be scavenged by cloud droplets. Scavenged radicals could initiate extensive homogeneous aqueous-phase chemistry involving OH, H_2O_2, and other ionic forms of the generalized formula H_xO_y as well as various halogen and trace element species. No data exist on the concentration of any of the active photochemical species in the interstitial air. Con-

centrations of interstitial photochemically active trace gases could be contrasted to measurements in contiguous noncloud air. These early experiments are diagnostic, as it is unlikely that all key photochemical species will be measurable in the near future, and the experiments could be carried out in conjunction with the *Photochemical Transformations Study*. Species of potential interest include NO, NO_2, O_3, SO_2, H_2O_2, CH_3OOH, OH, HO_2, and certain halogen species.

These diagnostic studies could be carried out with one or two aircraft measuring in-cloud and simultaneously collecting cloud water and rainwater (without reference to any large-scale flow field).

Phase II When the appropriate instruments, experimental methods, and theoretical concepts have been developed from the Phase I diagnostic studies, it will be possible to undertake more detailed experiments in selected cloud systems. Such studies are essential to progress in cloud chemistry and should be of great value in planning subsequent field operations in increasingly more complex meteorological systems. They also would serve as a test of laboratory kinetic studies, particularly those performed under experimental conditions different from those expected in clouds. In terms of the physics of clouds, unglaciated wave clouds are perhaps the least complex, with orographic stratus, marine stratus, and isolated cumulus clouds being increasingly more complex. The choice of specific cloud systems for study should also be related to source functions for the chemical species in the cycles being investigated.

Cloud water, aerosol, gas, and rain sampling experiments would be conducted in these relatively simple cloud systems. Although the detailed experiments cannot be defined at this time, an example can be suggested on the basis of current knowledge of atmospheric cloud chemistry. Orographic clouds downwind of biologically active ocean water offer a reasonably simple cloud physical situation in a potentially important source region for certain chemical species. Careful utilization of meteorological information would be required to provide an accurate description of the temporal history of an air parcel as it is processed through the orographic cloud. Possible studies could include the following:

1. The oxidation of $(CH_3)_2S$ and SO_2 in the gas phase and within cloud water itself;
2. The role of organic substances in cloud water chemistry;
3. The incorporation of aerosol particles into the cloud droplets;
4. The removal of sulfur species in rain.

It would also be useful to attempt a mass-balance calculation for sulfur species during these studies, perhaps as part of a Lagrangian observation scheme.

Phase III After completion of studies in less complex cloud systems, investigations could ultimately proceed to much more complex systems, including extratropical cyclones, either marine or continental, and their associated frontal systems. A significant fraction of the precipitation in midlatitude regions is related to such systems, and we suggest a study with particular emphasis on warm frontal precipitation. Warm frontal precipitation associated with cyclonic storms results, in principle, when warm, moist air ascends over cold air (cold sector) north of the warm front. The rising warm sector air cools until condensation occurs, effectively scavenging trace gases and aerosol particles from the warm air mass by in-cloud processes. Most of the precipitation occurs north of the surface front (in the northern hemisphere) and falls through the colder air below the cloud, from which trace gases and aerosol particles are also scavenged. Precipitation also results from convective instability, which often develops in the warm sector. Air trajectory analysis at several levels can help establish the general flow pattern into the complex storm system.

Detailed meteorological and cloud physical background information would be required in such a study. This phase of the *Wet Removal Experiment* should be developed in conjunction with comprehensive meteorological research and support programs if at all possible. An example of such a program in a continental region is the proposed *National Stormscale Operational and Research Meteorology Program* (STORM).[1] One emphasis of STORM is the study of the evolution of cyclonic storm systems.

One of the primary requirements of a study of warm frontal precipitation would be to direct and position the participating aircraft fleet in the following general flow regimes:

- in dry air in the warm sector before it ascends.
- in nonprecipitating clouds ahead of the warm front.
- in precipitating clouds involving the overriding air.
- just below cloud base and below the cloud penetration aircraft.
- in dry air west of the front (cold sector).

Close coordination with forecasters would also be required for mobile collection of ground-level precipita-

[1] *The National Storm Program, Framework for a Plan,* University Corporation for Atmospheric Research, Boulder, Colorado, 1982, 21 pp.

tion coinciding with the aircraft samples taken aloft in the same region. Surface-based Doppler radar would be required to document flow fields in the vicinity of the selected warm frontal system, as well as a surface-based calibrated PPI radar for documenting spatial and temporal variability of precipitation intensity. The success of this experiment would rest largely on the ability to follow the air as it is being cycled through the storm system, an extremely difficult task at present.

Information on chemical conversion rates for species in the various chemical cycles may be deduced from measured differences in the chemical composition of this air before and after cloud encounter and from the chemical composition of cloud and precipitation elements. The removal rate would be obtained, in principle, from the measured composition of cloud and precipitation water obtained at ground level and aloft, from the measured composition of inflowing air, from the liquid water content of clouds, and from the precipitation rate.

Requirements for chemical measurements would be as follows in cloud-free air and interstitial air:

- gases: Various species in the sulfur, nitrogen, carbon, and halogen cycles.
- aerosol: $SO_4^=$, NO_3^-, NH_4^+, Ca, Mg, K, Na, Cl, carbonaceous material, transition metals, and acidity.

Requirements in cloud water and precipitation would be as follows:

- $SO_4^=$, NO_3^-, halides, H^+, NH_4^+, Ca^{++}, Mg^{++}, Na^+, K^+, H_2O_2, CH_2O, carbonaceous material, and transition metals.

Additional variables that should be measured include aerosol particle and cloud droplet size distributions and liquid water content. Liquid sample measurements could be obtained with current instrument technology.

Such a study would require approximately five medium-sized aircraft equipped with cloud physics and dynamics sensors, with cloud and precipitation samplers, and with sufficient payload to accommodate the chemical and aerosol particle sensors and samplers. Small trailers equipped with replicate sets of chemical sensors and samplers and with several sequential precipitation collectors would also be required. A review of existing research aircraft (see Part II, Chapter 9) confirms that adequate platforms are available for this study. These platforms include well-instrumented meteorological aircraft with adequate space for tropospheric chemistry measurements. The timetable for a Phase III warm frontal precipitation study would depend to some extent on the schedule of major meteorological field experiments on mesoscale weather systems.

Dry Removal Experimental Program

Dry deposition includes the turbulent and diffusive transfer of trace gases and aerosol particles from the air to the underlying surface, and the gravitational settling of large particles. The processes that control dry deposition are normally associated with the nature of the surface itself, or with characteristics of the neighboring media. Detailed knowledge of the factors controlling dry deposition of key chemical species is still rudimentary. No large field effort can be offered as a panacea. The nature of the problem requires close attention, at least initially, to small-scale factors in process-related field studies rather than large-scale integrated research programs.

Flux determinations would provide detailed knowledge of appropriate deposition velocities and their controlling properties. These deposition velocities could be used directly in numerical models, but they could not be used to evaluate dry deposition fluxes from field data unless suitable concentration data were available. Such data must be obtained sufficiently close to the surface so that surface-based formulae can be used to interpret them. Simple but reliable sampling methods need to be developed for this purpose. Methods analogous to high-volume filtration for airborne particles appear to offer special promise. Such methods are already in operation in some networks (e.g., in Canada, Scandinavia, and over the Pacific), but these methods need improvement to permit routine and inexpensive operation on a large scale.

Special attention must be given to deposition at sea, where flux determinations are rare. Rates of exchange between the atmosphere and the ocean are major unknowns in many geochemical cycles. For example, the atmosphere may be the primary transport path for a number of trace elements, including heavy metals, found in the open ocean, but efforts to determine deposition rates for these substances are complicated by resuspension processes at the sea-air interface. Similarly, air-sea exchange plays a major role in the biogeochemical cycle of iodine, but little information is available on the conversion rates of iodine species in the marine troposphere or their exchange rates with the ocean.

Existing knowledge of air-sea exchange of trace gases is based primarily on extrapolation of laboratory studies conducted over calm, clean water with some guidance derived from the use of chambers in light wind conditions at sea. It is suspected, however, that exchange rates are greatly accelerated in strong winds when effects of breaking waves cannot be disregarded. Recently developed aircraft and surface tower methods for determining dry deposition rates of specific trace gases must be extended to oceanic situations so that moderate- to high-wind conditions are addressed adequately. Furthermore, it is likely that air-sea exchange of trace gases will be moderated by biological activity in the surface waters. The influence of these biological factors must be investigated.

Techniques for Determining Fluxes There is no generally accepted method suitable for routinely monitoring dry deposition fluxes. However, suitable techniques are available for application in intensive field studies. These techniques include micrometeorological gradient and covariance methods from towers, aircraft covariance methods, direct measurement of the accumulation of material on exposed natural surfaces (especially of leaves and snow), and mass budget studies conducted over a closely monitored research area (such as a calibrated watershed). These intensive measurements are usually intended to identify controlling processes and to quantify those that are important.

Aircraft and tower studies of dry deposition frequently make use of the covariance (or eddy-correlation) method of flux determination in which output from a fast-response chemical sensor is multiplied by the vertical wind component measured with fast-response co-located instruments. The average of this product over a period long enough to obtain a statistically significant estimate gives the vertical flux of the chemical species through a horizontal plane at the sensor height.

Aircraft covariance measurements require chemical sensors with at least 10-Hz frequency response; tower studies can be conducted with instruments of at least 1-Hz frequency response. In general, instruments that meet the requirements for aircraft eddy-correlation measurements will also satisfy the requirements for tower operation. In some cases, however, eddy-correlation methods with their requirements for rapid response sensors can be replaced by gradient techniques. The latter place great demands on accurate measurement of mean concentration differences between several levels near the surface instead of high-frequency measurements of concentration fluctuations. Except in a few instances, suitable instruments for flux determination by the eddy-correlation technique are not yet available. Therefore dry deposition studies remain at the mercy of sensing technology. Techniques for estimating chemical fluxes are discussed in Part II, Chapter 8, where measurements of various terms that contribute to the large-scale mass budget of a trace species are considered.

Experimental Approach We recommend the continuation and intensification of laboratory, theoretical, and experimental research programs to identify and quantify dry deposition processes, with particular emphasis on the development of fast-response and high-precision chemical sensors. When the appropriate instrumentation is available, we recommend consideration of a large-scale study of dry deposition at sea.

Conducting experimental studies at sea is a demanding task that should be attempted only after appropriate new instruments have been thoroughly tested over land. Once the necessary experimental facilities are developed, considerable benefit would derive from testing the various components in an integrated fashion in a large-scale "box-budget" experiment over the ocean. Such a large-scale field study would serve not only as a target for instrument development but also as a means for exploring the links between such interrelated aspects of oceanic geochemical cycling as the importance of flux divergence terms, the rate of chemical transformation in oceanic air, and eventually the role of deep convective processes and precipitation in redistributing and scavenging tropospheric trace species. However, such an experimental box-budget study would not be warranted before the development of accurate models for evaluating meteorological transport and dispersion within remote areas of the troposphere, over distances on the order of at least 1000 km. We propose the following studies leading to experimental investigations of surface exchange at sea.

Phase I Through the use of meteorological towers at appropriate land sites, evaluate surface fluxes for trace gases, including SO_2, NO, NO_2, HNO_3, NH_3, H_2O, H_2O_2, O_3, and measurable species in the halogen and carbon cycles, as well as species present on aerosol particles. Studies should be conducted over several types of surfaces at different seasons to develop confidence and experience with the measurement techniques, as well as to investigate the importance of seasonally varying biological and meteorological factors. Such studies permit direct comparison of alternative methods of measurement and should be expanded to include research aircraft equipped with comparable sensors.

Phase II Once fast-response sensors for several key species in the sulfur, nitrogen, carbon, and halogen cycles are available, prototype tower experiments should be conducted over the ocean, possibly by selecting a suitable island or existing tower within some selected experimental area. Fly-by missions with instrumented aircraft should be conducted both for comparison with tower-based results and to assess spatial variability.

Phase III Extension of this program beyond the process-oriented and instrument development case studies would occur only after careful evaluation of the integrated box-budget experiment approach. Figure 3.3, using the sulfur cycle as an example, draws attention to some of the physical and chemical constraints and meteorological factors that must be borne in mind. At this time, such a study cannot be discussed in detail because the required experimental capabilities do not exist. However, it is appropriate to bear in mind the wide range of phenomena to be considered in any such experiment because they are all closely interrelated, and any one could influence the interpretation of experimental programs designed primarily to address another.

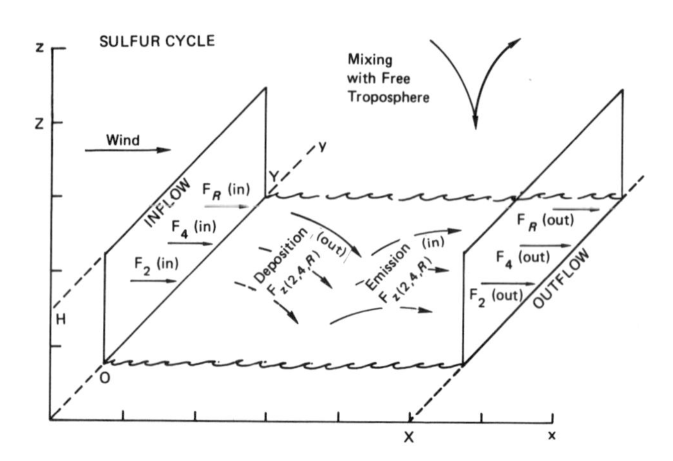

FIGURE 3.3 A simplified schematic representation of the design of a box-budget experiment involving reduced and oxidized sulfur species. Horizontal fluxes F and vertical fluxes F_z have subscripts indicating reduced sulfur species (R), SO_2 (2), and $SO_4^=$ (4). H is the depth of the marine mixed layer, somewhat less than the height Z to which measurements must be made. The horizontal dimension of the study volume is about 1000 km, with Z in the range of 2 to 5 km. Note that fluxes at the top and longitudinal sides of the volume are not shown.

Future Experiments

Models that describe the fate of species in the various tropospheric chemical cycles incorporate parameters that must be obtained from experiments such as those described above. As new instrumentation and theoretical concepts are developed, investigations of conversion, redistribution, and removal can be expanded both scientifically and geographically. Ultimately, we would explore all chemical cycles and the relationships among them within the cloud systems embedded in the major flow regimes (global circulation) of the earth. This exploration would yield information required for development, completion, and verification of global *Tropospheric Chemistry Systems Models* (TCSMs).

MODELING THE TROPOSPHERIC CHEMICAL SYSTEM

Over the last decade, tropospheric chemistry research has progressed largely through the development of new instruments and the concomitant pioneering measurements of previously undetected species. It is now evolving to a more mature science with the design of large-scale field programs devoted to the systematic collection of required data. Modeling has also always been a part of tropospheric chemical studies. However, in contrast to experimental chemistry, global chemical modeling has mostly been carried out by a few individuals whose primary goal has been to leap to new levels of understanding by using simple models tailored to display the mechanistic role of the new species of interest. Regional air-quality problems, on the other hand, because of their more immediate practical concerns, have forced other modelers to be more empirical and to model the data at hand without going into much detail on the physical processes involved. Almost all past efforts in modeling global problems were accomplished by one or two scientists working together for one or two years.

It is expected that such modeling will continue profitably not only for introducing new concepts but also for illuminating individual subprocesses within the overall global tropospheric chemical system. However, to achieve the objectives of the proposed *Global Tropospheric Chemistry Program*, it will also be necessary to have a more ambitious long-term perspective toward model development. That is, for the successful application of the field program data and to advance scientific understanding, it will be necessary to develop comprehensive models of the overall tropospheric chemical system. Such models should necessarily include the meterological processes that transport and in other ways interact with the chemical species.

Experience in closely related areas shows that the development of such models would involve considerable investments in time and computer resources over a long term. Further, it is unlikely that such model development can be achieved within existing institutional arrangements, such as individual short-term grant programs. Thus **we recommend the initiation of one or more global *Tropospheric Chemistry Systems Models* (TCSMs) with stable long-term support.**

Considerable modeling effort will also be devoted to the various individual processes related to tropospheric chemistry. These processes are required for the TCSM to have a sound physical basis. The field programs that have been proposed as part of the *Global Tropospheric Chemistry Program* are intended in part to provide an observational basis for developing the critical models required for developing the TCSM. **We recommend that additional modeling efforts be initiated to develop the critical submodels required for the TCSM.**

Models for Biological and Surface Sources and Sinks

Biological and surface source models fall into three categories: (1) global empirical models, (2) mechanistic models of biological processes, and (3) micrometeorological and oceanic models of surface transport processes.

The observational efforts in the *Biological Sources of Atmospheric Chemicals Study* will provide measurements at individual field sites. Initial exploratory efforts will identify the ecological communities that provide significant emissions, but as a second stage it will be necessary to obtain sufficient observations to determine annual average emissions at various sites. Variability with environmental parameters such as temperature, solar radiation, moisture, soil pH, and Eh will also be obtained. However, due to the great variety and small-scale structure of biological systems, it will always be very difficult to collect sufficient data to permit straightforward numerical averaging to establish regional and global average emissions. Rather, more sophisticated approaches should be developed to interpolate and extrapolate the available observations to all the nonsampled areas and thereby to derive regional and global budgets. We refer to such models as global empirical models.

It will also be necessary to develop models of the detailed biological mechanisms and processes responsible for the measured emissions. These will range from models of soil or oceanic biochemical processes to models of whole-leaf physiology.

Boundary layer and surface transport models are required to describe the movement of gases and aerosols between ocean or land surfaces and the atmosphere. In the case of the oceanic processes, such models require consideration of oceanic as well as atmospheric boundary layers and the effects at the ocean interface of wave breaking, i.e., the movement of air bubbles on the ocean side and spray droplets on the atmospheric side.

In the case of land surfaces, it is necessary to model, in conjunction with observational studies, the detailed physical mechanisms that together determine the deposition of a given species to a given surface. It has been found convenient to model these processes as "resistances" or "conductances." A gas molecule being transferred from the atmosphere to a surface first passes through the atmospheric mixed layer above the vegetation canopy or other roughness elements. Thus it must be transferred to the air within the canopy or other roughness elements and from there to surfaces of deposition. These surfaces in turn may not instantly capture the molecule but rather, as in the case of ocean surfaces, provide additional resistance to the transfer of the depositing species to its ultimate sink. Many gases of interest have sinks in the internal cavities of leaves so that the biomechanics of leaf stomata must be modeled, possibly in terms of leaf resistance.

Models for Global Distributions and Long-Range Transport

Here we outline some of the research studies that one or more TCSMs would carry out in support of the *Global Distributions and Long-Range Transport Study* and, more generally, for modeling exploration of the tropospheric chemical system. Two classes of investigations would be carried out with TCSMs. First would be studies intended to validate and possibly develop the capability of models to simulate long-range transport and global distributions and the variability of long-lived chemical species. These studies would compare model results with the data sets including measured variances obtained through the *Global Tropospheric Chemistry Sampling Network* and, if necessary, would develop the model improvements required for satisfactory validation. Most of the modeling studies would be carried out in a climatological framework, but detailed event studies would also be performed in conjunction with intensive periods of field data collection. Such investigations could also be carried out with more simple chemistry if needed to simulate the sources and sinks of the long-lived species.

The second class of studies would emphasize the simulation of species with tropospheric lifetimes of days to about a year. At present, the sources and sinks of these species are not sufficiently well known for such studies to be used to test model transports. Rather, these studies, assuming adequate transport submodels, would explore the role of meteorological processes in determining the spatial distributions and temporal variability of these species. Such modeling studies, for example, could address the question of the importance of continental pollution sources of sulfur and odd nitrogen for atmospheric distributions at remote sites. A second extremely important example is that of O_3 in the global troposphere. TCSM studies are needed to compute how the marine troposphere responds to continental sources of nitrogen oxides and hydrocarbons (which, together, produce O_3) and to the export of O_3 itself from continents.

The exploration of such questions would help improve interpretation of the data on many of the species to be monitored in the *Global Distributions and Long-Range Transport Study*. Species of special current interest include NO, NO_2, HNO_3, CO, SO_2, O_3, sulfate-nitrate aerosol particles, and continental soil aerosol particles. The soil aerosol is of interest not only because of its optical-radiative effects but also as a source of Ca, which would neutralize acidity and thereby raise the pH of airborne droplets.

Models for Photochemical Transformations

The questions of photochemical transformation involve the concentration and interrelationships of H_xO_y, N_xO_y, and O_3. The immediate connections among these species occur so rapidly that the effects of transport are negligible; field observations can hence be tested by simple box models. However, the longer-lived species, including O_3, that ultimately determine the concentration of the fast-radical families must be generated by the TCSM. Global models are also needed to provide meteorologically realistic descriptions of the radiation fluxes that drive the fast photochemistry.

Models for Conversion, Redistribution, and Removal Processes

Conversion, redistribution, and removal involve both dry and wet processes. Dry deposition at surfaces has already been discussed. Besides the physics of surface removal, it is necessary to model the meteorological transport through the planetary boundary layer and the surface mixed layer. Models for the wet processes also need to be improved. Clouds and precipitation play important roles in the removal, transport, and transformation of species in element cycles. For instance, wet removal is probably one of the most effective sinks for nitrogen, sulfur, and inorganic halogen compounds. Important species such as SO_2, N_2O_5, and perhaps NO_3 may go through fast aqueous transformation in cloud droplets. Furthermore, cloud convection may be an efficient vertical transport mechanism for trace gases and aerosol particles.

In order to evaluate these processes quantitatively, it is necessary to develop a cloud-removal model that includes detailed treatments of the physical and chemical mechanisms involved. The cloud model would be a submodel of the TCSM. Physical aspects of the cloud model would include the parameterization of radiation, condensation, evaporation, stochastic coalescence and breakup, and precipitation development. Chemical aspects of the model would include both homogeneous gas-phase and liquid-phase reactions as well as heterogeneous reactions. In the clean atmosphere, chemical species treated within the cloud model should include at least O_3, odd-nitrogen species, hydrogen radicals, H_2O_2, sulfur species, CO, and CH_4 and its oxidation products. In the polluted atmosphere, nonmethane hydrocarbons and their oxidation products, metals such as Mn and Fe, and graphitic carbon should also be considered.

Summary

Modeling should be a major component of the *Global Tropospheric Chemistry Program*. It is clear that, in addition to current modeling efforts, there is also a need to focus on the development of tropospheric chemistry systems models. The meteorological processes in these models would best be provided through specially designed atmospheric general circulation models (GCMs) that not only account for large-scale tracer transports in the free atmosphere but also incorporate transport through the planetary boundary layer and through cloud processes. These models also require physically based submodels, a good description of land surfaces, and adequate treatment of the solar radiation that drives tropospheric photochemistry. With proper support, it should be feasible to develop TCSMs successfully because well-posed questions are apparent, suitable computers exist, and the substantial experience of scientists with global meteorological models and with complex air pollution models can be tapped.

The three-dimensional distribution of chemical species should be represented with a spatial and temporal resolution comparable to that of the meteorological variables. The distribution of these species is determined on the one hand by meteorological transport and source and removal processes, and on the other hand by wet and dry chemical transformations.

Besides the development of the TCSM, there should be additional new modeling programs devoted to the treatment of crucial subprocesses that will be needed for the TCSM to have a sound physical basis. We anticipate a two-way interaction with the other components of the *Global Tropospheric Chemistry Program* (i.e., field programs and laboratory studies of kinetics, photochemical data, and heterogeneous equilibria and reaction rates). As shown previously in Figure 3.1, these components would provide the data needed for the development and testing of the TCSM. The TCSM and the submodels would in turn provide guidance and direction for the planning of the field programs and laboratory studies.

INSTRUMENT AND PLATFORM REQUIREMENTS

Field and Laboratory Instrumentation

To attain the long-term goals and objectives of the *Global Tropospheric Chemistry Program*, sensitive instrumentation is required for both the measurement of chemical species and their fluxes in the remote troposphere and the elucidation of critical reaction mechanisms and rates in the laboratory. Although currently available instrumentation is adequate to initiate some of the exploratory phases of major field studies in the proposed *Global Tropospheric Chemistry Program*, the instrumentation is not adequate to carry out the detailed research program outlined in this report. Currently available instrumentation ranges from low-technology, low-cost sensors and collection systems with limited accuracy and sensitivity to high-technology, delicate, accurate, but costly bench-type instruments that still require considerable development and intercalibration before they can be deployed in the field. In addition, there are as yet no instruments available for the measurement of certain critical species in the global troposphere, such as HO_2. A summary of currently available measurement techniques for critical species in tropospheric chemical cycles is presented in Chapter 9.

While it is feasible today to make concentration measurements for most of the major species in tropospheric chemical cycles within the planetary boundary layer, many of these measurements cannot be made in the free troposphere, where concentrations are considerably lower. In addition, relatively few instruments are capable of making in situ measurements with a frequency greater than one measurement per second. Absolute calibrations, instrument intercomparisons, and other quality control procedures during all research efforts in the *Global Tropospheric Chemistry Program* are needed. Specific requirements for instrument development have been described in the individual research program sections. **We recommend that a vigorous program of instrument development, testing, and intercalibration be undertaken immediately and that it continue throughout the *Global Tropospheric Chemistry Program*.**

As new and more sensitive instruments are developed, the requirements for accurate calibration and standardization of these instruments will become more difficult to achieve. At the present time the National Bureau of Standards produces no Standard Reference Material gas standards with concentrations below several parts per million. The *Global Tropospheric Chemistry Program* will require accurate analytical standards for many trace gas species in the part per thousand to part per billion range. **We urge the initiation of a program to develop accurate trace gas standards in this concentration range.**

A strong, active program of laboratory measurements of chemical reaction kinetics will be required in the *Global Tropospheric Chemistry Program*. The program will require the continuing commitment of chemical kineticists and the further development of laboratory instrumentation systems for the investigation of the mechanisms and rates of the gas- and liquid-phase reactions critical to an understanding of tropospheric chemical cycles. **We recommend that an increased effort be initiated on laboratory studies of the rates and pathways of fundamental chemical reactions in tropospheric chemical systems.**

Platforms

Aircraft

There is a wide variety of aircraft platforms currently available in the United States from government, university, and private operators. A detailed compilation of the specifications and characteristics of these aircraft is now being developed by the National Center for Atmospheric Research. A brief summary of research aircraft platforms in the United States is presented in Part II, Chapter 9. Available platforms range from single- and two-engine aircraft with limited range and space for scientific equipment (more than 20 such aircraft) to long-range, four-engine turbo-jet and turbo-prop transports. Currently, three jet aircraft and five turbo-prop aircraft are being used in various aspects of tropospheric chemistry research. In some cases, the aircraft platform is available as an unmodified vehicle and in others as a complete aircraft measuring system, often dedicated for extensive periods of time to meteorological and atmospheric chemistry studies. We believe that the current aircraft fleet number and type is adequate to undertake the *Global Tropospheric Chemistry Program*, assuming ready access to these aircraft. Certain improvements and modifications to some aircraft and the meteorological support equipment aboard them will undoubtedly be necessary.

Ships

A large number of dedicated oceanographic ships are now active in the United States registry. They are operated by academic institutions through coordination with the University National Oceanographic Laboratory System (UNOLS), the Navy, NOAA, the Coast Guard, and other institutions. There are currently more than 30 active ships from academic institutions and 35 operated by the federal government. These ships vary widely in capability and in the geographic areas they cover. A summary of the specifications of academic oceanographic ships in the United States is given in Part II, Chapter 9. There is no oceanographic vessel designed specifically to carry out tropospheric chemistry research or dedicated to this area of research. Therefore, oceanographic sampling platforms will be a compromise between the needs of the atmospheric chemistry community and the missions for which the ships were designed. Sufficient ships are available to undertake the proposed field research at sea in the *Global Tropospheric Chemistry Program*.

Satellites

Ideally, spaceborne remote sensors could provide near-global measurements and thus could satisfy the ultimate goal of obtaining a three-dimensional distribution of certain atmospheric trace constituents. The importance and future potential of spaceborne instrumentation for investigating the chemistry of the troposphere are highlighted in a forthcoming report of the National Research Council.[2] Eventually, this approach should provide the tropospheric chemistry community with the opportunity to iterate a variety of distribution measurements with evolving mathematical models of the troposphere. An assessment of the capability of current remote sensor technology for performing measurements in the global troposphere reveals that three classes of remote sensors have demonstrated unique capabilities in meeting some of the measurement needs. The first class includes imaging spectroradiometers currently being used in earth observation satellite systems. A second class includes passive remote sensors that measure spectral emission or absorption of atmospheric molecules by using external sources of radiation. A third class includes active remote sensors in which lasers are used in a manner similar to an active radar system. Through a combination of scattering by aerosol particles and molecules in the atmosphere and selective absorption by atmospheric molecules, these sensors should ultimately be able to provide range-resolved measurements of aerosol particles and many tropospheric molecular species. However, significant technological advances, relative to both the species that can be detected and spatial resolution, are necessary to satisfy the foreseeable needs of the *Global Tropospheric Chemistry Program*. A review of current spaceborne sensor capabilities is presented in Part II, Chapter 9. **We suggest that a study be undertaken to define the role of satellite measurements in the *Global Tropospheric Chemistry Program*.**

Regardless of the technology developed for the remote detection of tropospheric constituents, it will always be necessary to perform complementary point measurements using in situ sensors from surface-based or airborne platforms (ground truth). Thus it will be necessary to maintain in-depth research and development programs for both types of sensors.

[2] *A Strategy for Earth Sciences from Space in the 1980's and 1990's. Part II: Atmosphere and Interactions with the Solid Earth, Oceans, and Biota.* Committee on Earth Sciences, Space Science Board, National Academy Press, Washington, D.C., 1984 (in press).

INTERNATIONAL COOPERATION

The *Global Tropospheric Chemistry Program* will be international in scope. The research plans outlined on the preceding pages recognize the necessity for investigating these critical biogeochemical cycles throughout the many and varied physical and chemical regimes found in the global troposphere. To attain a detailed and comprehensive understanding of global tropospheric chemistry will require cooperative efforts by many nations in addition to the United States. The resources and commitment of the international scientific and political community will be vital to the success of the *Global Tropospheric Chemistry Program*. For this reason, **we recommend that the United States join in a cooperative international effort to commit these resources and that it do so with confidence that the international community is both ready and willing to join in this initiative.**

Many members of the international community of atmospheric chemists were contacted as this report was being developed, and many provided thoughtful and valuable comments. Copies of the report will be distributed widely abroad. It is clear that to achieve the maximum benefit from international cooperation in such a global-scale study, any plan of action must be discussed and developed in an open international forum. The program proposed in this document is only a start toward the development of a truly international effort. Although we have attempted to outline the major types of investigations necessary to achieve an in-depth understanding of global tropospheric chemistry, many additional studies will be required in the future.

We suggest that a forum for discussion of an international *Global Tropospheric Chemistry Program* could be provided for the international scientific community through the International Council of Scientific Unions (ICSU) and some of its member bodies. The three most appropriate organizations in ICSU would include these:

1. The International Union of Geodesy and Geophysics (IUGG) and its member organization, the International Association of Meteorology and Atmospheric Physics (IAMAP);
2. The International Union of Pure and Applied Chemistry (IUPAC); and
3. The Scientific Committee on Problems of the Environment (SCOPE).

We further suggest that one possible focal point for this forum within ICSU could be the IAMAP Commission on Atmospheric Chemistry and Global Pollution. Commission members are active atmospheric chemists from many nations who are concerned about global-scale problems.

A second possibility is to form an Interunion Commission with representation from IUGG (IAMAP), IUPAC, and SCOPE. Within IUPAC, the appropriate committee would probably be the Commission on Atmospheric Chemistry of the Applied Chemistry Division. SCOPE has wide-ranging interests and concerns relative to biogeochemical cycles in the atmosphere and other compartments of the environment. Any or all three of these groups could initiate discussions and formulate programs relative to the international science aspects of a *Global Tropospheric Chemistry Program*.

A third possible approach to the development of international cooperation in a *Global Tropospheric Chemistry Program* would be the formation of a special committee by the International Council of Scientific Unions with appropriate representation from the nations and scientific unions concerned with tropospheric chemistry research.

The strength and ultimate success of any international *Global Tropospheric Chemistry Program* will depend upon the quality of science and scientists involved. The best research groups in the world will be required. However, routine global observations of meteorological data will also be needed. Some of these meteorological observations can be obtained by the international scientific research groups involved in the *Global Tropospheric Chemistry Program*, and others can be obtained from currently existing observation stations. The capability to provide this type of data has been ably demonstrated by the member countries of the World Meteorological Organization. It is expected that maximum use of these facilities will be made in appropriate geographical regions by the various investigations undertaken as part of a *Global Tropospheric Chemistry Program*. Close cooperation between chemists and meteorologists will be a critical factor in the success of such a global-scale program.

4 Global Tropospheric Chemistry—A Call to Action

In the past 12 years, the National Research Council has issued at least 10 reports concerned wholly or partly with atmospheric chemistry. Most of these reports have focused on identified atmospheric environmental problems and ways to alleviate damage to the environment. For example, the issues arising from fossil fuel combustion, stratospheric ozone perturbations due to supersonic aircraft, and man-made chlorofluorocarbons and airborne particles and their effects have been studied and exposed in these reports. A recent report[1] on acid deposition in North America adds to this impressive body of literature. The latter report was commissioned and prepared in an attempt to discern, from available information, scientific conclusions that could lead to formulation of public policy.

By contrast and in response to the charge provided to our panel (see Preface to this report), the present report looks ahead to future research. We conclude that a global study of tropospheric chemistry is needed to provide answers to major questions about the chemistry of the world's atmosphere and the effects of this chemistry on the physical state of the atmosphere. This conclusion arose from an assessment of the current state of atmospheric chemistry knowledge (see Part II of this report). Further, we attempt to outline an overall scientific strategy to allow the identified objectives to be attained economically. Although the scientific strategy adopts the intellectual framework of geochemical and biogeochemical cycling of chemical elements, the proposed research program has a strong heuristic character. In many respects, the proposed research program is similar to programs envisioned earlier in National Aeronautics and Space Administration reports and in less formal discussions involving U.S. and European scientists and National Science Foundation staff.

The focus on the *global* troposphere is required scientifically. It does not preclude attention to existing questions of smaller-scale air pollution. Indeed, we believe that the knowledge to be gained from the research we propose will permit much sounder assessments of many pollution issues and eventually more effective protection and management of the world's natural resources.

The major observational elements of the proposed *Global Tropospheric Chemistry Program* are outlined in Figure 4.1. The four major field studies—biological sources, global distribution and long-range transport, photochemical transformation, and conversion and removal—are illustrated along with their component experiments. These field studies, combined with data on

[1]*Acid Deposition: Atmospheric Processes in Eastern North America, A Review of Current Scientific Understanding*, Committee on Atmospheric Transport and Chemical Transformation in Acid Precipitation, National Academy Press, Washington, D.C., 1983, 375 pp.

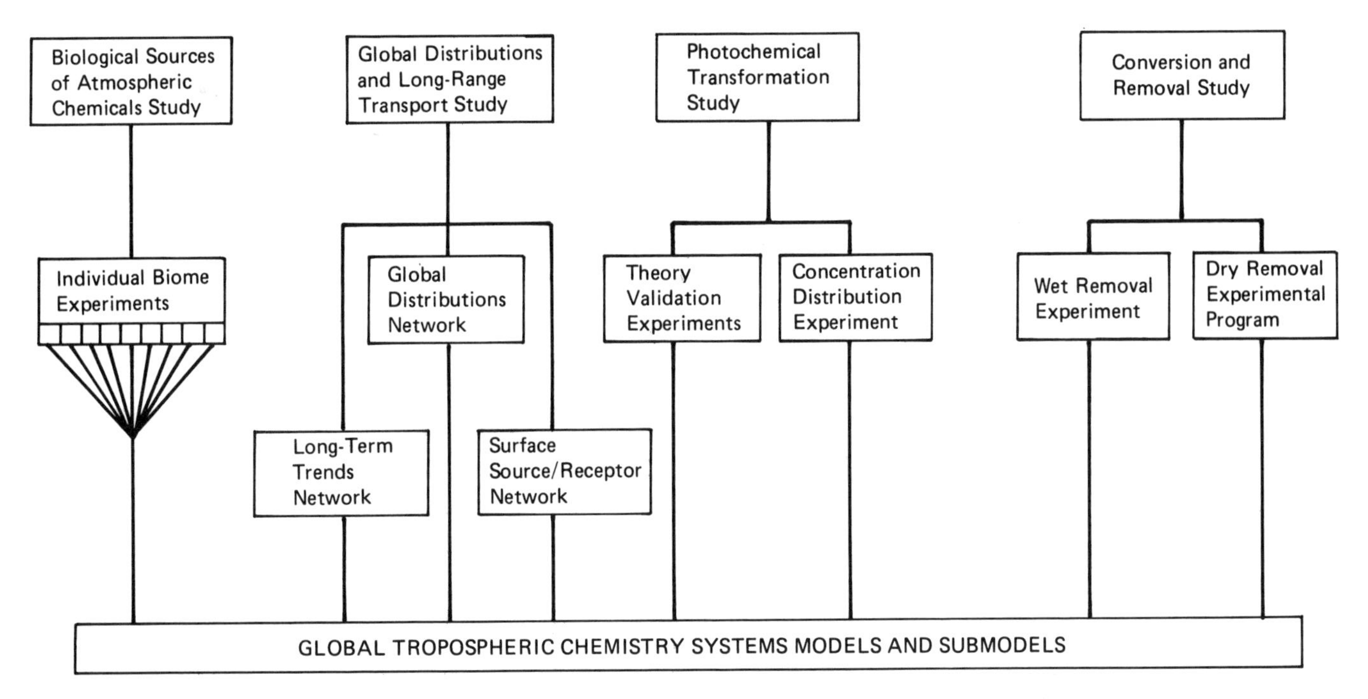

FIGURE 4.1 Major observational elements in the proposed Global Tropospheric Chemistry Program.

the rates and mechanisms of atmospheric chemical reactions determined in the laboratory, will provide data for the development of *Global Tropospheric Chemistry Systems Models*. These global TCSMs are necessary if we are to obtain both a comprehensive understanding of global tropospheric cycles and a predictive capability in the future.

Because the proposed program is one of basic research, the focus is on questions and concepts rather than on a detailed plan of execution. Accordingly, a detailed estimate of program cost is not attempted. It is clear, however, that incremental funding of several tens of millions of dollars per year for a decade or more would be required to support the proposed investigations. Finally, it should be emphasized that we have outlined only the U.S. national component of a necessarily international research program—for which there appears to be significant and growing support.

PART II

ASSESSMENTS OF CURRENT UNDERSTANDING

Part I of this report, "A Plan For Action," presents a coordinated national blueprint for scientific investigations of biogeochemical cycles in the global troposphere. Part II, "Assessments of Current Understanding," presents much of the background information from which the proposed *Global Tropospheric Chemistry Program* described in Part I was developed. Brief authored reviews are presented in Part II, and they evaluate both the present knowledge and gaps in the understanding of tropospheric chemical cycles and processes. Four review papers discuss the sources, transport, transformation, and removal of trace species in the troposphere. These are followed by a review of the role of modeling in understanding tropospheric chemical processes. A general review of tropospheric biogeochemical cycles is given followed by individual reviews of the water, ozone, nitrogen, sulfur, carbon, halogen, trace element, and aerosol particle cycles. Instrumentation development needs for use of mass-balance techniques is reviewed, and an instrument and platform survey completes the main part of Part II. Brief bibliographies are given at the end of each review paper. These bibliographies are not intended to be comprehensive but simply to serve as representative reference lists from which further information can be obtained for each topic.

Appendix A presents a brief review of current tropospheric chemistry research in the United States. In recognition of the future importance of spaceborne remote sensing techniques in tropospheric chemistry research, a brief review of this subject is presented in Appendix B. A matrix approach was used by our panel to systematically evaluate present knowledge of the primary species in several of the major elemental cycles in the troposphere. These matrices are presented in Appendix C.

5 Critical Processes Affecting the Distribution of Chemical Species

BIOLOGICAL AND SURFACE SOURCES

BY R. CICERONE, C. C. DELWICHE, R. HARRISS, AND R. DICKINSON

THE IMPORTANCE OF BIOLOGICAL SOURCES

In recent years we have seen increasingly clear manifestations of the important, even dominant, roles of biological processes as sources of atmospheric chemicals. While it has long been recognized that the sheer variety and adaptivity of the biosphere permit a wide range of phenomena, the quantitative size of biological influences on the atmosphere (examples below) has amazed many atmospheric chemists and climate experts. More biologically oriented scientists have long recognized the great potential of atmosphere-biosphere exchange processes; it has even been proposed that much of our chemical and physical environment is under biological control. For example, Lovelock has postulated the existence of an encompassing living feedback system through which the biosphere regulates the physical environment by and for itself as external stimuli change—the Gaia hypothesis.

There are both general and specific indications for the importance of biology in affecting or controlling the chemical composition of the atmosphere. Rough indications may be obtained from examining the state of disequilibrium of the atmosphere's chemical concentrations of gases that would exist if the earth's oceans and atmosphere were in perfect thermodynamic equilibrium (TE) or with only external inputs of solar, galactic, and electrical energy. With calculations such as these, Lovelock and Margulis found concentrations of N_2, O_2, CH_4, N_2O, NH_3, and CH_3I many orders of magnitude lower than in the actual atmosphere. Large departures from pure TE or an analogous abiological system in the atmosphere's composition are one indication of the influences of the biosphere. From a more empirical consideration, a comparison with Venus and Mars, planets with no life, suggests that there would be several orders of magnitude less O_2 on earth without life.

The importance of biological sources is indicated strongly by other empirical data and analyses. For example, the ^{14}C content of atmospheric CH_4 was measured to be about 80 percent of that in recent wood. These data, mostly from W. F. Libby, were gathered in 1949-1960, before nuclear explosive devices were tested widely in the atmosphere, and they show that at least 80 percent of atmospheric CH_4 then was derived from recent organic material and not from old carbon, e.g., primordial CH_4 or fossil fuels. Although the burning of biomass could also yield CH_4 high in ^{14}C, it is likely that ruminating animals and termites, rice paddies, and shallow inland waters are the principal contemporary CH_4 sources. Also, from a geological and geochemical analysis the mostly biological origin of atmospheric O_2 and the important biological role in controlling CO_2 levels can be deduced.

Other strong evidence exists for the importance of biological processes in the cycles of many elements whose volatile forms pass through and affect the atmosphere. In the nitrogen cycle, biological and industrial fixation of N_2 from the atmosphere leads to subsequent return of NH_3, N_2O, N_2, and possibly NO to the atmosphere. The importance of biogenic sulfur compounds to the atmosphere and to the global sulfur cycle was recognized long ago. Principal compounds of interest are dimethyl sulfide, dimethyl disulfide, H_2S, COS, and possibly others. Methylated materials of several other elements also appear to constitute globally important biological sources. These include CH_3Cl, CH_3I, CH_3Br, and several methyl-metal compounds, e.g., those of mercury, arsenic, and probably others. While biomass burning is a source of CH_3Cl, the oceans appear to be a more important source. The physical Cl^--I^- exchange reaction in seawater postulated in the early 1970s may not be compatible with recent measurements indicating the simultaneous existence of CH_3I and CH_3Cl in seawater. In any case, CH_3Cl is the only significant natural source of chlorine atoms to the stratosphere, and its origin, probably biogenic, requires study. Over land, much of the water transferred from the surface passes through vegetation.

Regardless of whether the biosphere acts as an integrated, almost purposeful, system in its release and regulation of atmospheric gases, or whether individual species and biomes act independently with inadvertent results, the biologically released materials are important in atmospheric chemistry and climate. Available data show that biogenic sulfur compounds can make appreciable contributions to sulfate deposition measured in certain regions, and probably to dry SO_2 deposition; it is important to define these contributions and regions more clearly. Similarly, field survey data and studies on individual plants show that the natural emissions of vegetation represent potentially significant hydrocarbon sources for the atmosphere. Often these are isoprene and terpenes (comprised largely of isoprene-like units). These biogenic hydrocarbons react photochemically and along with NO_x gases can lead to photochemical production of O_3. There is also evidence that the direct emission of natural hydrocarbons and the burning of biomass can generate significant amounts of organic acids in rainfall; tropical forest areas are especially interesting in this regard.

Processes and raw materials that produce and/or control O_3 concentrations in the background troposphere are not only central to tropospheric chemistry but are also of importance to climate. In the relatively clean, unpolluted troposphere, O_3 is a major source of reactive free radicals as well as a key reactant itself. Its climatic role arises from its 9.6-μm band absorption (in the atmospheric wavelength window); in the troposphere, this band is considerably pressure-broadened. In the photochemical control of tropospheric O_3, biogenic gases such as CH_4, CO, many hydrocarbons, N_2O, and possibly NH_3 are important players, N_2O through its stratospheric production of odd-nitrogen oxides that flow downward into the upper troposphere, and NH_3 as a possible NO_x source and as an NO_x sink.

From a climatic point of view, the potential warming effects of several biogenic and anthropogenic trace gases are startling. Upward trends in the concentrations of tropospheric O_3, CH_4, N_2O, CH_3CCl_3, CCl_2F_2, and CCl_3F are of particular concern, although other species also need attention. The effects of these gases will add to those of CO_2. For O_3, the extent of influence of biogenic input, summarized above, is a key question. For CH_4 and N_2O, biological sources are known to dominate the global budgets. For CH_4 and CH_3CCl_3, whose primary sinks are tropospheric OH reactions, any understanding and ability to predict future trends will require knowledge of the natural and human-controlled sources and of global patterns of OH and their controlling processes. For the fluorocarbons CCl_2F_2 and CCl_3F, it will be important to understand how tropospheric chemistry will respond to the stratospheric changes they cause, e.g., more ultraviolet radiation reaching the tropopause and higher O_3 concentrations in the lower stratosphere and upper troposphere.

Finally, there is another large role for biological processes in atmospheric chemistry, that of surface receptors for depositing gases and particles. There is evidence that vegetated continental areas are major sinks for O_3, HNO_3, SO_2, and possibly NH_3, for example. Further, the effectiveness of these surface sinks is largely controlled by dynamic responses of the involved plants. In certain regions and seasons, the deposition of atmospheric gases and particles delivers important nutrients and, at other places and times, pollutants and toxins for plant life and soils.

There has been an increasing awareness by the scientific community of the critical role played by biosphere-atmosphere interactions in biogeochemical cycles in general and in tropospheric chemical cycles in particular. A number of recent documents discuss these interactions and their implications in considerable detail. Examples include the National Research Council report *Atmosphere-Biosphere Interaction: Toward a Better Understanding of the Ecological Consequences of Fossil Fuel Combustion* (1981); the NASA Technical Memorandum 85629, *Global Biology Research Progress—Program Plan* (1983); and several publications by the Scientific Committee on Problems of the Environment (SCOPE), including *Some Perspectives of the Major Biogeochemical Cycles* (1981), and *The Major Biogeochemical Cycles and Their Interactions* (1983).

THE NATURE OF BIOLOGICAL SOURCES

Those atmospheric constituents to which biological sources are major contributors (nitrogen, oxygen, CO_2, N_2O, compounds of sulfur, halogens, and others) are products of the energy reactions of one or another organism. The process of photosynthesis and its reversal (respiration) are the most familiar expression of this fact, but the close similarity of processes yielding the other constituents is not always appreciated. An examination of the interrelationships between these biological oxidations and reductions from the viewpoint of their driving forces and modulating influences is in order.

First, although no one has yet offered a completely convincing explanation of the driving force behind the phenomenon called "life," there is much evidence of its potency. Virtually any chemical reaction that can take place in an aqueous system, that yields energy in excess of 40 kilojoules (kJ) per mole for two-electron transfer, and for which the required reactants have been available in reasonable abundance over the evolutionary period of the earth has been exploited. Aside from photosynthesis, no physical source of energy has been so utilized. The importance of this concept from the standpoint of the atmospheric chemist is that it implies that the present composition of the atmosphere is closely coupled to the biological system, and that any alteration by physical phenomena or human activity may be countered by a biological response giving the appearance of "resistance" to the change. As a single example from many possible, the annual fixation of CO_2 in photosynthesis (and the equivalent release of CO_2 in oxidative reactions) is about 10 percent of the atmospheric carbon pool. This speaks for an exceedingly tightly coupled cyclic exchange.

Any increase in atmospheric CO_2 levels would be expected to result in a countering increase in photosynthesis with the fixation of more CO_2. This is known to happen, but because there are so many other factors influencing photosynthetic production, the relationship is not linear.

Of the elements from biological sources appearing in the atmosphere, the ones most actively cycled are those having several oxidation states within the range of stability of water, and for which at least one reasonably stable gaseous form exists. Under the reducing conditions of anoxic environments (reducing because some biologically oxidizable compound is present and the influx of atmospheric oxygen is limited by a diffusion barrier or other means), compounds of these elements serve as "electron acceptors" for the biological oxidation of other, more reduced, compounds. In the process, gaseous compounds are released, some of them reaching the atmosphere.

Typical reactions are as follows:

1. Denitrification:

   a. $[HCHO] + 0.8\ NO_3^- + 0.8\ H^+ \rightarrow CO_2 + 1.4\ H_2O + 0.4\ N_2$

   or

   b. $[HCHO] + NO_3^- + H^+ \rightarrow CO_2 + 1.5\ H_2O + 0.5\ N_2O$

2. Sulfate reduction:

   $[HCHO] + 0.5\ SO_4^= + H^+ \rightarrow CO_2 + H_2O + 0.5\ H_2S$

3. Hydrogen production:

   $[HCHO] + H_2O \rightarrow CO_2 + 2\ H_2$

4. Methane production:

   $[HCHO] \rightarrow 0.5\ CO_2 + 0.5\ CH_4$

In all the above, [HCHO] represents carbohydrates, although the organic substrate can be any of a variety of materials. The reactions are simplified in their representation, with none of the intermediates shown. The point is that each reaction yields one or more volatile compounds to the atmosphere.

A cursory treatment of the subject, as given here, is sufficient to demonstrate the variety of reactions possible and to suggest some of the implications discussed below.

The operation of the various biomes contributing to the whole of this biological process is dynamic, influenced by all of the parameters that drive it. Most of these ecosystems (indeed, all from an absolute standpoint) are limited by one or more of their constituents. Available organic substrate, as we have seen, is a major limitation, but in most systems, one or more mineral elements required for life are also limiting. Compounds of nitrogen and sulfur are among the more notorious products of modern industry, and, although one tends to classify these emissions as "pollutants," they probably also are "fertilizers" for some species and in some locations. Unquestionably, human activities have altered the biosphere, but it is difficult to evaluate that alteration as quantitatively as desired.

Concentration of atmospheric constituents on an areal basis is part of the question. Immediately downwind of a point source, the concentration of an element or compound can be lethal for some organisms. On the basis of the considerations offered above, one can assume that any alteration of the concentration of a particular element can only result in an alteration of the associated biological population. The significance of this alteration is more difficult to interpret.

The excess of CO_2 injected into the atmosphere over the rate at which it can be sequestered by various carbon sinks is a good example of the complexities of this sort of question. Concern over the possible effects that increased atmospheric CO_2 levels may have on climate is tempered by the uncertainty regarding the consequences of these effects. It is not known what would have been "normal" secular trends in the absence of this excess CO_2. Intuitively, it is often assumed that any change is undesirable, but firmer information is required for planning purposes.

An interesting feature of the atmospheric carbon cycle is the role of CH_4. Available data indicate that there has been a significant increase in the concentration of CH_4 in the atmosphere since the industrial revolution, and ^{14}C data imply that 80 percent or more of atmospheric CH_4 was recently in living matter. The concentration increase of CO_2 is better documented. The fact that the increase in atmospheric concentration of CO_2 is less than would be expected on the basis of known processes for the removal of carbon from the atmosphere has led to some debate. The carbon of atmospheric CH_4 contains less ^{14}C than that of atmospheric CO_2. Biological sources are largely "modern," and as noted above, they are large. The quantity of fossil carbon from natural venting and fossil fuel burning does not appear to be sufficient to explain the deficit of ^{14}C in atmospheric CH_4. There is a discrimination against ^{14}C in the CH_4 formation reaction, but this can explain only part of the deficiency in ^{14}C.

Recent suggestions that large quantities of CH_4 are coming from magmatic sources could explain some of this "fossil" CH_4 but not all. Thus there are debates within the debates, all of them emphasizing the need for more information.

The biological production of CH_4 is a marginal thing from an energetic standpoint. The rather elegant exposure of details of the process by H. A. Barker in 1941 has proved to be even more complex. What was thought to be the conversion of ethanol to acetate, with a concomitant reduction of bicarbonate ion to CH_4, has turned out to be a coupling of reactions by two interdependent organisms, one oxidizing ethanol to acetate with the production of hydrogen, the other forming CH_4 from hydrogen and the bicarbonate. Although there are two separate organisms involved, the removal of hydrogen by the methanogen utilizing hydrogen is required to provide the energy gradient for life support of the hydrogen producer.

Because of the close constraints placed on energy yield (and therefore growth) by the concentration of hydrogen gas in reducing systems, anything that reduces hydrogen concentration will accelerate oxidation of available organic substrate. On the other hand, hydrogen consumption (and CH_4 production) depends on CO_2 concentration. Thus an increase in atmospheric CO_2 levels will correspondingly affect diffusion rates from CO_2 sources (anoxic environments) and could stimulate CH_4 production. This, in turn, could accelerate the oxidation of available carbon compounds (some of them fossil) and result in an increased production of CH_4, some of it deficient in ^{14}C relative to atmospheric CO_2. These energy relationships are shown below.

Fermentation of ethanol to acetate and hydrogen:

$$CH_3CH_2OH + H_2O \rightarrow CH_3COO^- + H^+ + 2\,H_2.$$

At pH 7 and with other reactant concentrations standard, this reaction yields only 5.3 kJ of energy, insufficient to support life. With the partial pressure of hydrogen at 1.0×10^{-3} atm, the energy yield is about 39.4 kJ, adequate for life support if properly coupled to synthetic reactions.

Oxidation of hydrogen with CO_2 as an electron acceptor:

$$4H_2 + CO_2 \rightarrow 2H_2O + CH_4.$$

The standard free energy for this reaction as written is about -140 kJ. By taking the 1×10^{-3} atm concentration of hydrogen suggested by the former reaction, a partial pressure for CO_2 of 3.2×10^{-4} atm, and a partial pressure for CH_4 of 1.6×10^{-6} atm, the energy yield becomes 85.0 kJ.

As written above, there are four molecules of hydrogen involved. The exact pathway of the reaction is not known, so it is not possible to identify the point at which energy is extracted. For each mole of water produced (a two-electron process), there is a yield of 42.5 kJ of energy, and it is difficult to visualize any other energy-coupling reactions. Because there is no room in the energy figures for increase in the CH_4/CO_2 ratio, the generalization probably is permitted that atmospheric CO_2 concentrations should influence biological CH_4 production. The close parallel in their rates of increase in the atmosphere may well be related to this interdependence of biological processes. The increase in CH_4 in the atmosphere would then be explained at least in part by the mobilization of organic matter in anoxic zones, some of which is "fossil" on the comparatively short time scale (thousands of years) of carbon half-life.

The data to test the significance of processes such as this are lacking, and the extent to which a concentration feedback such as this can explain present inconsistencies in the data is not known. The example does serve to demonstrate the complex interaction of biological and other factors in establishing atmospheric composition, and the challenge to unravel the processes at play.

GLOBALLY IMPORTANT BIOMES

Given the discussions above on the apparent importance of biological sources and on the nature of the biological processes that release materials to the atmosphere, it is now necessary to review some of the characteristics of world biomes and to formulate criteria for evaluating their potential importance. Before examining data on various distinct biomes, we present the following criteria that permit identification of biomes that require research relative to their potential role as significant sources of atmospheric chemical species:

1. Biomes covering large geographic areas;
2. Biomes with high gross primary productivity rates;
3. Biomes with fast cycling rates for nutrients;
4. Biomes with anoxic sites;
5. Biomes with fast rates of change of the local population compared to the time scale for natural succession (30 to 70 years);
6. Biomes where processes can trigger irreversible changes (e.g., desertification or climatic change);
7. Biomes of special importance to human life (e.g., agricultural areas);
8. Biomes with unique characteristics (due, for example, to toxicological, meteorological, or successional considerations);
9. Biomes or processes that are poorly understood and that satisfy some of the criteria above.

We will refer to these criteria frequently in the discussion that follows.

Schemes to classify world biomes vary somewhat depending on the purpose of the classification, need for detail, and other reasons. Many of the available data and compilations have grown from research on the global carbon cycle and from needs of individual researchers to extrapolate data from isolated, in situ measurements into regional or global estimates. A further complication results from the distinction between gross and net primary productivity of the biomes. The former is the rate of photosynthetic carbon fixation; the latter is this rate minus the rate of respiration, i.e., the rate of carbon storage. Table 5.1 presents one compilation of data on terrestrial biomes, their sizes, net primary productivities, and phytomasses. This particular compilation accounts for nearly all terrestrial surfaces, or about 30 percent of the total global surface. For our purposes at present, the most important entries in Table 5.1 are the geographical areas covered by the individual biomes.

To identify biomes of special interest in atmospheric chemistry, Table 5.2 lists about twenty specific, although informally classified, biomes and one process, biomass burning, as the rows of a matrix. The columns of this matrix are biogenic gases, individual species such as CH_4, and groups of species such as methylated metals (CH_3M) and organohalides (RX). A measure of the scientific interest in the emissions of the listed biogenic gases from the listed biomes is assigned, considering the criteria outlined above and the available data. An "X" indicates reason to expect significant emissions, and a circled "X" indicates strong reason (or directly applicable, available data) to expect a particularly significant biome-emission relationship. In the remainder of this section, we focus attention on several biomes and processes that are potentially significant as sources for tropospheric chemical species.

Tundra and Other Northern Environments

The "tundra" biome and the boreal forest at a lower latitude cover about 14 percent of the land area of the globe, most of it in the northern hemisphere. Total photosynthetic productivity of this area, although less than many environments on an area basis, still is large (an estimated 10 percent of all land area, based on the figures of Whitaker). Underlain by permafrost, much of it poorly drained, this area could be a large contributor to the reduced compounds delivered to the atmosphere by the biosphere (CH_4, reduced sulfur compounds, and the products of denitrification).

Because of its secondary economic interest and its inaccessibility, this area has not been studied intensively, and its significance in the budget of atmospheric constituents is poorly known.

A research program to obtain needed information on fluxes from tundra regions should be flexible, starting with exploratory studies. The results of these preliminary investigations will then guide further program development. Because of the two-phased nature of the tundra research, the exploratory studies should be initiated as early as possible on a modest scale, with the extent and nature of future studies left flexible. Aside from the physical (logistic) problems involved in investigations of this environment, there are geopolitical constraints. Initial studies can be performed in North America and in the Scandinavian countries, but because of the large area involved on the Eurasian continent, cooperative participation by Soviet scientists should be sought.

Because these frequently water-logged environments are expected to yield significant quantities of reduced carbon, nitrogen, sulfur, and in some cases, halides, the species of interest will be the products of denitrification and sulfate reduction: CH_4, NH_3, halides (in the vicinity of the ocean), various hydrocarbons, and other reduced carbon species in forested areas. Sulfur may be a limiting nutrient in many of these areas, so the sulfur components may be low or absent except within the range of ocean spray delivery. Analysis of precipitation may be desirable in later phases, but because of costs and logistic problems, precipitation sampling should not be attempted during the first 2 years except where facilities of cooperating institutions provide the opportunity.

Initial investigations at three sites are proposed. Site selection is based upon a compromise of factors including the likelihood that these sites will yield information representative of significantly large areas, factors of cost and logistics, and the probable availability of cooperating individuals and institutions. The suggested sites include:

TABLE 5.1 Surface Areas, Net Primary Productivity, and Phytomass of Terrestrial Ecosystems of the Biosphere[a]

Ecosystem Type	Surface Area (10^{12} m^2)	NPP DM (g^2/yr)	Total Production DM (10^{15} g)	Living Phytomass DM (10^3 g/m^2)
1. Forests	31.3		48.68	
Tropical humid	10	2300	23	42
Tropical seasonal	4.5	1600	7.2	25
Mangrove	0.3	1000	0.3	30
Temperate evergreen/coniferous	3	1500	4.5	30
Temperate deciduous/mixed	3	1300	3.9	28
Boreal coniferous (closed)	6.5	850	5.53	25
Boreal coniferous (open)	2.5	650	1.63	17
Forest plantations	1.5	1750	2.62	20
2. Temperate woodlands (various)	2	1500	3	18
3. Chaparral, maquis, brushland	2.5	800	2	7
4. Savanna	22.5		39.35	
Low tree/shrub savanna	6	2100	12.6	7.5
Grass-dominated savanna	6	2300	13.8	2.2
Dry savanna thorn forest	3.5	1300	4.55	15
Dry thorn shrubs	7	1200	8.4	5
5. Temperate grasslands	12.5		9.75	
Temperate moist grassland	5	1200	6	2.1
Temperate dry grassland	7.5	500	3.75	1.3
6. Tundra arctic/alpine	9.5		2.12	
Polar desert	1.5	25	0.04	0.15
High arctic/alpine	3.6	150	0.54	0.75
Low arctic/alpine	4.4	350	1.54	2.3
7. Desert and semidesert shrub	21		3	
Scrub dominated	9	200	1.8	1.1
Irreversible degraded	12	100	1.2	0.55
8. Extreme deserts	9		0.13	
Sandy hot and dry	8	10	0.08	0.06
Sandy cold and dry	1	50	0.05	0.3
9. Perpetual ice	15.5	0	0	0
10. Lakes and streams	2	400	0.8	0.02
11. Swamps and marshes	2		7.25	
Temperate	0.5	2500	1.25	7.5
Tropical	1.5	4000	6	15
12. Bogs, unexploited peatlands	1.5	1000	1.5	5
13. Cultivated land	16		15.05	
Temperate annuals	6	1200	7.2	0.1
Temperate perennials	0.5	1500	0.75	5
Tropical annuals	9	700	6.3	0.06
Tropical perennials	0.5	1600	0.8	6
14. Human area	2[b]	500	0.4	4
Total	149.3	895	133.0	3.75

[a]Annual average values.
[b]Of which 40 percent (or 0.8×10^{12} m^2) productive.

SOURCE: Adapted from Ajtay et al. (1979).

TABLE 5.2 Twenty-two Biomes, Sites, and Processes and Twelve Gaseous Species or Groups

	NO_x	RCN	NH_3	N_2O	CH_4	CO	RS	H_2S	COS	RX	CH_3M	NMHC
Sterile ocean				X	X					X	X	
Productive ocean	X		X	(X)	X	(X)	(X)		(X)	X	X	
Tropic wet	X			(X)	X	X	X	X	X		X	(X)
Tropic dry				X	(X)	X					X	
Desert 1												
Desert 2												
Desert 3 productive			X	X		X						
Wet subarctic	X	X	X	X	X	X	X	X	X		X	
Dry subarctic						X						X
Tundra	X	X		X	X	X	X	(X)	X	X	X	
Tropic agriculture			X	X	X	X					X	
Temperate agriculture	X	X	(X)	(X)	X	X					X	
Rice agriculture			(X)	(X)	(X)	X	X	X	X		(X)	
Temperate evergreen			X			X						(X)
Temperate mixed			X			X						X
Temperate grassland			X	X	1(X)	X					X	
Wetlands	X		(X)	X	(X)	X	(X)	(X)	(X)		X	
Inland waters			X	X		X						
Sewage sources	X		(X)	X	X	X	X	X	X		X	
Feedlots	X		(X)	X	X		X	X	X		X	
Coastal shelf			X	X	X	(X)	(X)	(X)	(X)	(X)	X	
Biomass burning	(X)	X	X	X	X	(X)			X	X		(X)

NOTES: An "X" indicates that there is some reason to expect a significant source; a circled X indicates especially strong interest or evidence. The symbol "R" represents an organic group, RX means a methyl halide, and CH_3M means a methylated metal. "1" includes termites and ruminants.

1. Alaskan arctic seaboard, vicinity of Point Barrow or Prudhoe Bay;
2. Interior Alaska, vicinity of Fairbanks;
3. Hudson Bay area, vicinity of Churchill.

Species to be measured in the initial studies should include CH_4, N_2O, H_2S, dimethyl sulfide [$(CH_3)_2S$], CO, and volatile halides. In later studies, such species as NH_3 and volatile metals (e.g., mercury, arsenic, and selenium) should be measured.

Ideally, gradient measurements should be made for flux determinations. During this initial phase, portable equipment with sufficient sensitivity will probably not be available, so bulk samples should be collected for analysis at cooperating laboratories.

Samples of opportunity should be collected, preferably during the spring ice breakup and during midsummer. Where possible, samples should be collected on a regular schedule throughout the year.

Temperate Forests

Observations on temperate forests will be of greatest value if accomplished at sites where ongoing research and monitoring programs will provide supportive information. A number of these are available within the contiguous United States. They provide representative sites for the forest types of interest.

Forest Type	Possible Sites
Pacific Northwest coniferous	Oregon State—collaboration with OSU School of Forestry
Sierran mixed conifer	Sequoia National Park (California)—collaboration with UCSB and UCB scientists
Southern coniferous	Tennessee—collaboration with ORNL scientists
Mixed hardwood	Hubbard Brook (New Hampshire)

Analytical Protocol

This portion of the study could be done at different levels of intensity, but for the information needed, a rather elaborate and detailed (including micrometeorological information) approach is most desirable. This approach would provide boundary layer gradient information on volatiles of interest, which, in turn, would make possible the estimation of emission and absorption rates. Instruments of the resolution and sensitivity needed for such a study exist, but they have not been applied specifically to any study of this sort. Initial application will emphasize the development of appropriate procedures and the calibration of instruments and procedures, possibly at only one or two of the possible sites. As procedures are refined and important data are obtained, the direction and intensity of program development will become known.

Both laboratory and field studies of processes are desirable, and they involve both the canopy trapping and the gradient measurements cited above. Procedures for canopy trapping and bench analysis are at a reasonable state of development.

Molecular Species of Interest

NH_3, CO, and NMHCs (including terpenes and other hydrocarbons) are of most interest. N_2O, NO_x, CH_4, and reduced sulfur compounds probably are not present in significant quantities, but should be examined.

Tropical Areas (Forests and Savannas)

Tropical continental areas, both wet forested regions and drier sites such as savannas, are probably major sources of a variety of trace gases important in tropospheric chemistry. The potential importance of these regions stems from their geographical size, biological productivity, anoxic environments (in wet areas), and high turnover rates (largely by insects in dry areas). According to the criteria we have adopted in identifying biomes of potential importance, measurements of biological emissions from both tropical wet and dry areas deserve a high priority.

In tropical forests, investigations are needed on the fluxes of various nonmethane hydrocarbons, CH_4, CO, N_2O, and volatile sulfur-containing gases. Volatile species containing metals (probably methylated) should also be sought, and the potential for regional CO production from hydrocarbon oxidation must be explored through measurements and photochemical modeling. A variety of specialized approaches must be utilized: airborne instruments must be deployed to determine the concentrations and fluxes of these gases above the forest canopy, sampling of emitted species from individual trees and plants is required, and airborne studies of the photochemical and cloud-mediated transformations in the forest plume must be undertaken. The latter studies would address questions related to the production and destruction of photochemical oxidants, e.g., O_3, peroxides, NO_x, and CO. Similarly, investigations focused on cloud-water chemistry in these regions will result in a more complete understanding of the origins of a variety of organics and acids in precipitation collected in heavily forested areas remote from populated or industrialized centers. Because biological systems such as forests can act as sinks as well as sources for trace species in the troposphere, a final recommendation is for measurements of deposition of these species to tropical forests—particularly of potential nutrients such as NH_3 or ammonium ion (NH_4^+), NO_x or wet nitrate ion (NO_3^-), SO_2 or wet sulfate ion ($SO_4^=$), O_3, CO_2 (to deduce exchange rates), and trace metals.

The measurement of chemical fluxes from and to the tropical forests will be difficult. Base facilities must be established and local scientists with similar interests must be involved. Before a large, coordinated expedition is undertaken in the tropics, methods for measuring concentrations and fluxes should be tested in more accessible forests, for example, in North America. Because qualitatively different emissions are likely to distinguish tropical from temperate forests, investigations in both regions will be required.

Dry tropical areas also display high cycling rates for nutrients. Although they store less material in their shrubs and grasses than is found in the wood of tropical forests, the biological material of savannas has a shorter lifetime and has higher nitrogen/carbon and sulfur/carbon ratios than hardwood. From the rapid turnover rates, the chemical composition of the material, and the present data that suggest a large role for herbivorous insects, we conclude that significant volatile emissions are likely from certain dry tropical areas. In particular, there is potential for large emissions of CH_4, CO, CO_2, nonmethane hydrocarbons, N_2O, and possibly methylated metals. Initial measurements should focus on sites and processes that concentrate nutrients (termite colonies, for example), but the large land areas involved leave room for significant emissions from lower intensity sources distributed over large areas.

The most difficult task will be to estimate total emission rates from tropical forests. The general methods mentioned above and the use of meteorological towers are envisioned. About six full-time scientists and technicians would be needed. Measurement of chemical deposition to the tropical forests would require sustained observations over a period of at least several months, both near the top of the forest canopy and at ground level for useful interpretation. In the dry tropical areas, wet versus dry season differences may be marked. Two separate seasonal investigations or one longer investigation stretching through the wet and dry seasons would be required. Four full-time scientists would be needed for this study.

Tropical Areas

Two distinct kinds of tropical biomes are identified for concentrated research on biological sources important in tropospheric chemistry. These will be termed "wet tropics" and "dry tropics." The potential importance of each of these environments and relevant investigations in each are outlined below. Separate discussions are presented for biomass burning and rice agriculture below.

Wet Tropical Areas

Wet tropical areas are usually covered by tropical forests. These are evergreen or partly evergreen, they are frostfree and have average temperatures of 24°C or higher and rainfall rates of 100 mm or more per month for 2 of every 3 years. About 1.6×10^7 km^2 of the earth's surface enjoys this climatic range, and (0.9 to 1.1) $\times 10^7$ km^2 is actually covered by tropical forests now. These tropical forests contain about 60 percent of the global biomass, but 1 to 3 percent of it is permanently deforested annually, usually for agriculture, timber harvests, cattle grazing, and firewood. There are perhaps several million species of life in these forests, but only about 500,000 have been described. Barring irreversible climatic changes, it is possible that similar floristic species could be regenerated in perhaps 50 years after cutting. Approximately 80 percent of this biome occurs in only nine nations, five in South America and four in Southeast Asia. These forests cycle nutrients rapidly through microrhizial (root) systems, and there are few inorganic reserves in their highly leached soils. Other indicators of the potential importance of tropical forests include their high net primary productivity, their potential trigger effects of large changes, scientists' relative ignorance about them, and the existence of many anoxic sites. Several climatic effects could be triggered by changes in tropical forests because they mediate regional hydrologic cycles. Cleared areas are more susceptible to prolonged droughts and to much more erosion and flooding in wet periods than forested areas. Also, these forests store CO_2 and present a darker surface to sunlight than cleared areas.

Potentially large emissions of gaseous hydrocarbons, N_2O, CH_4, reduced sulfur compounds, and possibly methylated metals and CO (see below) can be released from tropical forests. This potential arises from the prevalence of anoxic sites in saturated soils and, during nocturnal hours, in shallow surface waters. The largest potential source of CO could be the oxidation of biogenic hydrocarbons like isoprene (C_5H_8), although several steps in the gas-phase oxidation of this and other hydrocarbons are not clear at this time. Effects of these emissions on the acidity of regional watersheds and rainfall, principally through oxidation of organics to formic and acetic acids, are likely and need investigation.

Dry Tropical Areas

In dry tropical areas there are likely to be globally important biological sources of key gases. For example, in tropical savannas there is generally high net primary productivity although little of the fixed carbon enters into long-lived tissue (wood). The indigenous grasses and shrubs are short-lived and are recycled by animals, largely insects. The chief species among these are termites and ants; in some areas, termite mass densities per unit area exceed the highest mass densities of grazing animals anywhere. The potential for large emissions of CH_4, CO_2, CO, and perhaps other species from termites, whose digestion is fermentative, is quite large. Also, more speculatively, emissions of N_2O and NO_x gases are possible because nitrogen-fixing bacteria are known to live in termite guts. The roles of these creatures and of the relatively unexciting but extensive dry tropical areas in nature's atmospheric chemistry are ripe for investigation.

Coastal Marsh, Estuary, and Continental Shelf Environments

Coastal ecosystems are characterized by high biological productivity, active chemical and physical exchange, and transport driven by freshwater runoff, tidal forces, and wind-driven circulation. Existing data on emission of reduced sulfur gases, CH_4, and N_2O indicate that specific habitats in the coastal environment may be intense sources. Thus, though the areal extent of these sources may be relatively small, their contributions to the atmospheric budgets of certain reduced gases may be considerable.

An additional consideration for placing priority on these environments is that higher fluxes and concentrations of many chemical species of interest place less demand on instruments (detection limits, response time, etc.). For gases such as N_2O, CH_4, CO, COS, CS_2, $(CH_3)_2S$, SO_2, and a few others, it is reasonable to propose initiating immediately a field research program emphasizing basic processes of gas production from these source areas and their exchange with the atmosphere. For more reactive chemical species such as NH_3, NO, and volatile metals, existing technology is probably inadequate for quantitative biosphere-atmosphere exchange studies even in intense source areas. For almost all reduced gas species, measurement technology is currently inadequate for studying very low-level sources and sinks.

The primary objective of these studies is to develop an improved quantitative understanding of the processes that control the production and consumption of biogenic gases in coastal wetland and aquatic environments and their exchange with the lower troposphere. When this information is available, it should be possible to extrapolate to regional and global flux estimates with supporting data from remote sensing and other geographical and meteorological data bases.

The proposed studies will require measurement of a wide range of biological, physical, and meteorological variables at a selected set of coastal habitats. Five specific habitats of importance are (1) salt marshes, (2) mangrove swamps, (3) sea grass, (4) kelp, and (5) exposed mud and soil surfaces. Variables that can influence trace gas production and consumption in soils and sediments and their exchange with the atmosphere include biological community composition and dynamics, nutrient quantities, inputs of energy from different sources, removal of wastes by tidal forces and river runoff, temperature, salinity, pH-Eh, sediment physical properties, and wind stress on water and soil surfaces.

Previous studies have demonstrated high variability in time and space. A quasi-continuous research program at each coastal habitat of interest will be required to accomplish the objectives of this experiment. One efficient way of conducting studies on biogenic gas sources in coastal environments would be to use NSF-LTER (Long-Term Ecological Research) sites for the proposed measurement program. Ongoing activities at these sites would provide supporting biological, meteorological, and geochemical data required for the comprehensive gas production and exchange studies proposed here.

Point Sources

There is a class of potentially important biological and surface sources that are either intense and spatially small, or qualitatively distinctive in their emissions. Examples include lightning, industrial emissions such as combustion or waste plumes, volcanoes, and animal feedlots. A further source is biomass burning, an activity that is widespread in certain regions at sites whose exact locations vary from year to year.

Biomass Burning

Although not a biome but a process, biomass burning is included in the present discussion because of its great potential importance and distinctive characteristics. Qualitatively, biomass burning may be regarded as a type of nonindustrial pollution. Many types of biomass burning combine to yield a large total of biomass burned annually. For example, biomass burning is used to clear tropical forests for agriculture, to prepare forested areas for settlements, and to dispose of agricultural wastes (e.g., sugar cane). Large quantities of biomass are burned as fuel in industries, for individual human needs, and in wildfires. Recent estimates of the annual global area involved in biomass burning range from 3 to 7×10^6 km^2, with estimates of the total biomass burned ranging from 4400 to 7000 Tg/yr. Although total biomass burning quantities are probably uncertain to within a factor of 2 or 3 and vary from year to year, they are almost certainly significant in the global atmospheric carbon cycle and probably in other cycles as well, e.g., oxygen and nitrogen. Ecologically, pronounced changes accompany deforestation through biomass burning, e.g., in flora, soil structure, and surface hydrology. Much biomass burning for deforesting occurs in areas that are not well characterized ecologically. Surface albedo values, surface winds, and turbulence are also affected. Certain types of biomass burning also produce charcoal, thus effectively constituting a CO_2 sink.

From a physical and chemical point of view, biomass burning is a high-temperature process that is dramatic both in quality and in quantity. Large amounts of material are transformed, mobilized, and volatilized quickly. Partially combusted particles become airborne, and a wide spectrum of gases are produced in the flames and through the process of smoldering. Gases containing carbon, hydrogen, oxygen, nitrogen, sulfur, halogens, phosphorus, and trace metals are involved. Many of the gaseous species so produced are highly reactive photochemically, but stable gases like CO_2, CH_4, N_2O, and CH_3Cl are also generated. The photochemically active species are known to give rise to rapid O_3 production and probably yield other photochemical smoglike species, e.g., peroxyacetyl nitrate ($CH_3COO_2NO_2$ (or PAN)). Not surprisingly, a number of carcinogenic substances are also produced in biomass burning; gases include benzene (C_6H_6) and toluene (C_7H_8), and related airborne solids are certain to be found. Further, it now appears that several oxygenated hydrocarbons from biomass burning yield organic acids in sufficient quantity to acidify precipitation and groundwaters regionally.

There is a great deal of fundamental research to be done on the atmospheric chemistry effects of biomass burning. The full spectrum of compounds injected into the atmosphere in this way needs description. Quantitative production rates of CH_4, C_2H_6, and other alkanes, N_2O, CH_3Cl, aldehydes, ketones, nitrogen oxides, NH_3, HCN, CH_3CN, oxides of sulfur, CO, CO_2, H_2, and several other species must be determined. Methods need to be developed to quantify the various production and atmospheric injection rates; simply ratioing each species to CO_2 and deducing CO_2 emission rates might not suffice. Fortunately, some existing data suggest that the relative yields of some key gases (alkanes, alkenes, aldehydes, ketones) in temperate and tropical forest fires are not terribly different, so we can reasonably propose to concentrate initial research in temperate latitude areas where logistics are less of a concern. For example, there are controlled (or prescribed) forest fires managed by experts in Georgia and Oregon that might be suitable for several studies. Obviously, certain distinctive features of tropical forests (e.g., smoldering in damp areas), tropical agricultural products (e.g., sugar), and tropical grasses (e.g., the high cyanide content of sorghum) will eventually require targeted research expeditions. Global considerations such as for long-lived gases (CH_4, CO_2, N_2O, CH_3Cl, and COS) will also demand that the size of areas (biomass) being burned annually be quantified region by region.

Lightning

The most dramatic types of atmospheric electrical discharge undoubtedly produce certain gases and destroy others in their intense pulses of thermal and optical energy. Spectroscopic and chemical analyses have detected many interesting products of lightning. The single most pressing question today concerns the amount of fixed nitrogen (as NO) that is produced annually by atmospheric electrical discharges. Estimates of this quantity vary widely; they cover the range from major, i.e., comparable to combustion production of NO_x, to smaller but significant fractions of this quantity. Key uncertainties are whether laboratory discharges simulate the full range of atmospheric electrical phenomena adequately and the actual frequency of atmospheric discharges. Microdischarges near pointed biological surfaces (e.g., pine needles) might also have important consequences.

Volcanoes

By mass fraction, the principal emissions from volcanoes are H_2O and CO_2. A variety of other gaseous and particulate substances is also emitted in quantities that are potentially significant to the regional and global atmosphere. These include SO_2, several volatile heavy metals, and ash particles. Through large explosive events, volcanic inputs and impacts can be enormous if short-lived. By their nature, these effects defy prediction, and once they occur, they defy averaging—that is, it is not easy or particularly meaningful to compare volcanic emissions to annual averages from other sources.

Animal Feedlots

Volatile losses of fixed nitrogen (principally as NH_3, R-NH_2, and possibly N_2O and NO) and organic sulfur compounds from animal feedlots are appreciable, at least when expressed as a fraction of the feedlot's animal waste and possibly in absolute terms. In the United States, perhaps 40 percent of all fertilizer nitrogen is consumed by cattle, and the average length of time that cattle spend in feedlots or other concentrated populations is high. It is likely that regional impacts of these emissions are significant, especially for levels of gaseous NH_3, particulate NH_4^+, and NH_4^+ in precipitation. If so, the atmospheric transport of NH_3 represents an airborne fertilizer distribution system.

Industrial Emissions

This group includes combustion products, wastes from chemical production processes, mineral, gas and oil exploration and refining, burning or processing of wastes, and losses of solvents. As sources of atmospheric gases and particles, these processes can be distinctive both in kind and size, for example, by emitting unnatural substances or natural ones in quantities that are somehow comparable to natural cycling rates. For several clearly important gases like NO_x species and SO_2, emission inventories are available for most industrialized countries, and some of these have been prepared with good spatial resolution (100 km $\times$ 100 km). Changing industrial intensities, processes, and practices dictate that these emission inventories will require updating.

Rice Agriculture

The potential importance of rice paddies as sources of atmospheric chemicals might not seem obvious. Several lines of objective reasoning and preliminary field data combine to argue strongly in this direction, however. First, a number of our criteria (expressed above) indicate that rice agriculture represents a globally significant biome. These include physical area, reasonably high primary productivity rates, and the anoxic character of rice paddy soils. As a principal staple in world food diets, rice is extremely important in world agriculture. Accordingly, large areas are cultivated—1.3×10^6 km^2 in the late 1960s and 1.45×10^6 km^2 in 1979. Further, in many countries new emphasis has been placed on multiple cropping through improved irrigation. Two crops per year is becoming the norm in tropical areas. One practical result of this is that rice paddy soils are underwater for perhaps 8 months instead of 4 months annually, and the more negative redox potential of water-covered soils allows more reduced gases like CH_4, NH_3, and N_2O to form. Indeed, the rice paddy environment is ideal for the evolution of a number of volatile species containing nutrient elements. The soils are oxygen-poor (strongly reducing), and they are nutrient-rich, often through fertilization.

Direct indications of the impact of rice paddies on the global atmosphere have centered on CH_4. In the early 1960s, a Japanese scientist showed that paddy soils, when cultured in the laboratory, released copious quantities of CH_4, especially at the elevated temperatures that characterize midsummer temperate-latitude and tropical soils. Extrapolations of these results implied that rice paddies were responsible for perhaps one-third of all atmospheric CH_4. Later field studies of CH_4 emission rates continue to suggest the potential global importance of rice agriculture, but they find several complicating factors that require attention. For example, the emitted flux of CH_4 depends on the amount of nitrogen fertilizer (and possibly on the type of fertilizer) used. The principal mode of CH_4 emission is direct transmission from rice roots upward through the (hollow) plant, and less by diffusive or bubble transport across the water-air interface (although all three modes are active). Relatively more CH_4 is released late in the growing season. Other factors are doubtless involved, e.g., soil organic content. Reliable global source estimates must recognize these complexities.

Volatile nitrogen emissions are also thought to be products of rice agriculture. This is mostly due to evidence that rice utilizes fertilizer nitrogen relatively inefficiently, 50 percent or less. On the basis of observations of low leaching rates, one concludes that nitrogen is lost in large amounts, probably as NH_3 and/or N_2O, because of the reducing nature of waterlogged, water-covered paddy soils. Nitric oxide emissions would not be inconceivable. Direct field studies are needed to determine the principal nitrogen-bearing volatile species. If it is NH_3, the effects would be regional; if emitted as N_2O, then a global impact is possible. Once again, several complexities must be recognized: variations in fertilizer types, application protocols, soil types, and so on. While preliminary data suggest that ethane (C_2H_6) emissions from rice plants are not significant, those of isoprene (C_5H_8) might be. Methylated metals could be released easily by rice, and volatile phosphorous compounds deserve some attention as well.

Productive Oceans

The oceans are a large-area, low-intensity source of reduced sulfur compounds to the lower troposphere. Preliminary studies have identified a number of compounds in seawater that are presumed to be of biogenic origin including H_2S, COS, CS_2, dimethyl sulfide (DMS), dimethyl disulfide (DMDS), and dimethylsulfoxide (DMSO). Dimethyl sulfide appears to be one of the most abundant species, contributing an estimated flux of 34 to 56 TgS/yr to the marine boundary layer. Once in the marine boundary layer, DMS is probably oxidized by photochemical processes to produce SO_2 with intermediates such as DMSO and methane sulfonic acid. Qualitatively, the concentration of reduced sulfur compounds in surface seawater correlates with indicators of algal biomass. However, in one study no relationship was found between DMS in surface water and in the overlying atmosphere. A research effort to investigate the sulfur cycle in productive areas of the ocean should be initiated to elucidate sources of reduced sulfur in the ocean and their role in the global sulfur cycle and budget.

A program to investigate the sulfur cycle in productive areas of the open oceans should include a multidisciplinary team to study in situ biogenic production of sulfur species in the water column, fluxes across the sea-air interface, and chemical processes in the marine boundary layer determining transport and fate. Most of the research would be conducted from a major research vessel dedicated to the project, with simultaneous aircraft overflights to obtain information on the vertical distribution of sulfur species in the troposphere and rates of exchange between the boundary layer and free troposphere. Initial sites for these studies might include the ocean area off the west coast of the United States, continental shelf waters off the southeastern United States, and a major upwelling area.

ISSUES OF CLOSURE

Preceding discussions on biological and surface sources and following sections of this report on selected tropospheric chemical cycles illustrate the complexity of quantifying the role of biological processes and other surface sources in global atmospheric chemistry. Sufficient data are available on temporal and spatial variability in emissions of biogenic gases such as N_2O, CH_4, and $(CH_3)_2S$ to raise the issue of limits of understanding and predictability. For example, although it is established that emissions from soils are a significant source of global tropospheric N_2O, measured emission rates vary by at least a factor of 400. Nitrous oxide is produced during both denitrification and nitrification processes with emissions to the atmosphere influenced by a wide range of biochemical, soil physiochemical, climatic, and land-use variables. Emissions of N_2O may be episodic, with significant variations on time scales as short as hours and space scales of meters (e.g., see Figure 5.1). Similar ranges of variability may be typical of CH_4 emissions from diverse biological sources such as rice paddies, natural wetlands, and termite mounds (e.g., see Figure 5.2). The complexity and variability of emissions of biogenic gases from soil derive from nonlinear interactions of the wide range of environmental factors that control the metabolism of the source organisms and the physical process of gas exchange between the soil and atmosphere. Biogenic gas emissions from aquatic environments (e.g., $(CH_3)_2S$ emissions from the ocean) are

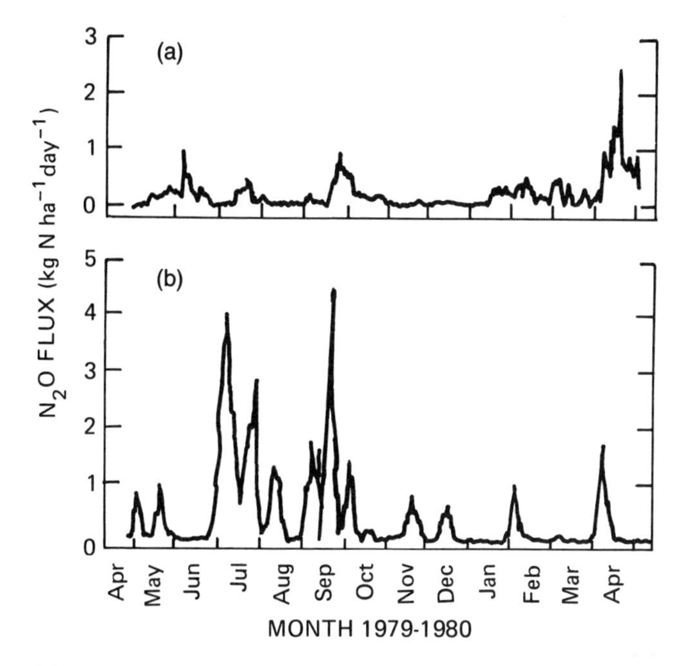

FIGURE 5.1 Emissions of nitrous oxide from (a) corn field and (b) fallow field. Note the episodic character of the emissions (data from Duxbury et al., 1982).

less well studied, but can be expected to be equally variable in time and space. The problem of "patchiness" in the distribution of marine bacteria and phytoplankton, which are two groups thought to be most involved in the production of reduced sulfur compounds in seawater, is a long-standing one in aquatic ecology.

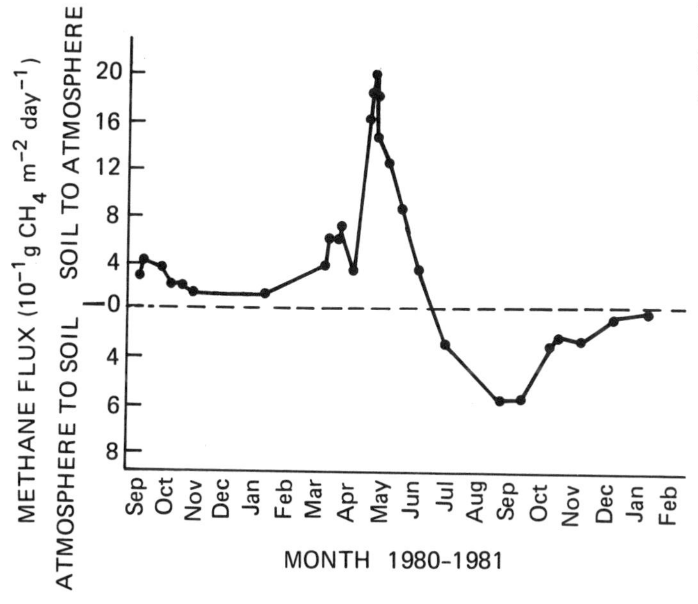

FIGURE 5.2(a) Methane emissions from organic soils in the Great Dismal Swamp, Virginia (data from Harriss et al., 1982).

Most research to date on surface source strengths has focused on "hot spots" (i.e., CH_4 emissions fron anoxic sediments, mineral aerosols derived from intense dust storms). Inadequacies in flux measurement technology and techniques have limited studies of low-intensity, large-area sources (e.g., CH_4 flux from slightly supersaturated open ocean surface layers and NH_3 flux from nonagricultural soils). Recent developments in high-sensitivity, fast response detectors, aircraft micrometeorological measurement capabilities, and ground-based flux chamber design offer new opportunities for studies of sources of atmospheric O_3, N_2O, CO, and CH_4. For most C, N, and S species, accurate flux measurements are not currently possible.

To achieve the goal of a better understanding of global tropospheric chemistry requires that both the magnitude of surface sources at their origin and the subsequent transport from the planetary boundary layer to the free troposphere be determined. For certain reactive reduced gas species, removal processes in the boundary layer may restrict the significant influence of a particular species to regional or hemispheric scales (e.g., H_2S, CO, and C_5H_8). Processes driving exchange from the boundary layer to the free troposphere are largely decoupled from processes controlling emissions from the surface; thus an experimental design to quantify fluxes of gas or aerosol species from surface sources to the free troposphere must take into account a very complex set of potentially nonlinear, episodic variables.

Several problems raise the issues of the limits of predictability and of scientists' ability to extrapolate from field measurements. These problems are inadequate methodologies for flux measurements and nonlinear interaction of most biological, geochemical, and meteorological variables that determine flux rates at the surface and through the boundary layer.

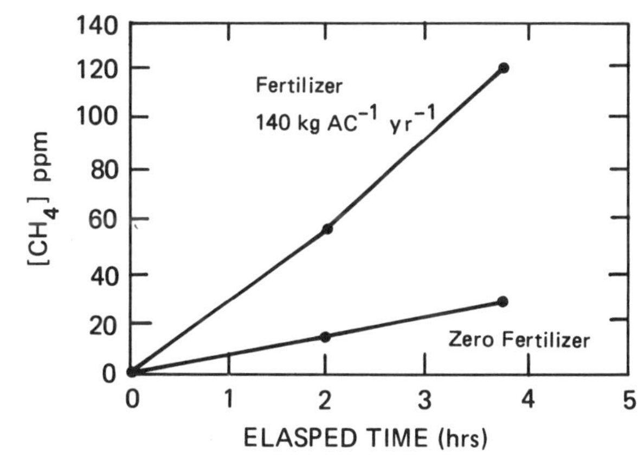

FIGURE 5.2(b) Emissions of methane from fertilized and unfertilized rice plants (data from Cicerone and Shetter, 1981).

Prediction of global emissions of most biogenic gases from surface sources may be somewhat analogous to weather prediction—short-term, local-to-regional processes are amenable to fundamental understanding through long-term studies with a combination of in situ and remote sensing techniques; long-term, global predictions will evolve through the development of a statistical chemical climatology of the troposphere.

BIBLIOGRAPHY

Ahrens, L. H. (1979). *Origin and Distribution of the Elements.* Pergamon, New York, 537 pp.

Ajtay, E. L., P. Ketner, and T. Duvigenaud (1979). Terrestrial primary reduction and phytomass, in *The Global Carbon Cycle.* Wiley, New York.

Andreae, M. O., and H. Raemdonck (1983). Dimethyl sulfide in the surface ocean and the marine atmosphere: a global view. *Science 221*:744–747.

Bonsang, B., B.-C. Nguyen, A. Gaudry, and G. Lambert (1980). Sulfate enrichment in marine aerosols owing to biogenic gaseous sulfur compounds. *J. Geophys. Res. 85*:7410–7416.

Bremmer, J. M., and A. M. Blackmer (1978). Nitrous oxide: emission from soils during nitrification of fertilizer nitrogen. *Science 199*:295–296.

Broda, E. (1975). *The Evolution of the Bioenergetic Process.* Pergamon, Oxford, 211 pp.

Broecker, W. S., T. Takahashi, H. M. Simpson, and T.-H. Peng (1979). Fate of fossil fuel carbon dioxide and the global carbon budget. *Science 206*:409–418.

Burns, R. C., and R. W. F. Hardy (1975). *Nitrogen Fixation in Bacteria and Higher Plants.* Springer-Verlag, New York, 189 pp.

Chapman, V. J. (1977). *Wet Coastal Ecosystems.* Elsevier, Amsterdam.

Cicerone, R. J., and J. D. Shetter (1981). Sources of atmospheric methane: measurement in rice paddies and a discussion. *J. Geophys. Res. 86*:7203–7209.

Crutzen, P. J., L. E. Heidt, J. P. Krasnec, W. H. Pollock, and W. Seiler (1979). Biomass burning as a source of atmospheric gases CO, H_2, N_2O, NO, CH_3Cl and COS. *Nature 282*:253–256.

Delwiche, C. C. (1981), (ed.). *Denitrification, Nitrification and Atmospheric Nitrous Oxide.* Wiley, New York, 286 pp.

Delwiche, C. C., and G. E. Likens (1977). Biological response to fossil fuel combustion products, in *Global Chemical Cycles and Their Alterations by Man*, Werner Stumm, ed. Dahlem Konferenzen, Berlin, pp. 73–88.

Duxbury, J. M., D. R. Vauldin, R. E. Terry, and R. L. Tate (1982). Emissions of nitrous oxide from soils. *Nature 298*:462–464.

Ehhalt, D. H., and U. Schmidt (1978). Sources and sinks of atmospheric methane. *Pure Appl. Geophys. 116*:452–464.

Friend, J. P. (1973). The global sulfur cycle, in *Chemistry of the Lower Atmosphere*, S. I. Rasool, ed. Plenum, New York, 335 pp.

Garrels, R. M., A. Lerman, and F. T. Mackenzie (1976). Controls of atmospheric O_2 and CO_2: past, present and future. *Amer. Sci. 64*:306–315.

Harriss, R., D. I. Sebacher, and F. T. Day, Jr. (1982). Methane flux in the great dismal swamp. *Nature 297*:673–674.

Holland, H. D. (1978). *The Chemistry of the Atmosphere and Oceans.* Wiley, New York, 351 pp.

Hutchinson, G. L., A. R. Mosier, and C. E. Andre (1982). Ammonia and amine emissions from a large cattle feedlot. *J. Environ. Qual. 11*:288–293.

Jorgensen, B. B. (1977). The sulfur cycle of a coastal marine sediment (Limfjorden, Denmark). *Limnol. Oceanogr. 22*:814–832.

Junge, C. E. (1972). The cycle of atmospheric gases—natural and man-made. *Quart. J. Roy. Meteorol. Soc. 98*:711–729.

Kellogg, W. W., R. D. Cadle, E. R. Allen, A. L. Lazrus, and E. A. Martell (1972). The sulfur cycle: man's contributions are compared to natural sources of sulfur compounds in the atmosphere and oceans. *Science 175*:587–596.

Kvenvolden, K. A., ed. (1974). *Geochemistry and the Origin of Life.* Dowden, Hutchinson & Ross, Stroudsburg, Pa., 422 pp.

Li, Y.-H. (1972). Geochemical mass balance among lithosphere, hydrosphere, and atmosphere. *Amer. J. Sci. 272*:119–137.

Lovelock, J. E., R. J. Maggs, and R. A. Rasmussen (1972). Atmospheric dimethyl sulfide and the natural sulfur cycle. *Nature 237*:452–453.

Mann, K. H. (1982). *Ecology of Coastal Waters: A Systems Approach.* University of California Press, Berkeley, Calif., 322 pp.

Margulis, L., and J. E. Lovelock (1978). The biota as ancient and modern modulator of the earth's atmosphere. *Pure Appl. Geophys. 116*:239–243.

Martell, E. A. (1963). On the inventory of artificial tritium and its occurrence in atmospheric methane. *J. Geophys. Res. 68*:3759–3769.

Ponnamperuma, C. (1977). *Chemical Evolution of the Early Precambrian.* Academic, New York, 221 pp.

Rasmussen, R. A. (1974). Emission of biogenic hydrogen sulfide. *Tellus 26*:254–260.

Rasmussen, R. A., L. E. Rasmussen, M. A. K. Khalil, and R. W. Dalluge (1980). Concentration distribution of methyl chloride in the atmosphere. *J. Geophys. Res. 85*:7350–7356.

Redfield, A. C. (1958). The biological control of chemical factors in the environment. *Amer. Sci. 46*:205–221.

Singh, H. B., L. J. Salas, and R. E. Stiles (1983). Methyl halides in and over the eastern Pacific (40°N-32°S). *J. Geophys. Res. 88*:3684–3690.

Söderlund, R., and B. H. Svensson (1976). The global nitrogen cycle, in *Nitrogen, Phosphorus and Sulphur—Global Cycles.* SCOPE Report 7. B. H. Svensson and R. Söderlund, eds. *Ecol. Bull. Stockholm 22*:23–72.

Walker, J. C. G. (1977). *Evolution of the Atmosphere.* Macmillan, New York, 318 pp.

Whittle, K. J. (1977). Marine organisms and their contribution to organic matter in the oceans. *Mar. Chem. 5*:381–411.

GLOBAL DISTRIBUTIONS AND LONG-RANGE TRANSPORT

BY J. M. PROSPERO AND H. LEVY II

The ultimate objective of the Global Tropospheric Chemistry Program is to understand the global chemical cycles of the troposphere to such a degree that the response of the atmosphere to natural or man-made perturbations can be predicted. In order to do this, a thorough knowledge of the chemistry of the species in question and of the physical behavior of the atmosphere is needed. At the simplest level, the linkage to atmospheric conditions is implied in any chemical process that involves solar radiation or water in any of its states. The linkage between meteorology and atmospheric chemistry becomes more explicit when the variable of time becomes a factor in the problem. Because the atmosphere is never still, the time dependence of chemical processes will also invoke a space dependence.

Meteorological processes are intimately linked with chemical processes over a wide range of time and space scales. In the early days of atmospheric chemistry, when the subject was synonymous with pollution chemistry, field experiments combining chemistry and meteorology were carried out on a scale commensurate with that of the perceived problem; consequently efforts focused on studies of pollutant dispersion from relatively localized sources such as smoke stacks or industrial complexes. Later, pollution began to be viewed as a regional problem, and experiments were designed to study pollutant dispersion on a regional scale. A number of experiments of this type were carried out with some degree of success.

It has come to be realized that the scale of the pollution problem is much greater than that of a particular urban region. For example, the area affected by acid rain encompasses the entire northeastern United States and much of eastern Canada. Some of the major industrial sources for the sulfur and nitrogen oxides that are responsible for the increased acidity are located in the central United States. A similar situation was recognized much earlier in Europe, where the acid rain problem in some countries is clearly attributable to a major extent to materials injected into the atmosphere in another country. Thus the subject of atmospheric chemistry in general, and atmospheric pollution in particular, must be viewed as one that transcends national boundaries. Transport studies must be planned accordingly.

Before the long-range transport of chemical species in the atmosphere can be understood and predicted, a good understanding of global meteorology is needed. Fortunately, tremendous progress has been made in this area over the last decade or so. A number of models have been developed that have been used (or could be used) for predicting transport. At this time a number of general circulation (climatological) transport models exist that have been applied to a limited range of tropospheric chemistry studies with mixed results. Unfortunately, there is no chemical data set that can be used to validate these models. Thus the future development of these and other models is severely hampered. It is because of this need that we established our first objective:

1. To obtain data on the distribution of selected chemical species so as to be able to characterize important meteorological transport processes and to validate and improve the ability of models to simulate the long-range transport, global distribution, and temporal variability of selected chemical species.

This first objective is primarily meteorological in nature. The second objective focuses on the needs of the atmospheric chemists studying the various cycles:

2. To obtain data on the distribution of those chemical species that play an important role in the major chemical cycles of the troposphere.

The third objective derives from a concern about the possible impact that some chemical species might have on weather and climate. To this end we recommend measurements:

3. To establish, quantify, and explain long-term trends in the concentration and distribution of environmentally important trace gases, especially those that are radiatively active.

These three objectives could be met by establishing a global network comprising three different types of subnetworks: the Global Distributions Network; the Surface Source/Receptor Network; and the Long-Term Trends Network. The configuration of these networks and the species to be measured in each are discussed in the Global Distributions and Long-Range Transport experiment description in Part I, Chapter 3 of this report. At this time we will discuss only the broader philosophical aspects of setting up networks of this sort. Although the general subject of atmospheric models is fundamental to our topic, it will not be discussed here, as it is adequately covered later in Part II, Chapter 6. Likewise, we will not deal with any of the chemistry of the species that are designated in the network protocol because this subject is covered in Part II, Chapter 7 of this report.

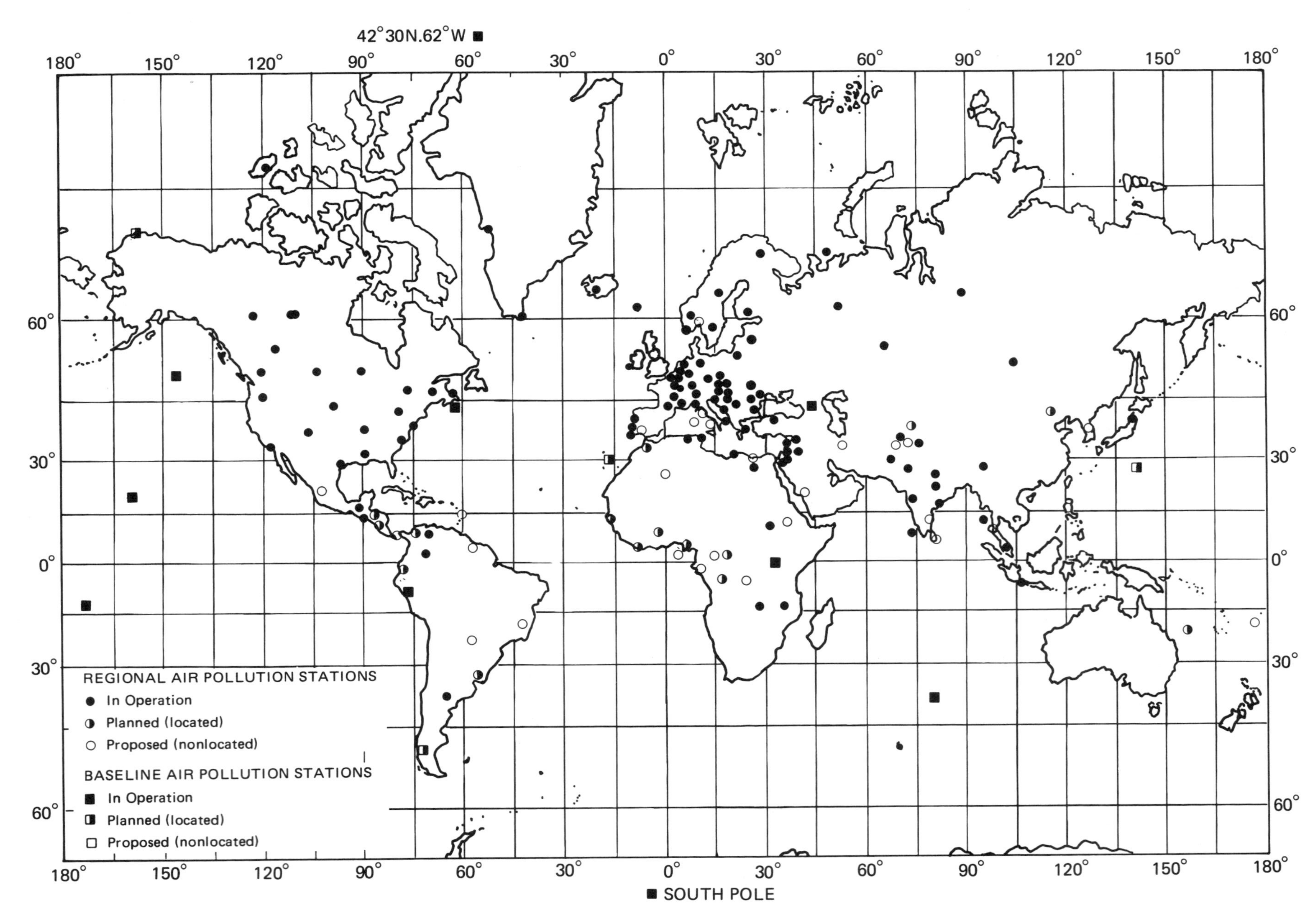

FIGURE 5.3 Present stage of WMO network for monitoring background air pollution.

MONITORING CONCEPTS

The word "monitor" is defined as "to watch, observe, or check especially for a special purpose." Thus a network of stations engaged in a routine program of sampling could be rightfully described as a monitoring program. Monitoring is an activity that in the past has been viewed with a certain amount of disdain by many scientists. The poor reputation that monitoring had earned for itself was due perhaps to the fact that some monitoring programs were poorly conceived and badly executed; nonetheless, they were permitted to continue. This poor image persists in some circles despite the fact that some of the most exciting developments in atmospheric chemistry in recent years have been derived from programs that involved a monitoring effort. The most outstanding example is that of CO_2. The excellent and vitally important record that exists for CO_2 concentrations in the atmosphere is primarily due to the efforts of one person, David Keeling, who persisted in his work despite the initial indifference of many in the scientific community and despite the sometimes flagging funding. It is relevant to our thesis that the CO_2 record is important not only for what it says about the carbon cycle but also for what it says about meteorological and oceanographic processes.

An excellent example of a monitoring program that has had a major impact on knowledge of the atmospheric circulation is the Department of Energy's (formerly Atomic Energy Commission's) high-altitude sampling program for radioactive materials. This program was initiated in response to the intensive nuclear weapons testing of the 1950s and 1960s. In effect, the individual nuclear detonations injected a point-source tracer pulse into the stratosphere. Circulation patterns could be determined by monitoring the subsequent dispersion of the tracer. By monitoring the total budget of radioactive materials in the stratosphere as a function of time, especially after the prolonged interruption and subsequent cessation of atmospheric testing, it was possible to calculate the residence time for stratospheric materials. As a consequence of this program of systematic aircraft and balloon flights, important stratospheric transport processes were identified and quantified. Without these data it would have been extremely difficult to make any realistic evaluation of the impact of anthropogenic emissions on stratospheric chemistry such as those assessments made for supersonic-transport emissions, halocarbons, and other materials. The current stratospheric chemistry models are, in effect, based on meteorology derived from these radionuclide tracer measurements.

To provide further support for the concept of monitoring for the sake of investigating processes, we cite an example from the field of oceanography. For over a decade, a number of programs have focused on the study of the distribution of ^{14}C and ^{3}H (tritium) in the oceans. These radionuclides were produced in very high yields during the period of intense weapons testing. Much of this material was subsequently deposited in the oceans in the high latitudes. After injection into the sea surface, the nuclides follow the movement of the water masses. The distribution of these radioactive materials has been measured periodically and systematically in a series of programs that have been carried out over the years. As a result of this work, there is now a much improved picture of transport and mixing processes in the oceans, and detailed ocean-mixing models have been developed. This new knowledge of oceanographic processes has been vitally important in developing the models of the global CO_2 cycle that are critical to assessing the impact of CO_2 on climate.

Thus it is clear that carefully planned and executed monitoring efforts can produce data that can be used to develop new insights into chemical and physical processes. The Global Tropospheric Chemistry Program network will, in effect, use chemical tracers in much the same way as radionuclides were used in the cited examples.

There are a number of monitoring networks currently in existence. The most extensive and ambitious is carried out under the United Nations Environment Program (UNEP), one of whose tasks is global environmental assessment. As a part of the task of information gathering, a Global Environmental Monitoring System (GEMS) was established. Some of the activities carried out under GEMS are as follows:

1. Health-related monitoring of pollutants in air, food, water, and human tissues (e.g., SO_2, heavy metals, chlorinated hydrocarbons);
2. Climate-related monitoring of variables that could effect climate change (e.g., CO_2, atmospheric turbidity, glacier masses, albedo);
3. Monitoring of long-range transport of pollutants (e.g., sulfur and nitrogen oxides and their transformation products).

In pursuit of these objectives, UNEP supports the Background Air Pollution Monitoring Network (BAPMoN), which is operated under the World Meteorological Organization (WMO). The BAPMoN network consists of three types of stations: regional, continental, and baseline (or global). Regional stations are located in rural areas far from population centers so as to be minimally affected by anthropogenic emissions. Continental stations are situated in places where no significant changes in land use practices are anticipated for a considerable time. Baseline stations are located in remote pristine areas such as islands and mountains. The distribution of BAPMoN stations is shown in Figure 5.3.

At regional stations the program consists of sampling of wet precipitation by automatic rain gauges and the measurement of atmospheric turbidity by sun photometers. At a limited number of locations, measurements of suspended particles are made by means of high-volume air samplers. Generally, the precipitation (and suspended particle) samples are grouped and analyzed on a monthly basis. However, some national programs analyze on a shorter term basis. In the United States, there are 12 BAPMoN stations, which are operated by NOAA. These stations (and those in Canada) operate on a weekly schedule. The United States and Canadian stations do not sample suspended particles. Baseline and continental stations, in addition to following the same program as the regional stations, also measure CO_2 and, in some cases, other constituents such as SO_2 and NO_x. Baseline stations are usually operated as research stations by the managing country. Some stations, such as those in the U.S. Global Monitoring for Climate Change (GMCC) program, also monitor other trace gases (such as the halocarbons, CO, and O_3).

As of 1981, the WMO network consisted of 100 regional, 12 continental, and 12 baseline stations operating in 49 countries (Figure 5.3). It is hoped that the network will ultimately include about 160 stations in 90 countries.

The WMO network is an impressive achievement considering the limited funding available to support its operation. In 1981, the entire GEMS program was budgeted at $2 million a year.

However, from the standpoint of the objectives of the Global Tropospheric Chemistry Program in general and the Global Distribution/Long-Range Transport Program in particular, the WMO network leaves much to be desired. First, the major emphasis in the network protocol is placed on pollution. Consequently, many species that are important to atmospheric chemical processes are ignored. Second, according to the network protocol, the chemical analysis activity focuses primarily on precipitation; suspended particle sampling is carried out at only a select subset of stations as a result of national initiatives. Gas sampling (other than for CO_2) is even more limited and is usually carried out only at baseline stations. Third, the BAPMoN protocol is based on a monthly sampling frequency; consequently, these data could not be used for developing and validating event-transport models. Last, the BAPMoN network is concentrated on continental areas in the northern hemisphere. There are only seven operational stations in the southern hemisphere, and of these only four are located south of 15°S: Cape Grim, Tasmania; Lopez, Argentina; Amsterdam Island; South Pole. This unbalanced distribution of stations is a major deficiency from the standpoint of the Global Tropospheric Chemistry Program objectives.

There are a number of other networks operating in the United States. The most widespread is the National Air Surveillance Network (NASN), operated by the Environmental Protection Agency. The primary responsibility for the operation of this network resides with the states; the primary purpose of the network is to monitor air quality and to verify compliance with the national air quality standards. A number of species measured in the NASN program are relevant to the Global Tropospheric Chemistry Program: SO_2, CO, NO_2, O_3, hydrocarbons, and total suspended particles. The basic device for measuring total suspended particles is the high-volume (hi vol) filter sampler; as of the late 1970s, there were several thousand of these in operation at NASN sites. The number of stations measuring the various gaseous species exceeds several hundred, with over a thousand measuring SO_2 routinely. Most NASN stations are situated in urban areas; however, a significant number are located in suburban and remote areas.

Some of the data obtained from the NASN program will be very useful for modeling purposes especially from the standpoint of providing information on atmospheric source strengths. Unfortunately, the data obtained from the high-volume particle samplers are almost useless for any purpose other than determining total suspended particle concentrations. First, sampling is too infrequent (one day in six), and second, the filter sampling medium generates artifacts that preclude meaningful chemical analysis for important species such as $SO_4^=$, NO_3^-, and other trace materials.

A rather extensive network of precipitation-sampling stations has evolved in the United States over the past few years as a consequence of increased concerns about acid precipitation. These studies are carried out as a part of the National Acid Precipitation Assessment Program. The national network is actually a composite of many smaller networks, each of which is operated by various agencies (EPA, NOAA, the Department of Energy, and the Department of Agriculture) and university groups. The operational protocol does not call for any concurrent air-sampling studies, although some work of this nature is carried out as a part of individual research efforts.

The WMO BAPMoN program and other related pollution-monitoring programs have been criticized for their strategy and their tactics. The deficiencies in these programs are to a certain extent inherent in the terms of reference mandated by the governing bodies. Such programs evolved in the crisis atmosphere that was generated by the environmental movement in the 1960s. As a consequence of the demands to take action on these issues, programs were instituted that have produced tremendous quantities of data. Unfortunately, many of the data are difficult to interpret in terms of chemical and physical processes and of environmental impact.

The shortcomings of these national and international monitoring programs are now generally recognized. In the United States, a National Research Council study group (Committee on National Statistics, 1977) on environmental monitoring has recommended changes in the EPA's programs, which they found to be deficient on a number of grounds: they were not firmly based on scientific principles; they gave insufficient attention to discovering or anticipating pollutants; their efforts were often fragmented and uncoordinated. Similar study groups in other countries have reached similar conclusions. As a consequence of these perceived shortcomings, a number of changes are imminent in some of these programs. Most significant from the standpoint of the objectives of the Global Tropospheric Chemistry Program is a move to implement high-volume air sampling for suspended particles at BAPMoN stations and to place all sampling on a weekly rather than monthly schedule.

There are a number of monitoring programs of more limited scope that have been relatively successful. The NOAA GMCC program and the Atmospheric Lifetime Experiment (ALE, sponsored by the Chemical Manufacturers Association) have monitored concentrations of a number of halocarbons including fluorocarbon-11 and fluorocarbon-12 since 1977 and 1978, respectively. Methyl chloroform has been monitored at a number of locations since 1979. These data clearly show that the concentrations of the halocarbons have been increasing as a result of man-made emissions. Likewise, extended time records of measurements for a number of other trace species are being developed; most notable are the data for N_2O, CH_4, O_3, and CO. Although the quality and extent of these data sets are variable and the interpretation is in some cases debatable, it is clear that temporal and areal variations are indeed occurring and that some changes can be attributed to natural processes and others to human activities.

A number of surface time-series data sets are also being developed for airborne particulate matter. Many of these data sets show large variations that can be attributed to large-scale transport phenomena. For example, the concentration of sulfate and other species in aerosols in the Arctic has been found to increase sharply with the arrival of air masses from Europe and the industrialized areas of Asia; these increases are also associated with the occurrence of widespread haze in the Arctic. More than 15 years of mineral aerosol measurements made over the western tropical North Atlantic show clear seasonal trends; the seasonal maximum concentrations are about 100 times greater than the seasonal minima. These mineral aerosols are derived from sources in Africa some 5000 km distant. Long-term trends in concentration have been related to drought in Africa. Likewise, mineral aerosol measurements in a network of seven SEAREX stations in the North Pacific show similar seasonal variations that are attributable to mineral dust transported from Asian sources that are over 8000 km away. In both the Asian and African dust studies, sharp day-to-day variations in dust can be related to specific synoptic events; in some cases, the dust can be traced to specific dust sources. Aerosol data such as these could be used to develop event-type meteorological transport models.

NETWORK DESIGN

In order to properly design a global network, it is necessary to have a fairly good idea of the temporal and areal variability of the concentration of the species in question. With a few exceptions, the data base for species of interest to the Global Tropospheric Chemistry Program is utterly inadequate. The situation is especially desperate with regard to the question of vertical distribution of trace species.

To further complicate matters, there is also a dearth of meteorological data over large areas of the globe. Especially troubling is the sparsity of upper atmosphere data over most of the oceans and much of the continents. Thus some types of meteorological studies (especially event studies) that require the input of real meteorological data will be handicapped.

The other problem is logistical. The earth is large, and field operations are difficult, time consuming, and expensive. It is unrealistic to expect to be able to dispatch simultaneously many large teams of highly trained scientists and technicians to many different parts of the globe and to keep them there for extended periods of time. Even if such resources were available, it is not at all clear what species these teams should measure, where they should measure them, and at what frequency. Indeed, many of these questions will be addressed by the individual research programs that are focused on specific processes; consequently, this essential information will not be available for some years to come.

Faced with these gaps in the knowledge of atmospheric chemistry and meteorology and because of the problems with logistics, we took a pragmatic approach. We surveyed the list of important species in the sections on cycles in Chapter 7 and also in Appendix C and asked the following question: Of the species that are relevant to our objectives, which are relatively easily measurable at ambient concentrations by using existing technology?

It is difficult to accurately define the criterion of being

"relatively easily measurable." To give some indication of what we mean, we shall cite a few examples. Current network operations in SEAREX in the North Pacific have shown that it is possible for untrained personnel to routinely collect filter samples that can be successfully analyzed for a number of species such as excess $SO_4^=$, NO_3^-, mineral particles, sea-salt aerosol, CH_4, sulfonic acid, ^{210}Pb, and some trace elements such as vanadium. On the other hand, field experience in this same region has led the SEAREX investigators to conclude that it is impossible to use these same sampling procedures to make measurements of trace species such as lead, mercury, and chlorinated hydrocarbons—the possibilities of contamination during the sampling operation are so overwhelmingly great that the data would be unreliable and hence useless.

Another example of a relatively easy sampling and analysis procedure is that for some of the halocarbons, which can be "grab sampled" in the field by using flasks, and subsequently analyzed in the laboratory by means of gas chromatographic procedures. Although this sampling procedure does require the use of specially constructed and treated flasks and some operator training, the procedure has been used more or less routinely by a number of investigators. Another example is O_3, which can be easily measured at ground level with off-the-shelf equipment. However, it is a much more difficult task to obtain vertical profiles of O_3. Although ozonesondes are readily available, their use is not so straightforward, and trained technicians are necessary, although they need not be scientists.

The types of measurements that we categorically exclude from our network protocol are those that require a highly trained scientist who has detailed knowledge of the chemistry of the species being measured and the idiosyncracies of the measurement technique—an example would be the current measurement of NO at concentrations typically found in remote areas. Finally, there is the type of measurement where the technique is still under development such as that for the hydroxyl radical. At this time, such measurements are completely out of the question for any sort of routine network field operation.

In order to circumvent the problems of a limited data base and difficult logistics, we propose a program that calls for a dynamic plan of growth for the networks. We would start with a few stations sited in grossly different environmental regimes. These stations would serve several purposes. They would yield data that would serve as initial values for testing models; in turn, the output from the initial model tests would be used to further refine the sampling protocol and to provide guidance in the placement of new stations in the network. Also, the start-up stations would serve as operational environmental test sites for the instrumentation and measurement protocol. Finally, the operation of these stations would serve to develop the logistical support base for the network.

An assumption that is central to the proposed program is that selected stations in existing networks such as BAPMoN would evolve along with or as a part of the Global Tropospheric Chemistry Program global network. The most likely mechanism would be for cooperating nations to augment the operations at suitable existing stations. An essential requirement would be that the protocol is identical to the Global Tropospheric Chemistry Program protocol and that analytical techniques are completely validated for accuracy and precision.

TEMPORAL CONSIDERATIONS

The question of the frequency of sampling was not addressed in a very explicit manner in the description of the proposed program in Part I. The sampling frequency will be a function of a number of factors including the lifetime of the species in the atmosphere, the rate of change of the mean concentration of the species in the atmosphere (assuming that a concentration trend is being monitored), and the magnitude of the transient concentration changes that are generated by source injections and by transport or removal processes. As an example, consider N_2O, which has a rather uniform global distribution (except for urban areas). The mean concentration of N_2O is about 300 ppbv; concentrations at the surface in the northern hemisphere are about 0.5 ppbv greater than those in the southern hemisphere. Moreover, the concentration of N_2O appears to be increasing at a rate of about 0.2 percent per year. Clearly, it should not be necessary to measure N_2O very often or at many different stations in order to characterize these trends. At the other extreme, aerosol particles have a residence time of 1 to 2 weeks. Thus the concentration of aerosol particles can change by orders of magnitude over the period of 1 day as the synoptic situation changes; for example, when a cloud of particles is advected to a sampling site. Consequently, in order to characterize an aerosol event in context with the meteorology, a relatively high sampling frequency is required, ranging from 1 day to 1 week. It is because of their short residence time that aerosols can serve as good tracers for use in event models.

To illustrate the difficulty in defining a sampling protocol at the outset of a program, we cite the example of CO_2, which is a relatively long-lived species. The concentration of CO_2 in the atmosphere is increasing steadily at the rate of about 0.3 percent per year because of human activities. However, the concentration of CO_2 varies seasonally, the magnitude of the variation being geographically dependent. For example, at Mauna Loa the annual variation is about 2 percent whereas at the South Pole it is about 0.5 percent. The annual variations are primarily attributable to the impact of the biosphere. Because of these variations, it is necessary to sample at a relatively high frequency in order to adequately characterize the inherent variance of the data. In the current CO_2 monitoring program, weekly CO_2 samples are collected.

Nonetheless, it is the variability of the concentration of the species that contains the information about the controlling processes. In the case of CO_2, this variability can be traced to the seasonal activity and distribution of biological sources and sinks and to the controlling meteorological and oceanographic processes. Indeed, the year-to-year fluctuations in the CO_2 growth rate have shown that the oceanographic processes in equatorial regions of the Pacific Ocean play a major role in determining the variability of atmospheric CO_2 concentrations. The CO_2 sampling program is an excellent example of how a carefully planned and executed monitoring program can provide great insights into important processes. The ultimate test of the knowledge of the CO_2 system (or any other chemical system) is the capability to use the collected data to construct a model that will predict future trends.

The way in which the CO_2 program was implemented is also a good example of the general approach that we espouse. Stations were established at a relatively slow rate, and sampling and analytical techniques were thoroughly tested. Finally, modeling has played a large role in the CO_2 program, and it has provided some important insights into the CO_2 system.

VERTICAL DISTRIBUTION MEASUREMENTS

The most difficult problem in the network operations will be that of obtaining vertical concentration profiles. In many cases, the boundary layer concentrations will not be representative of those in the free troposphere—the major sources or sinks may reside in the boundary layer, and the chemistry there could be different from that in the free troposphere. Long-lived gases are thought to be well mixed in the vertical. However, the medium-lived gases may require vertical profile measurements. Shorter-lived gases such a O_3, C_2Cl_4, and C_2HCl_3 will almost certainly require measurements of vertical profiles. Also, vertical distribution data will be necessary for many of the aerosol species.

Unfortunately, it is difficult and expensive to make vertical profile measurements on a routine basis. Other than the data obtained from a small number of ozonesonde stations, there is remarkably little data in the literature on vertical profiles. The most impressive and extensive data set on the distribution of a broad range of important atmospheric chemistry species was that obtained in a series of flights made in 1977 as part of the Global Atmospheric Measurement Experiment on Aerosols and Gases (GAMETAG). A team of 39 scientists from nine different institutions measured a large number of species on flights over the western United States and Canada and over large areas of the North and South Pacific. Profiles were made up to 6-km altitude.

GAMETAG was clearly a pioneering effort in the field of atmospheric chemistry. Aircraft operations of that type will be a mandatory part of the Global Tropospheric Chemistry Program as spelled out elsewhere in this report. However, it is completely unrealistic to expect to carry out flights of the GAMETAG type in conjunction with routine network operation—indeed, it would be impossible to do so because there are not enough aircraft and scientific personnel to do the work.

Instead, we propose a stepped approach to developing the knowledge of vertical distributions. The vertical distributions in a limited number of source and sink environments must be measured. The objective is to characterize the concentration profiles and their temporal and spatial variance and to relate these measurements to the sources (or sinks) and the local meteorology. Measurements of this type are already planned as primary components of the other major programs in this work. The data obtained from these studies will enable an appropriate protocol to be specified for obtaining useful vertical profile data on a global scale.

Another approach is to make a concurrent series of measurements above the boundary layer and at the surface using stations located on mountaintops and at ground level. Considerable care must be exercised in operating a sampling station on a mountain because the mountain generates its own meteorology, which, in turn, can affect the chemistry. Nonetheless, the program at the NOAA GMCC station on Mauna Loa, Hawaii, has shown that this problem can be circumvented. The station is located at an altitude of 3400 m, which is nominally above the marine boundary layer. However, it is clear that on many occasions, especially during the day, the station is definitely impacted by boundary layer air transported through the inversion by upslope winds. On the other hand, very extensive tests have shown that during downslope wind conditions the air at the station is derived from above the boundary layer.

There exists an extensive set of measurements made concurrently at a mountaintop and the surface. Mineral and sea-salt aerosol concentrations have been measured for the past several years at Mauna Loa by a group from the University of Maryland. Independently, a SEAREX group has been making measurements of some of the same species at a coastal site in Oahu. These two records reveal some very interesting similarities and differences, which, for the most part, appear to be interpretable on climatological and meteorological grounds. In another effort, a limited number of gases are being sampled concurrently at Mauna Loa and at Cape Kumukahi, Hawaii. These data suggest that a similar approach of concurrent mountain/surface measurements might be fruitful for other species as well. It would seem logical that an extended set of such measurements be initiated at the Mauna Loa GMCC and at a suitable surface site in the Hawaiian Islands. Some other candidate sites are available. A BAPMoN site is planned at the observatory on Tenerife in the Canary Islands. This station is at an altitude of 2370 m, which places it above the height of the nominal trade-wind inversion. There are other BAPMoN stations planned for mountain sites, some of which might be suitable for such studies.

While paired mountain/surface measurements will provide useful free troposphere/boundary layer data, extensive vertical profile data will require the use of aircraft (except for O_3 and water vapor, where sondes can be used).

What sort of aircraft program could we reasonably expect to carry out a routine network operation? First, it will be necessary to use locally based lightweight commercial aircraft. Second, we assume that only slight modifications could be made to the aircraft. Third, flight times would have to be relatively short, on the order of an hour or two; this means that one could spend only a relatively short time at each sampling level. Because of these limitations, it appears that only grab-sampling procedures would be possible for routine profiling work. Thus the protocol would be limited initially to the grab-sampling of gases and to those aerosol measurements that are feasible with low-volume samples.

Thus the vertical distribution studies within the Global Tropospheric Chemistry Program would consist of four types of activities:

1. Intensive short-term field experiments associated with the major Global Tropospheric Chemistry programs—these would involve surface experiments and highly instrumented aircraft.
2. Ozone and water vapor sondes launched from a limited number of bases on a fairly frequent schedule—several a week.
3. Mountain/surface stations carrying out the complete network protocol on a fairly frequent schedule (one or more samples a week).
4. Aircraft studies of grab-sample gases and particles.

A major task in developing the network protocol is to ascertain the optimum mix of these activities. A special effort must be made to relate the light aircraft work using grab samples to the ground-level program. We anticipate that this is an appropriate area for exploratory research in the early phases of the program. It might be appropriate for such a study to be carried out in the Hawaiian Islands in conjunction with the routine operations at Mauna Loa and a corresponding operation set up at sea level. For completeness, an ozonesonde program should be carried out concurrently with the aircraft and ground station study.

BIBLIOGRAPHY

Bodhaine, B. A., B. G. Mendonca, J. M. Harris, and J. M. Miller (1981). Seasonal variations in aerosols and atmospheric transmission at Mauna Loa. *J. Geophys. Res. 86*:7395–7398.

Broecker, W. S., and T.-H. Peng (1982). *Tracers in the Sea*. Lamont-Doherty Geological Observatory, Columbia University, Palisades, New York, 630 pp.

Cawse, P. A. (1982). Inorganic particulate matter in the atmosphere, in *Environmental Chemistry*, Vol. 2, H. J. M. Bowen, ed. Royal Society of Chemistry, London, pp. 1–69.

Committee on National Statistics (1977). *Environmental Monitoring*. National Academy of Sciences, Washington, D.C., 181 pp.

Davis, D. D. (1980). Project GAMETAG: an overview. *J. Geophys. Res. 85*:7285–7292.

Duce, R. A., et al. (1980). Long-range atmospheric transport of soil dust from Asia to the tropical North Pacific: temporal variability. *Science 209*:1522–1524.

Geophysics Study Committee (1977). *Energy and Climate, Studies in Geophysics*. National Research Council. National Academy of Sciences, Washington, D.C., 158 pp.

Mahlman, J. D., and W. J. Moxim (1978). Tracer simulations using a global general circulation model: results from a midlatitude instantaneous source experiment. *J. Atmos. Sci. 35*:1340–1374.

Marland, G., and R. M. Rotty (1979). Carbon dioxide and climate. *Rev. Geophys. Space Phys. 17*:1813–1824.

Newell, R. E. (1963). The general circulation of the atmosphere and its effects on the movement of trace substances. *J. Geophys. Res. 68*:3949–3962.

Oltmans, S. J. (1981). Surface ozone measurements in clean air. *J. Geophys. Res. 86*:1174–1180.

Parrington, J. R., W. H. Zoller, and N. K. Aras (1983). Asian dust: seasonal transport to the Hawaiian Islands. *Science 220*:195–197.

Pinto, J. P., Y. L. Yung, D. Rind, G. L. Russell, J. A. Lerner, J. E. Hansen, and S. Hameed (1983). A general circulation model study of atmospheric carbon monoxide. *J. Geophys. Res. 88*:3691–3702.

Prinn, R. G., P. G. Simmonds, R. A. Rasmussen, R. D. Rosen, F. N. Alyea, C. A. Cardelino, A. J. Crawford, D. M. Cunnold, P. J. Fraser, and J. E. Lovelock (1983). The atmospheric lifetime experiment, I: introduction, instrumentation, and overview. *J. Geophys. Res. 88*:8353–8368.

Prospero, J. M., and R. T. Nees (1977). Dust concentration in the atmosphere of the equatorial North Atlantic: possible relationship to the Sahelian drought. *Science 196*:1196–1198.

Rahn, K. A. (1981). Relative importance of North America and Eurasia as sources of Arctic aerosol. *Atmos. Environ. 15*:1447–1456.

Turco, R. P., R. C. Whitten, and O. B. Toon (1982). Stratospheric aerosols: observation and theory. *Rev. Geophys. Space Phys. 20*:233–279.

Uematsu, M., R. A. Duce, J. M. Prospero, L. Chen, J. T. Merrill, and R. L. McDonald (1983). Transport of mineral aerosol from Asia over the North Pacific Ocean. *J. Geophys. Res. 88*:5343–5352.

Wallén, C.-C. (1980). Monitoring potential agents of climatic change. *Ambio 9*:222–228.

Whitten, R. C. (ed.) (1982). *The Stratospheric Aerosol Layer*. Springer-Verlag, New York, 152 pp.

World Meteorological Organization, Environmental Pollution Monitoring Program (1981). *Summary Report on the Status of the WMO Background Air Pollution Monitoring Network as of April, 1981*.

HOMOGENEOUS AND HETEROGENEOUS TRANSFORMATIONS

BY D. DAVIS, H. NIKI, V. MOHNEN, AND S. LIU

As discussed in the section above by Cicerone et al., a variety of biological and geological processes results in the emission of trace gases into the troposphere. For the most part, the key elements (e.g., carbon, nitrogen, and sulfur) making up these trace gases are in reduced oxidation states. By contrast, when these elements are returned to the earth's surface, via precipitation and/or dry deposition, they most frequently are found in their thermodynamically stable oxidized forms. The atmospheric transformations that lead to the chemical oxidation of trace gases are complex and encompass homogeneous gas-phase, homogeneous aqueous-phase, as well as heterogeneous processes. Outlined in the text that follows are several of the more prominent features of each of these transformation types. The authors have drawn special attention to those outstanding questions that will require futher study in the near future.

HOMOGENEOUS GAS-PHASE CHEMISTRY

Cyclic Photochemical Transformations: H_xO_y

Most gas-phase oxidation processes are either directly or indirectly initiated as a result of the atmospheric absorption of ultraviolet solar radiation. One of the more important oxidizing agents formed from this absorption process is now believed to be the hydroxyl (OH) radical. Its pivotal role is illustrated in Figure 5.4. It is seen that OH-initiated reactions provide the major pathway for transforming a large number of tropospheric compounds into their oxidized forms. The OH species therefore plays a major role in controlling the chemical lifetimes of these "reduced" compounds.

The primary production of OH is initiated by the photolysis of O_3. Solar photons having wavelengths between 315 and 1200 nm dissociate O_3 to produce an oxygen atom in its ground electronic state:

$$O_3 + h\nu\,(1200 > \lambda > 315\,\text{nm}) \rightarrow O(^3P) + O_2. \quad (5.1)$$

The $O(^3P)$ atom combines rapidly with O_2 in a three-body reaction to reform O_3:

$$O(^3P) + O_2 + M \rightarrow O_3 + M\,(M = N_2, O_2). \quad (5.2)$$

Thus the sequence of reactions (5.1) and (5.2) is a null cycle with no net chemical effect. On the other hand, when O_3 absorbs a photon in the near-ultraviolet, with a wavelength shorter than 315 nm, an electronically excited oxygen atom is produced.

$$O_3 + h\nu\,(\lambda < 315\,\text{nm}) \rightarrow O(^1D) + O_2. \quad (5.3)$$

The $O(^1D) \rightarrow O(^3P)$ transition is forbidden and therefore $O(^1D)$ has a relatively long radiative lifetime (t = 110 s). In the troposphere, rather than relaxing radiatively, $O(^1D)$ most often collides with N_2 or O_2, i.e., reaction (5.4), and ultimately leads to the regeneration of O_3 via reaction (5.2) and another null cycle.

$$O(^1D) + M \rightarrow O(^3P) + M\,(M = N_2, O_2). \quad (5.4)$$

Occasionally, however, $O(^1D)$ collides with H_2O and causes the generation of two OH radicals:

$$O(^1D) + H_2O \rightarrow 2HO. \quad (5.5)$$

In anthropogenically unperturbed regions of the troposphere, OH reacts overwhelmingly with CO and CH_4; i.e.,

$$OH + CO \rightarrow CO_2 + H \quad (5.6)$$

$$OH + CH_4 \rightarrow CH_3 + H_2O. \quad (5.7)$$

The hydrogen atom and CH_3 radical rapidly combine with O_2 to form HO_2 and CH_3O_2 radicals, respectively. Even so, reactions (5.6) and (5.7) do not necessarily lead to removal of OH from the atmosphere, since both HO_2 and CH_3O_2 radicals can be partially converted back to OH via a complex series of chain reactions. This chemistry is illustrated in Figure 5.5. In this reaction scheme, the HO_2 radical formed by the reaction of HO and CO, regenerates OH via

$$HO_2 + NO \rightarrow NO_2 + OH \quad (5.8)$$

and

$$HO_2 + O_3 \rightarrow 2O_2 + OH. \quad (5.9)$$

Alternatively, the HO_2 species can result in radical chain termination via

$$HO_2 + OH \rightarrow H_2O + O_2 \quad (5.10)$$

or

$$HO_2 + HO_2 \rightarrow H_2O_2 + O_2. \quad (5.11)$$

However, a small fraction of the H_2O_2 formed in reaction (5.11) may be photolyzed to again regenerate two OH radicals, thereby serving as a temporary reservoir for OH. In addition to reactions (5.10) and (5.11), both OH and HO_2 may be removed by combination reactions with NO_2, i.e.,

$$HO + NO_2 + M \rightarrow HNO_3 + M \quad (5.12)$$

$$HO + NO_2 + M \rightarrow HONO_3 + M. \quad (5.13)$$

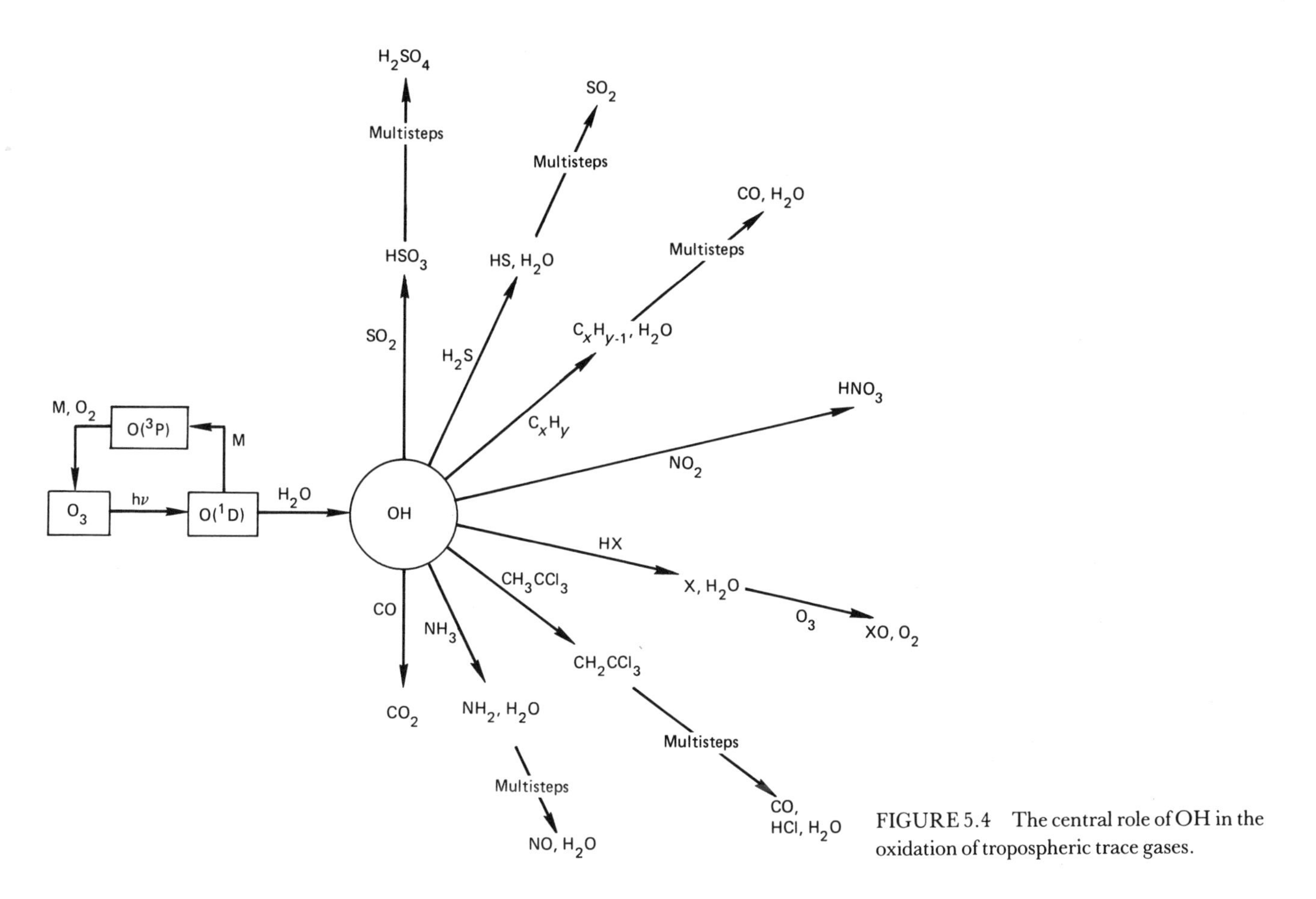

FIGURE 5.4 The central role of OH in the oxidation of tropospheric trace gases.

Depending on the temperature, the $HONO_3$ species can be relatively unstable, dissociating back to HO_2 and NO_2. Its lifetime at 300°K is 10 s, whereas at 250°K it is 10^4 s. At high altitude, reaction (5.13) may therefore provide a significant chain termination step. The HNO_3 formed in reaction (5.12) is very efficiently converted back to OH radicals via photolysis and typically is removed by wet or dry deposition processes.

The chemistry of the CH_3O_2 radical resulting from the OH-CH_4 reaction is quite complex, with the rate coefficients for several of the elementary reactions involved in this chemistry remaining unmeasured. One possible atmospheric degradation scheme for this species is that shown in Figure 5.6. As in the chemistry of HO_2 species, CH_3O_2 can react with either NO or HO_2, depending upon the [NO]/[HO_2] ratio:

$$CH_3O_2 + NO \rightarrow CH_3O + NO_2 \quad (5.14)$$

$$CH_3O_2 + HO_2 \rightarrow CH_3OOH + O_2. \quad (5.15)$$

Simple kinetic considerations indicate that reaction (5.14) becomes equal to reaction (5.15) for an [NO]/[HO_2] ratio of ~1.0 at 300°K (the uncertainty here is at least a factor of 3). As shown in Figure 5.6, the eventual fate of the CH_3O appears to be the formation of CO via the intermediate product, formaldehyde (CH_2O):

$$CH_3O + O_2 \rightarrow CH_2O + HO_2 \quad (5.16)$$

$$CH_2O + h\nu\,(\text{or OH}) \rightarrow CHO + H \quad (5.17)$$

$$CHO + O_2 \rightarrow CO + HO_2. \quad (5.18)$$

From this sequence of reactions, it is evident that the complete oxidation of the CH_3OO radical potentially can provide an additional source of HO_2 radicals to the H_xO_y system. However, in the presence of low NO levels, reaction (5.15) dominates, and CH_3OOH is formed. This species may be removed from the gas phase by heterogeneous processes or may undergo various homogeneous gas-phase reactions. At present, the relative importance of the various reaction pathways for CH_3OOH is unknown. Among the minor products that may be formed from further CH_3OOH chemistry are methanol (CH_3OH) and formic acid (HCOOH).

Complicating the H_xO_y fast-photochemical cycle still further is the possibility that other OH-initiated reactions also generate free radical species, some of which could feed back into the main H_xO_y cycle. One of the

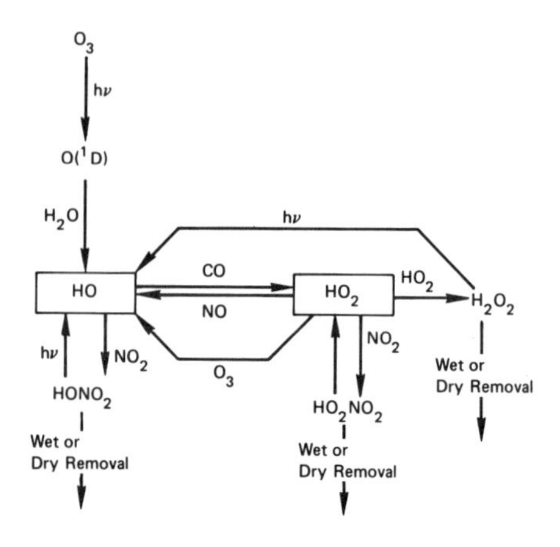

FIGURE 5.5 OH/HO_2 radical chain reactions.

most likely classes of compounds that might fit this role are the nonmethane hydrocarbons, NMHC (C_xH_y, where x and $y > 1$). In general, the atmospheric chemistry of the NMHCs is far more complex than that of CH_4. And, despite some recent progress, mechanisms describing this chemistry remain highly uncertain.

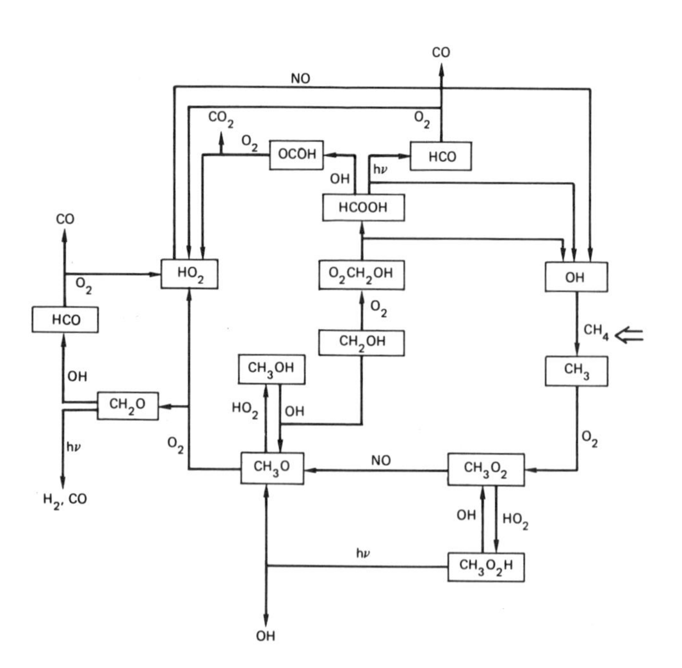

FIGURE 5.6 A possible tropospheric degradation scheme for CH_3O_2 radicals, formed from CH_4. The current lack of understanding of this chemistry defines one of the major uncertainties in the understanding of fast H_xO_y photochemistry.

Two potentially important types of NMHCs in the "remote" troposphere are isoprene (C_5H_8) and terpenes. Emitted by deciduous and coniferous trees, respectively, these compounds react rapidly with OH radicals. Their reactivity toward OH as well as O_3 suggests that they should be short-lived in the atmosphere; therefore, their concentrations will likely decrease rapidly as one moves away from their source regions. As noted previously, whether the oxidation of these compounds serves as a net sink or source for H_xO_y species can only be speculated on at the present time. Illustrating the potential complexity of this problem is the atmospheric degradation scheme for the compound C_5H_8, Figure 5.7. Two of the by-products from this chemistry that might affect the H_xO_y cycle are CO and O_3. For this impact to be realized, however, would require that significant fluxes of C_5H_8 be present such as those that might originate from a tropical rain forest.

Although it is perhaps understandable that the complex mechanisms involving NMHCs are not yet known,

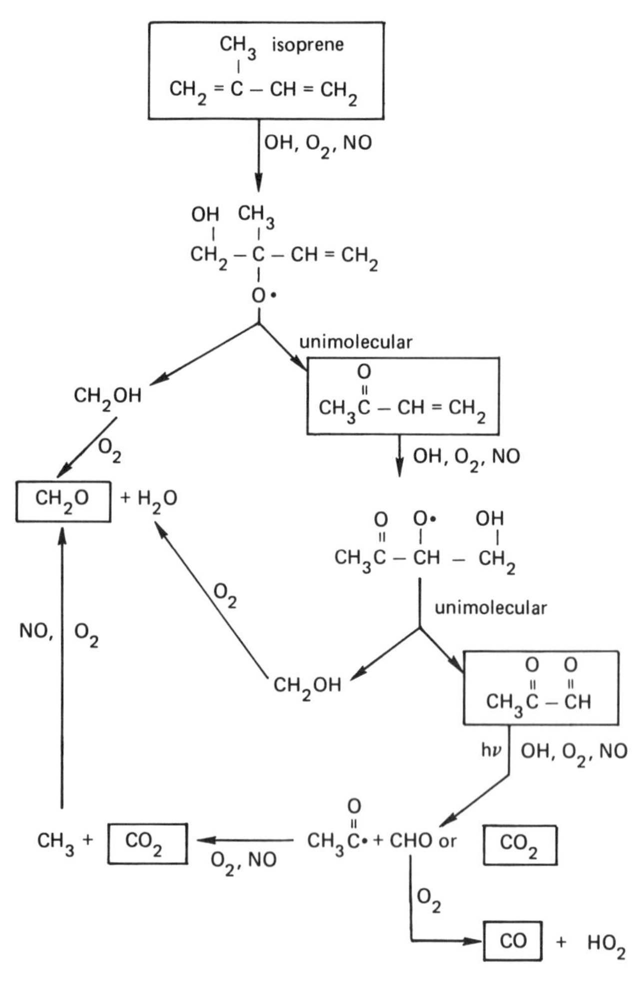

FIGURE 5.7 A possible reaction scheme for isoprene oxidation in the presence of NO_x.

the point should be stressed that even the simplest H_xO_y cyclic scheme has not yet been fully tested. Thus the central role of OH in atmospheric chemistry, even though strongly supported by laboratory kinetic measurements, has not been convincingly demonstrated. This state of affairs reflects the current absence of significant field data on many of the critical species involved in H_xO_y fast photochemistry. The absence of field data, in turn, has reflected an absence of available measurement technology for many of the critical species involved in this cycle.

Cyclic Photochemical Transformations: N_xO_y

Like the H_xO_y system, there are several N_xO_y species (e.g., NO, NO_2, NO_3, HONO, N_2O_5, and $HOONO_2$) that are believed to be present in the atmosphere at concentration levels corresponding to those predicted by photo-stationary-state chemistry. As previously noted, this N_xO_y cyclic system also has chemical coupling with the H_xO_y system. Figure 5.8 summarizes several of the key aspects of this cycle. It is seen that during daylight hours NO and O_3 are constantly produced from the photolysis of NO_2:

$$NO_2 + h\nu\,(285 \leq \lambda \leq 375\text{ nm}) \rightarrow NO + O(^3P) \quad (5.19)$$

followed by reaction (5.2). NO and O_3, in turn, continuously react to regenerate NO_2 via reaction (5.20):

$$NO + O_3 \rightarrow NO_2 + O_2, \quad (5.20)$$

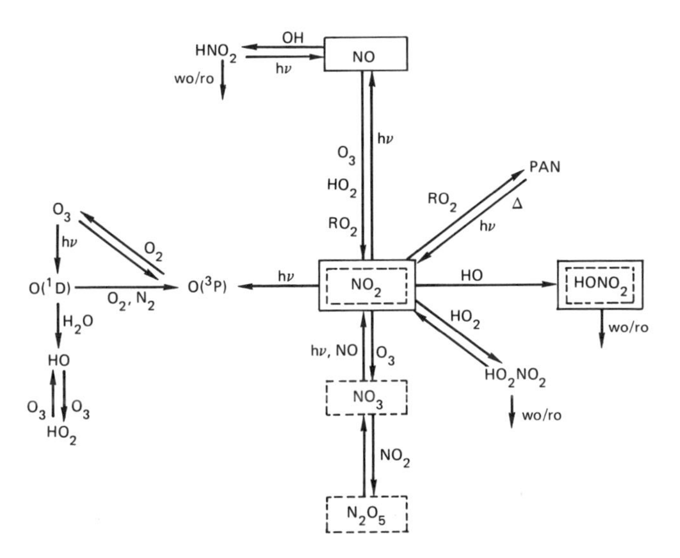

FIGURE 5.8 Major atmospheric reactions of N_xO_y species: Solid line boxes indicate major daytime nitrogen species; broken line boxes indicate significant nighttime nitrogen species. It is still uncertain whether PAN is a major species in the free troposphere. The notation wo/ro denotes washout/rainout process.

thus forming a null cycle in which an ultraviolet photon is converted into heat. In addition to reaction (5.20), both reactions (5.8) and (5.14) may also make significant contributions to the NO to NO_2 conversion process, provided high-solar-flux conditions are prevalent.

The NO_2 formed from reactions (5.20), (5.8), and (5.14) may be photolyzed to regenerate NO via reaction (5.19) or, alternatively, react with OH or HO_2 (Reactions (5.12) and (5.13)) to form nitric acid species. Both of the latter reactions can lead to the removal of N_xO_y from the atmosphere via wet or dry deposition. Yet another reaction possibility involving NO_2 is reaction with O_3:

$$NO_2 + O_3 \rightarrow NO_3 + O_2. \quad (5.21)$$

The NO_3 formed in reaction (5.21) may undergo photolysis (regenerating NO_2) or react with additional NO_2 (reaction (5.22)) and generate still another new N_xO_y species, N_2O_5.

$$NO_2 + NO_3 + M \rightarrow N_2O_5. \quad (5.22)$$

Although N_2O_5 may thermally decompose, it is now believed that some fraction of it reacts heterogeneously with H_2O, forming HNO_3.

During daylight hours, the dominant loss process for NO_3 is photolysis. Under nighttime conditions, reactions (5.21) and (5.22) become dominant. This nighttime chemistry thus predicts that NO_3 should be one of the major N_xO_y species in the atmosphere. Limited field observations of this species, however, indicate the levels to be far lower than those predicted from the chemistry shown in Figure 5.8. Possible NO_3 scavengers remain unidentified at this time.

A final uncertainty in the understanding of N_xO_y chemistry involves the coupling of this chemistry with complex hydrocarbon species known to react with NO_2 to form peroxyacetyl nitrate (PAN), a well-known air pollutant.

$$CH_3C(=O)OO\cdot + NO_2 \rightarrow CH_3C(=O)OONO_2. \quad (5.23)$$

PAN can thermally decompose with a lifetime of approximately 45 min at 300°K; but it has a lifetime of 200 days at 250°K. There is now some evidence, in fact, that suggests that this species could define a major reservoir for NO_x in the "remote" free troposphere.

Like the H_xO_y fast-photochemical cycle, there is now a great abundance of laboratory kinetic data that suggest that the fast N_xO_y cycle is one of the key photochemical cycles operating in the troposphere. Also like the H_xO_y cycle, no quantitative field test has yet been performed that has demonstrated this fact. Such tests need to validate basic relationships such as that shown in equation (5.24):

$$\frac{[NO]}{[NO_2]} = \frac{J_{23}}{k_{24}[O_3] + k_8[HO_2] + k_{14}[CH_3O_2]} \quad (5.24)$$

The absence of these tests to date again reflects the absence of appropriate field measurement technology in years past.

Ozone Transformations/Photochemical Sources and Sinks

As shown in Figure 5.9, the fast-photochemical coupling between NO, NO_2, and O_3 produces a null cycle in which an ultraviolet photon is converted into low-grade heat. When, however, NO is converted to NO_2 without the use of an O_3 molecule, there can be net production of O_3. The latter chemistry occurs when reaction (5.8) and/or reaction (5.14) becomes dominant over processes (5.11) and (5.15), respectively. These chemical conditions are found to be quite common in large urban population centers throughout the United States. As such, there are frequently times when the levels of photochemically generated O_3 far exceed those found in the natural environment.

Evaluating whether significant O_3 production occurs in the "remote" troposphere is a far more difficult task. As shown in Figure 5.9, the primary formation pathway for OH requires the consumption of one O_3 molecule. Whether this ultimately represents a net destruction pathway for O_3 depends on the subsequent chemistry of the HO_2 and CH_3O_2 radicals. Recall that these species are formed as a by-product from the reaction of OH with CH_4 and CO. Thus, if HO_2 and CH_3O_2 predominantly react to form H_2O_2 and CH_3OOH (reactions (5.11) and (5.15)), the formation of OH from O_3 defines an O_3 sink. Alternatively, the HO_2 species can react with O_3 itself, and this again would define a significant O_3 photochemical sink.

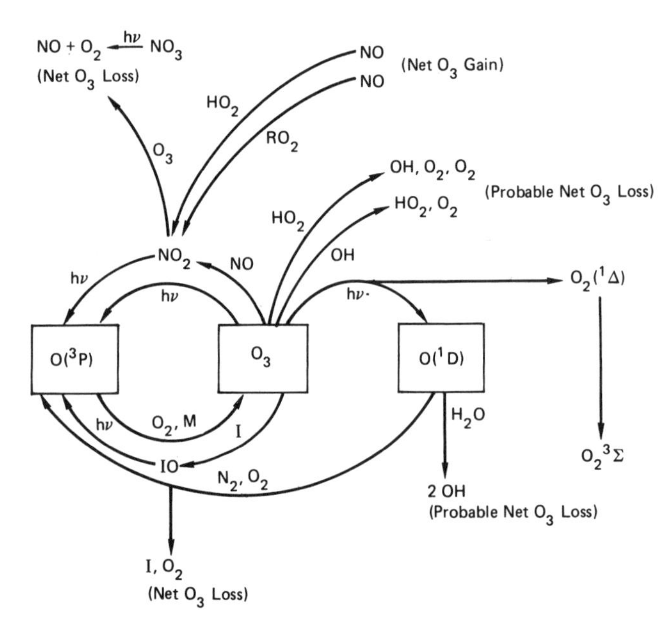

FIGURE 5.9 Photochemical production and destruction of O_3. Key oxygen species are shown in solid boxes.

Only when appreciable NO is present can reactions (5.8) and (5.14) (involving HO_2 and CH_3O_2) dominate (5.11) and (5.15), leading to a net photochemical source of O_3. Thus, whether the coupled H_xO_y/N_xO_y cycles define a net sink or source for O_3 in the "remote" troposphere stands as one of the major unanswered scientific questions in global tropospheric chemistry.

Noncyclic Transformations

Earlier, it was noted that the OH radical appears to be the principal species responsible for initiating the oxidation of numerous "reduced" trace gas species. Following the initiating step, most of the reaction products move systematically through one or multiple steps to a final oxidized product. This final product is then removed from the troposphere by wet or dry deposition. Illustrative of this reaction sequence is the degradation of complex NMHCs. The initial products resulting from the attack of OH or O_3 on the parent hydrocarbon compound are aldehydes, ketones, and/or organic acids. Still further chemistry involving these species may result in the final production of CO and CO_2. Alternatively, some of the larger organic radicals generated may undergo combination reactions or chain-polymerization-type reactions. Both of the latter type of reactions may result in the formation of organic aerosols. At present, the branching ratio for a given hydrocarbon undergoing complete degradation to form CO_2 versus its forming organic aerosols is still poorly understood.

Ammonia in the troposphere can react with OH to form NH_2 radicals. These radicals, in turn, may react with O_2 to form NH_2OO. The chemistry of this species is unknown. Possible final reaction products may involve N_2 or NO.

A very significant class of compounds for which noncyclic transformations are important are those that contain sulfur. For virtually all of these compounds (the one possible exception being COS), the initiating step involves reaction with OH. In the case of hydrogenated sulfur, the subsequent reactions of the initially formed sulfur radical species involve several possible reactants: O_2, O_3, NO, HO_2, or RO_2. In each case, this appears to lead to the formation of SO_2. The SO_2 species, in addition to being removed from the gas phase by heterogeneous processes, may react with the OH to form the intermediate $HOSO_2$ radical. The fate of the $HOSO_2$

radical is believed to be reaction with O_2 to form a peroxy radical or, alternatively, SO_3 and HO_2, i.e.,

$$H-O-\underset{\underset{O}{|}}{\overset{\overset{O}{|}}{S}}+O_2 \rightarrow \overset{O}{\diagdown}\underset{\diagup \atop O}{S}=O+HO_2.$$

The peroxy species may react with NO or HO_2 or perhaps hydrate as a result of collisions with H_2O. The only certainty in this chemistry now appears to be that, like SO_3, the final product is some form of sulfuric acid. The latter species is rapidly removed from the gas phase by various heterogeneous processes.

In each of the above systems (e.g., NMHC, NH_3, and the sulfur compounds), the mechanism following the initiating step—leading to the formation of a final oxidized product—is unknown. The absence of this kinetic information reflects, to a large degree, the absence of adequate methodologies to study the kinetics of polyatomic free radical species. New kinetic information will be essential to achieving an acceptable level of understanding of these important oxidative atmospheric pathways.

HOMOGENEOUS AQUEOUS-PHASE TRANSFORMATIONS

The aqueous phase is most frequently considered in the context of physical removal processes. Like the gas phase, however, this medium also encompasses extensive chemical transformations. And, like the gas phase, these chemical transformations are oxidative in their chemical nature and involve some of the same reactive agents i.e., OH, HO_2, and O_3, although the aqueous phase is far more complex in its chemistry than the gas phase. Not only are there a large number of single-step elementary-type reactions to contend with, but there are also numerous fast equilibria. Furthermore, this chemistry involves the reactions of neutral free radicals, free radical ions, *non*-free-radical ions, and nonradical, nonionic, reactive species such as H_2O_2 and O_3.

Making this chemistry still more complex is the fact that the aqueous phase is distributed in the atmosphere in the form of a broad spectrum of aqueous aerosols. The two most general classes of liquid aerosols may be identified as (1) those found in clouds or fogs, and (2) those present under clear air conditions. In the first case, the most important size range is 2 to 80 μm, although rain droplets up to a few millimeters can be found. The second category encompasses particles ranging from the size of critical clusters (10 angstroms) up to a few micrometers. The number density of aqueous aerosols as a function of size is also highly variable, being critically influenced by exact environmental conditions. How the chemistry of aqueous aerosols differs as a function of size is currently one of many poorly understood characteristics of these species.

Traditionally, the approach taken in unraveling this chemistry has involved studies of closed chemical reactor systems in which only two or perhaps three major chemical species are added to solution reactors. (In many respects, these studies have their analogue in gas-phase smog chamber investigations.) This has been particularly true of studies designed to elucidate the oxidation pathways of SO_2 and nitrogen oxides. In one of the most extensively investigated systems, involving aqueous SO_2 mixtures, added oxidizing agents have included O_2-saturated solutions with and without added metal ion catalysts, O_3-saturated solutions, and H_2O_2 solutions. The qualitative as well as semiquantitative data generated from these investigations have shown that each reaction system could potentially be important in the aqueous-phase oxidation of S(IV) to S(VI). All show some pH dependence, but of these the O_3 system appears to have a particularly high sensitivity to changes in pH level (see Figure 5.10).

For the most part, mechanistic details on the S(IV) to

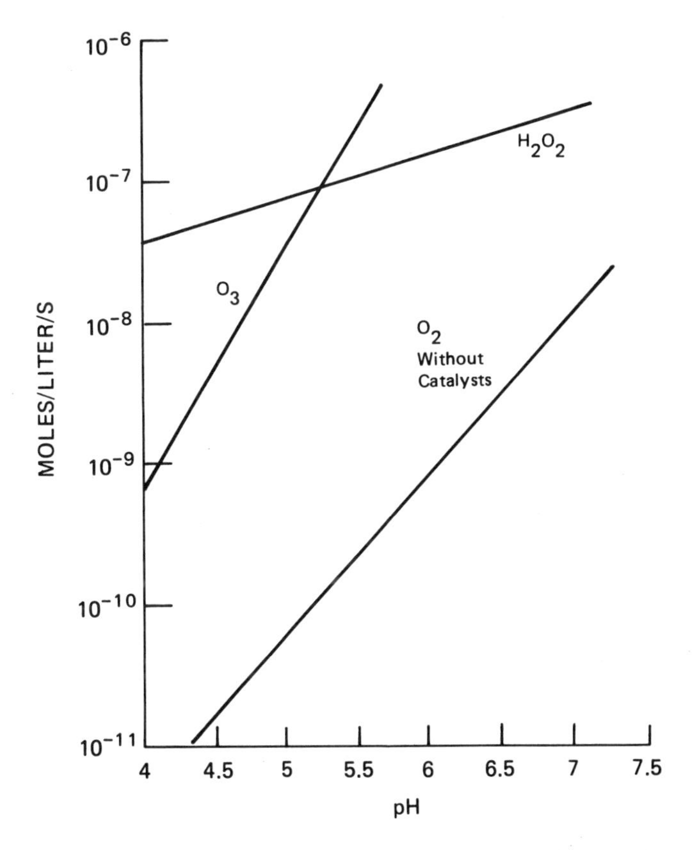

FIGURE 5.10 Conversion of S(IV) to S(VI). The pH dependence of the reaction rate is for the systems H_2O_2, O_3, and O_2.

S(VI) aqueous chemistry have been lacking. Thus it is unclear how the bulk chemical conversion rates measured for the O_3, O_2/metal ion catalyst, and H_2O_2 oxidizing systems may be combined for the case of a real aqueous aerosol environment. There is growing evidence, in fact, that many of the same reactive intermediates (especially free radicals) exist in all three systems. More recently, an integrated modeling approach has been attempted on these aqueous-phase systems. In this approach, as in modeling studies of homogeneous gas-phase chemistry, the entire reaction system is built up from a large number of elementary reactions. Sulfur, nitrogen, and/or carbon chemistries are taken to occur simultaneously and continuously in time. Illustrative of the latter approach is the chemical scheme shown in Figure 5.11. This chemical scheme portrays aqueous-phase chemistry as being made up of both numerous very fast equilibria and individual rate controlling elementary reactions. Chemical intermediates include both ionic and free-radical-type species. H_xO_y oxidizing agents in the system may result from aerosol scavenging of H_2O_2, O_3, OH, and HO_2 or by the in situ liquid-phase photolysis of the species H_2O_2 and O_3. The aqueous aerosol model may also be expanded to include the chemistries of NO_2 and NO_3 as well as soluble carbon species such as formaldehyde (CH_2O).

The use of building block elementary reactions to construct the complex chemistry of aqueous aerosols now appears to offer considerable potential. Facilitating this approach is the availability of a large volume of rate coefficients for elementary solution reactions. Most of these have been generated over the past 15 years by radiation chemists using, in particular, pulse radiolysis techniques. Even so, there remain numerous reactions of possible importance to this chemistry that still are without rate constants. Others, which have been measured, need to be reexamined with more advanced kinetic tools, especially with regard to establishing their temperature dependence.

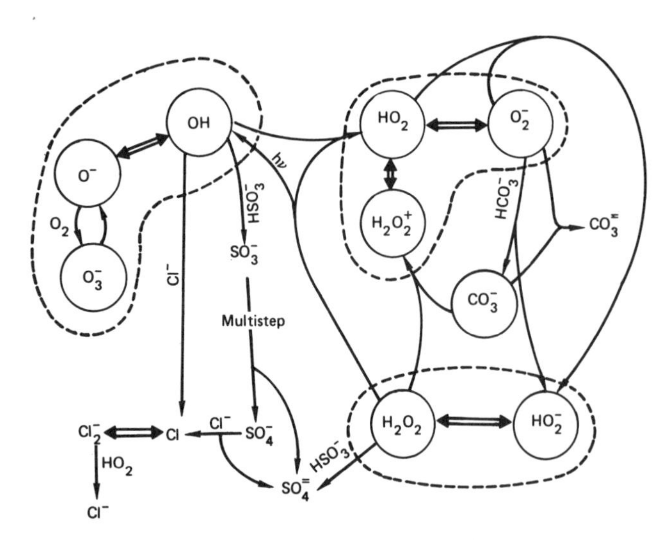

FIGURE 5.11 Primary chemical pathways for a cloud droplet containing reduced sulfur, carbonate, and Cl^- ion, and a source of reactive H_xO_y. The symbol ⟺ indicates fast equilibria, an → either an elementary reaction or multistep fast aqueous-phase process, and the dotted enclosures indicate various types of microchemistries taking place within the larger overall aqueous system.

In all cases, the question may be raised: Are rate coefficients measured in bulk liquid phase applicable to the broad spectrum of aqueous aerosols in the environment? Certainly, there would appear to be a need to investigate this chemistry under conditions where individual aerosol species could be studied as a function of time. Such studies will challenge the best technology, but must be viewed as a critical step in advancing the understanding of this science.

HETEROGENEOUS PROCESSES

Normally, a heterogeneous reaction implies one occurring at an interface between two phases, e.g., gas-liquid, gas-solid, or liquid-solid. Several interfaces may be involved in an overall process. Of course, solid-solid and liquid-liquid interfaces are also possible, but are not likely to be of great importance in atmospheric chemistry.

The study of interfaces, with their interesting and significant physical and chemical processes, is an old discipline that has recently reawakened. To appreciate the role of interfaces in many phenomena, one need only recognize that, in any multiphase system, communication between the bulk phases occurs through the surfaces that connect them. Even when these surfaces make up a small fraction of the total volume—as is usually the case for particles suspended in ambient air—they may have a dominant effect. A classical example can be seen in the phase transition of supercooled water droplets to frozen droplets (contact freezing) or ice crystals (sublimation freezing). This fundamental precipitation-producing process (Bergeron process) is believed to be induced and controlled to a significant extent by particles with very specific surface characteristics (ice nuclei).

As in all chemistry and solid-state physics, measurements at the atomic and molecular level lie at the heart of a satisfying description of surface structure and composition. Processes that occur at surfaces are described in terms of the time evolution of reactant, product, and intermediate structures. Without definition of surface structures in terms of equilibrium bond lengths and bond angles, as well as the potential energy functions that describe their variations, an adequate description of reactions at the molecular level is impossible. Because even the simplest heterogeneous reactions are very complex at the molecular level, understanding at this level requires the application of many complementary experimental and theoretical tools. As a result, attempts to resolve heterogeneous processes of importance to atmospheric chemistry are still quite rudimentary. An attempt to present a comprehensive picture of heterogeneous atmospheric chemistry can be found in various workshop documents listed in the bibliography.

Most reactions that occur on such surfaces are thought to be noncatalytic. These include chemical reactions in which both phases participate as consumable reactants, or physical processes involving either transport or growth, or both. The process of absorption or adsorption is a heterogeneous process. Heterogeneous catalytic processes normally imply the "conserved" participation of the interface material or a species adsorbed on it. Heterogeneous catalysis requires, among other things, the demonstration of "turnover" numbers far in excess of unity. The turnover number is essentially the number of repeated reactions conducted per unit time at a catalytic site.

Reactions can be heterogeneous overall but locally homogeneous, as represented by reaction within the bulk of an aerosol particle where reactants are transported in from the gas phase. These reactions might better be termed multiphase rather than heterogeneous because the reactants react in one phase, although some originate from another phase.

One special class of heterogeneous reactions is that referred to as gas-to-particle conversion. These reactions cause the transfer of a chemical species from the gas phase to an aerosol or liquid droplet suspended in the atmosphere, or they may cause formation of new particles.

Figure 5.12 shows a box diagram for gas-to-particle conversion processes including the following:

1. Homogeneous, homomolecular nucleation (the formation of a new stable liquid or solid ultrafine particle from a gas involving one gaseous species only);
2. Homogeneous, heteromolecular nucleation (formation of a new particle involving two or more gaseous species);
3. Heterogeneous heteromolecular condensation (growth of preexisting particles due to deposition of molecules from the gas phase).

The coexistence of homogeneous and heterogeneous reaction paths governing the distribution of key chemical species is shown in Figure 5.13.

Of the many gas-to-particle conversion processes believed to occur in the atmosphere, such as those depicted in Figure 5.13, one of the most interesting is the conversion of gas-phase SO_2 to sulfate. Because this process generates two hydrogen ions, it often is responsible for producing acid rain in regions containing high levels of SO_2.

Heterogeneous reactions may also be of importance in aqueous-phase atmospheric chemistry. The potential for transition metals, commonly found in atmospheric

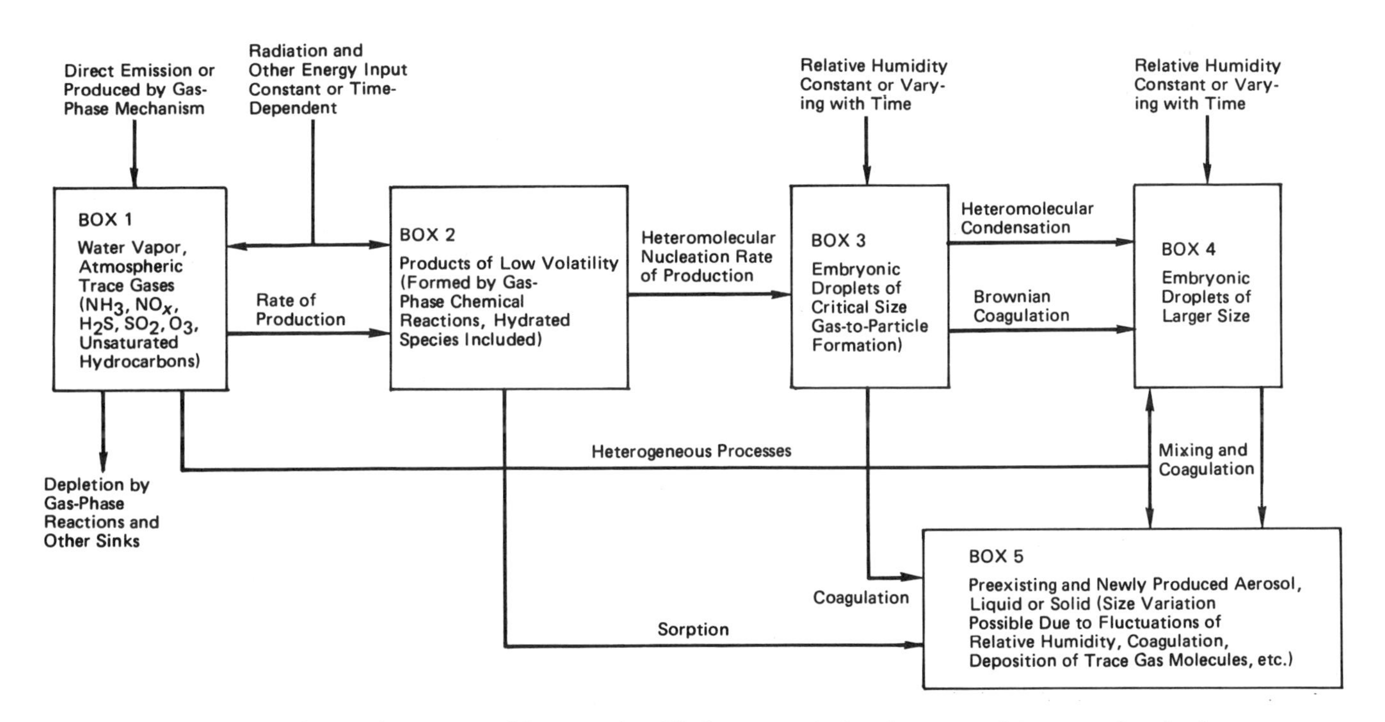

FIGURE 5.12 Box diagram for gas-to-particle conversion. The boxes contain the substance, and the arrows describe the process.

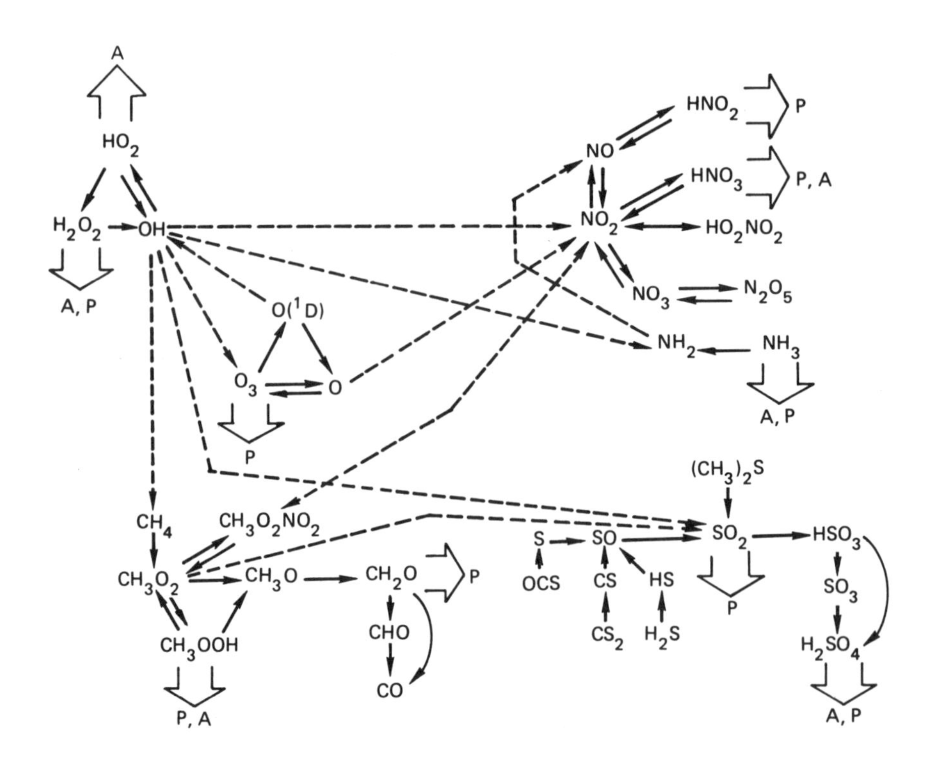

FIGURE 5.13 Gas-phase constituents and major reaction pathways (solid lines). Interactions between chemical families are indicated by dashed lines. Heavy (double) arrows show key heterogeneous pathways involving aerosols (A) and precipitation (P) (Turco et al., 1982).

aerosols and cloud droplets, to function as heterogeneous catalysts is relatively well known. Based on their abundance, chemical form, stable oxidation states, and bonding properties, one can speculate that iron, manganese, and perhaps copper are most likely to function in this way (especially in the case of rural or urban atmospheric environments). For an atmospheric reaction heterogeneously catalyzed by a transition metal, the rate-determining step is likely to be either reaction at the catalyst surface or permeation of the reactant through the organic film that is sometimes observed on aqueous atmospheric aerosols. Heterogeneous catalysis involving transition metals is likely to be of little consequence for species with rapid gas-phase or homogeneous liquid-phase reaction pathways, but may be significant for slower processes such as the aqueous-phase oxidation of SO_2.

Soot-catalyzed SO_2 oxidation may be another important mechanism for sulfate formation in the atmosphere. Soot is synonymous with primary carbonaceous particulate material. It appears that this material is present not only in urban atmospheres, but also in remote regions such as the Arctic. It is a chemically complex material consisting of an organic component and a component variously referred to as elemental, graphitic, or black carbon. Soot has properties similar to those of activated carbon, which is well known to be a catalytic surface active material.

The above discussion makes it quite apparent that the inclusion of heterogeneous processes is essential to achieving a complete understanding of the tropospheric cycles of sulfur, nitrogen, chlorine, carbon, and so on. However, because of the complexity of this chemistry, its quantification in existing models has not yet been satisfactorily accomplished. Thus both extensive laboratory and extensive field studies are needed.

BIBLIOGRAPHY

Homogeneous Gas-Phase Transformations

Atkinson, R., K. R. Darnall, A. C. Lloyd, A. M. Winer, and J. N. Pitts, Jr. (1979). Kinetics and mechanism of the reaction of the hydroxyl radical with organic compounds in the gas phase. *Adv. Photochem. 11*:375–488.

Chameides, W., and J. Walker (1973). A photochemical theory of tropospheric ozone. *J. Geophys. Res. 78*:8751.

Crutzen, P. J. (1983). Atmospheric interactions—homogeneous gas reactions of C, N, and S containing compounds. Chapter 3 in *The Major Biogeochemical Cycles and Their Interactions.* SCOPE 21, B. Bolin and R. Cook, eds. Wiley, New York, pp. 67–112.

Demerjian, K. L., and J. G. Calvert (1974). The mechanism of photochemical smog formation. *Adv. Environ. Sci. Technol. 4*:1–262.

Fishman, J., S. Solomon, and P. Crutzen (1980). Observational and theoretical evidence in support of a significant in situ photochemical source of tropospheric ozone. *Tellus 31*:432.

Levy, II, H. (1974). Photochemistry of the troposphere. *Adv. Photochem. 9*:5325–5332.

Liu, S. C., D. Kley, M. McFarland, J. D. Mahlman, and H. Levy, II (1980). On the origin of tropospheric ozone. *J. Geophys. Res. 85*:7546–7552.

Logan, J. A., M. J. Prather, S. C. Wofsy, and M. B. McElroy

(1981). Tropospheric chemistry: a global perspective. *J. Geophys. Res. 86*:7210–7254.

Seiler, W. (1974). The cycle of atmospheric CO. *Tellus 26*:116.

Seinfeld, J. H. (Chairman) (1981). *Report on the NASA Working Group on Tropospheric Program Planning*. NASA Reference Publication 1062.

Wofsy, S. C. (1976). Interactions of CH_4 and CO in earth's atmosphere. *Annu. Rev. Earth Planet. Sci. 4*:441–469.

Homogeneous Aqueous-Phase Transformations

Chameides, W., and D. D. Davis (1982). The free radical chemistry of cloud droplets and its impact upon the composition of rain. *J. Geophys. Res. 87*:4863–4877.

Farhataziz, and A. B. Ross (1977). Selected specific rates of reactions of transients from water in aqueous solution. III, Hydroxyl radical and perhydroxyl radical and their radical ions. NSRDS—NBS 59, special publication. Department of Commerce, National Bureau of Standards.

Graedel, T. E., and C. J. Weschler (1981). Chemistry within aqueous atmospheric aerosols and raindrops. *Rev. Geophys. Space Phys. 19*:505–539.

Heiko, B. G., A. L. Lazrus, G. L. Kok, S. M. Kunen, B. W. Grandrud, S. N. Gitlin, and P. D. Sperry (1982). Evidence for aqueous phase hydrogen peroxide synthesis in the troposphere. *J. Geophys. Res. 87*:3045–3051.

Junge, C. E., and T. A. Ryan (1958). Study of the SO_2 oxidation in solution and its role in atmospheric chemistry. *Quart. J. Roy. Meteorol. Soc. 84*:46–55.

Penkett, S. A., B. M. R. Jones, K. A. Brice, and A. E. J. Eggleton (1979). The importance of atmospheric ozone and hydrogen peroxide in oxidizing sulfur dioxide in cloud and rainwater. *Atmos. Environ. 13*:123–137.

Pruppacher, H. R., and J. D. Klett (1978). *Microphysics of Clouds and Precipitation*. Reidel, Boston, Mass., pp. 1-714.

Ross, A. B., and P. Neta (1979). Rate constants for reactions of inorganic radicals in aqueous solution. NSRDS—NBS 65. Department of Commerce, National Bureau of Standards, pp. 1–55.

Scott, W. D., and P. V. Hobbs (1967). The formation of sulphate in water droplets. *J. Atmos. Sci. 24*:54–57.

Stedman, D. H., W. L. Chameides, and R. J. Cicerone (1975). The vertical distribution of soluble gases in the troposphere. *Geophys. Res. Lett. 2*:333–336.

Taube, H., and W. C. Bray (1940). Chain reactions in aqueous solutions containing ozone, hydrogen peroxide and acid. *J. Amer. Chem. Soc. 62*:3357–3375.

Heterogeneous Processes

James, D. E., ed. (1979). *U.S. National Report, 1975–1978*, Seventeenth General Assembly International Union of Geodesy and Geophysics, Canberra, Australia, December 2–15, American Geophysical Union, Washington, D.C.

Kiang, C. S., D. Stauffer, V. A. Mohnen, J. Bricard, and D. Vigla (1973). Heteromolecular nucleation theory applied to gas-to-particle conversion. *Atmos. Environ. 7*:1279–1283.

Schryer, David R., ed. (1982). *Heterogeneous Atmospheric Chemistry*. Geophysical Monograph 26. American Geophysical Union, Washington, D.C.

Turco, R. P., O. B. Toon, R. C. Whitten, R. G. Keesee, and P. Hamill (1982). Importance of heterogeneous processes to tropospheric chemistry: studies with a one-dimensional model, in *Heterogeneous Atmospheric Chemistry*. Geophysical Monograph 26. David R. Schryer, ed. American Geophysical Union, Washington, D.C., pp. 231–240.

Vali, Gabor, ed. (1976). *Proceedings of the Third International Workshop on Ice Nucleus Measurements*, Subcommittee on Nucleation, International Commission on Cloud Physics, International Association of Meteorology and Atmospheric Physics, International Union of Geodesy and Geophysics, Laramie, Wyo., Jan. 1976.

WET AND DRY REMOVAL PROCESSES

BY B. HICKS, D. LENSCHOW, AND V. MOHNEN

In simple terms, the tropospheric concentrations of many species are determined by their rates of emission and removal. The species source is not usually a single term; it typically includes contributions from both natural and anthropogenic sources, as well as in situ production. Likewise, the removal rate is made up of both transformation and transport terms. However, deposition to the earth's surface constitutes the major sink for many tropospheric trace gases and aerosols. This section discusses removal at the earth's surface, which in many cases is the major factor limiting tropospheric trace gas concentrations.

The residence time of aerosol particles ranges from the order of a day in the atmospheric boundary layer (the lowest ~ 1000 m of the troposphere, which is closely coupled to the surface by convection and mechanical mixing) to more than a week in the upper troposphere. These residence times suggest that the physical removal processes are equivalent to chemical transformation rates of about 1 percent per hour.

It is convenient to differentiate between wet and dry deposition processes. The process by which falling hydrometeors (e.g., rain, snow, and sleet) carry atmospheric trace constituents to the surface is known as wet deposition. The processes of gravitational settling of particles and of turbulent transport (and subsequent impaction, interception, and absorption to exposed surfaces) of particles and gases to the surface are collectively known as dry deposition. There are several potentially important processes that do not fit neatly into either category. These include fog droplet interception, scavenging by spray droplets at sea, and processes associated with dewfall.

In practice it is sometimes not possible to apportion total deposition between wet and dry components. Moreover, in some circumstances it is clear that wet dominates dry, while in other cases the opposite appears to be true.

Such generalities should be modified according to the chemical and physical nature of the species under consideration. For example, submicrometer particles are poorly captured by falling raindrops and are inefficiently deposited by dry mechanisms. However, they can enter into the in-cloud nucleation, coagulation, and coalescence processes that precede precipitation. Soluble trace gases (such as HNO_3 vapor) are easily scavenged by falling raindrops and are rapidly adsorbed at exposed surfaces.

WET DEPOSITION

Wet deposition constitutes a very intermittent but highly efficient mechanism for transforming and eventually removing trace gases and aerosol particles from the troposphere. Aerosol particles act as nuclei for the condensation of water in warm clouds and for the generation of ice crystals in supercooled clouds. Subsequent coalescence and accretion lead to a wide range of droplet sizes, the largest of which initiate the precipitation process (e.g., snow and hail). The droplets collect other particles and gas as they fall, especially when passing through urban plumes or through a polluted boundary layer. The terms rainout and washout are sometimes used to differentiate between in-cloud and subcloud scavenging, but their use is dropping from favor.

Airborne particles are removed by falling raindrops below cloud base by much the same physical processes as cloud droplets scavenge particles within clouds. Scavenging efficiencies are related to particle size and chemical composition. In-cloud nucleation processes scavenge soluble, hygroscopic particles more easily than particles that do not have an affinity for water. Likewise, soluble and chemically reactive trace gases are more readily removed than less reactive species.

There has been considerable effort to document and model cloud scavenging systems. For example, a deep-rooted convective cell feeds on air from the boundary layer, which is normally the most polluted portion of the atmosphere, whereas some stratiform cloud systems form above the boundary layer and thus exist in a relatively cleaner environment. Scavenging characteristics of the two kinds of cloud systems will certainly be different; futhermore, the trace gases and aerosol particles accessible to them will differ. Because of the differences between scavenging within clouds by nucleation (and related cloud-physical processes) and subcloud scavenging by impaction and adsorption by falling hydrometeors, special care must be taken to interpret correctly the results of experimental case studies. The results of a study of particle washout by raindrops falling through a smokestack plume may not necessarily be applicable to the case of long-range transport and subsequent precipitation scavenging in remote regions.

The manner in which trace gases and aerosol particles are scavenged by clouds and by falling precipitation determines the preferred parameterization for inclusion in models. Steady precipitation falling through a polluted air mass will deplete pollutants at a rate proportional to the instantaneous concentration, so that an exponential decay of concentration with time will result. Measurements of the composition of precipitation through the duration of such a simple precipitation episode will display the same exponential time decay. Thus, for some short-term studies an exponential "scavenging rate" is used to relate precipitation quality to air chemistry, analogous in form to the decay constant of radioactive decay. This scavenging rate for small particles is typically of the order of 10^{-5} to 10^{-4} per second.

If average concentrations of chemical species in precipitation are of concern (with respect either to space or to time), then it is usual to assume a first-order linear relationship between concentrations of the species of interest in precipitation and their concentrations in air. The "scavenging ratio" defined in this way (i.e., the precipitation concentration divided by the air concentration) is expressed either on a volumetric or on a mass basis, and sometimes the precise definition is not made clear. Further confusion arises from the influence of meteorological factors, especially precipitation type and intensity, and the frequent uncertainty concerning the relative contributions of in-cloud and subcloud processes. The term "washout ratio" is frequently used synonymously with "scavenging ratio."

Early studies of radioactive fallout showed that in-cloud mechanisms result in highly efficient scavenging of many types of airborne gaseous and particulate material. Contemporary studies of precipitation acidity have shown that in-cloud reactions can be rapid, and that experimental determinations of scavenging ratios can be strongly affected by these reactions. Ambient SO_2 can interact with other chemical constituents in hydrometeors (e.g., H_2O_2 and nitrogen oxides) and can be deposited as sulfate. The role of clouds as sites for accelerated chemical reactions is a major emphasis of the research program described elsewhere in this report.

Evaporation of falling hydrometeors is sometimes sufficiently rapid that none of the precipitation leaving the cloud base reaches the ground. This process (virga) is a familiar example of cloud-related mechanisms for transforming material chemically and physically and for relocating it in the troposphere. The overall effect of clouds that do not rain is not well understood.

An illustration of the uncertainty regarding wet deposition processes is the case of SO_2 scavenging. It is known that SO_2 is absorbed in rain droplets at a rate that is strongly affected by the pH of the droplet. This absorption causes sulfur scavenging to be dependent on all other factors that influence precipitation acidity, many of which are not yet known. Temperature is acknowledged to have a strong influence on the rate at which dissolved SO_2 is oxidized; the results of scavenging studies carried out in winter must be expected to differ from those obtained in summer. Finally, it is certain that scavenging characteristics depend on the physical nature of the precipitation. Most studies to date have been of rain. Freezing rain, hail, and snow have yet to receive much attention.

Research conducted on the relationships between precipitation chemistry and air quality has often been hampered by the lack of chemical data at cloud height. There are obvious difficulties involved in using ground-level air chemistry observations as a basis for calculating scavenging ratios. Scavenging ratios for materials of surface origin are likely to be underestimates if ground-level air concentrations are used in their derivation, because air concentrations near the surface will generally be greater than those characteristic of the air from which the material is being scavenged by precipitation. Similarly, experimental evaluations of scavenging ratios for substances with sources in the upper troposphere will tend to be too high if ground-level air concentration data are used. Unless this source of error is eliminated by appropriate use of aircraft sampling or remote probing to measure chemical concentrations in the air that is being scavenged, there is little hope of resolving questions regarding the role of synoptic variables and cloud chemistry.

The mechanism for generating precipitation clearly affects precipitation quality. If rain falls through a polluted layer of air beneath cloud level, then a first-order dilution effect results. Thus the concentration of some soluble trace gas in rain sampled at ground level would tend to vary inversely with the amount of rain that fell. On the other hand, if air from the same polluted layer were drawn into an active orographic cloud scavenging material from a constant air stream and depositing it in a steady rain, then the concentrations in the rain would be far less influenced by the amount of rain that fell. In general, the relationship between precipitation chemistry and precipitation amount is indeed found to lie between the extremes corresponding to these two conceptual examples.

Just as the quality of rain depends on the quality of the air from which it falls, the total deposition of chemical species in precipitation is closely linked to the quantity of precipitation. Precipitation is a highly variable phenomenon that cannot be predicted with accuracy. The net deposition of chemicals associated with precipitation is more variable and even more difficult to predict. It seems unlikely that the capability to predict wet deposition at a single location on an event basis will ever be developed, since no organized prediction scheme can hope to reproduce the details of the random factors associated with the location and intensity of single storm cells. These deposition "footprints" have been the subject of some study; first as a consequence of concern about radioactive fallout, but most recently under the aegis of acid rain. Precipitation quality recorded during a single period of uninterrupted precipitation (an event) will display features corresponding to a cross section through the event. As a consequence, interpretation of the fine structure of time-sequence records observed at a single station is quite difficult, since it is often not possible to determine which part of the observed behavior is due to meteorological or air chemical processes and which is a result of the vagaries of the sampling cross section. However, the prediction of average patterns and of statistical variability (both with time and space) are achievable goals, provided appropriate information becomes available on the physical and chemical processes of importance.

Because the deposition "footprints" of different chemical compounds in single precipitation events tend to look alike, comparisons between deposition records of different chemical species must be expected to yield high correlation coefficients. Time records of sulfate deposition in rain at some specific site should be highly correlated with nitrate, for example, without the need to imagine some cause-and-effect relationship between these two species. In this regard, the determination of a low correlation coefficient may be as informative as detecting an unusually high value.

Recent emphasis on precipitation acidity has tended to divert attention from the basic questions of precipitation scavenging of particular trace gases and aerosol particles. High rainfall acidity does not necessarily mean very high concentrations of dissolved trace species in the rain, nor does a pH of 7 mean that the rain is completely free of dissolved material. Precipitation collected at remote sites is usually somewhat more acidic than expected solely on the basis of equilibrium with atmospheric CO_2 (pH about 5.6) as a result of background levels of nitrates and sulfates. The worrying feature of acid deposition over North America, for example, is not only its pH but also the concentrations of chemicals in the solution being deposited.

In contrast to the case of dry deposition, wet deposition rates can be monitored with existing techniques. Wet/dry collectors, which protect precipitation samples from contamination by dryfall processes during periods between rain events, became popular during the era of radioactive fallout studies and are now familiar instruments in most deposition measurement programs. The use of bulk collection devices is discouraged for studies of long-term wet deposition, because of the considerable uncertainty about the effect of dry deposition between precipitation events.

Collection of precipitation for chemical analysis of trace constituents, although conceptually simple, is susceptible to many problems. Many trace species of interest have extremely low concentrations, particularly in the remote regions that are often areas of concern when studying global biogeochemical cycles. Precipitation samples containing these species are easily contaminated during collection and subsequent sample handling. Problems with wall losses in the collection and storage vessel, biological activity in the samples, loss of volatile species, and so on, demand that the greatest care and preparation be taken before undertaking the seemingly simple task of collecting rain for chemical analysis.

FOG AND DEWFALL

Precipitation collection devices fail to provide representative data on deposition via fog interception and dewfall. Fog droplets can contain relatively high concentrations of pollutants; the physical and chemical processes involved are precisely those that contribute to the in-cloud component of normal precipitation scavenging. If fog forms in polluted air, significant deposition via fog droplet interception and deposition is likely. It is not obvious whether this process best fits under the general category of wet or dry deposition, and this uncertainty sometimes causes the process to be overlooked.

Studies of the acidity of cloud liquid water have shown that droplet interception by forest canopies can be a major route for acid deposition. Exceedingly low pH values have been reported, presumably in circumstances (such as high-altitude, stratiform clouds) in which there is minimal buffering and negligible access to the trace metals and NH_3 compounds that can serve as neutralizing agents.

It is clear that even uncontaminated fog droplets will cause dry deposition rates to be modified by wetting surfaces. Dewfall (and other processes that cause liquid water to form on exposed surfaces) will modify dry deposition rates in much the same manner, and for some chemical species net deposition rates can be significantly affected.

DRY DEPOSITION

Dry deposition rates are influenced strongly by the nature of the surface and by source characteristics. Surface emissions are held in closer contact with the ground than emissions released at greater altitudes, so that in the former case concentration loss by dry deposition would be expected to be greater. Consequently, dry deposition fluxes tend to be highest near sources, whereas the highest rates of wet deposition of the same substances may be found much further downwind.

Dry deposition rates are intimately related to atmospheric concentrations in the air near the surface. A first-order linear relationship is usually assumed. The coefficient of proportionality between atmospheric concentrations and dry deposition rates, which is known as the deposition velocity, clearly depends on the meteorological conditions, the chemical nature of the substance in question, and the nature of the surface on which it is being deposited.

The term "deposition velocity" suggests an analogy with gravitational settling that is usually incorrect. In most instances, deposition through the atmosphere is accomplished by turbulent mixing to within a very short distance of the final receptor surface, followed by diffusive transfer across a layer of near laminar flow immediately next to the surface. Turbulent transfer very near the surface is possibly influenced by the presence of small roughness features of the surface, electrostatic forces, and other mass and energy exchanges that are occurring. Discussion of the relationship between these (and other) potentially important factors is simplified by use of a resistance analogy, in which the inverse of the deposition velocity is viewed as a resistance to transfer in direct analogy with electrical resistance as described by Ohm's law. Individual resistances are associated with each process contributing to the dry deposition phenomenon, and these individual resistances are combined in a network whose structure reproduces the conceptual linkage between the various contributing mechanisms. A total resistance to transfer is then evaluated by using the electrical analog. The analogy is not perfect, however, it permits the processes involved in trace gas and aerosol particle deposition to be compared and combined in a logical manner.

Particles already deposited on a dry surface can be resuspended by wind gusts exceeding some critical value related to the size and density of the particle. Soil grains and particles of surface biological origin can be entrained in the lower atmosphere under some conditions. Suitable circumstances are not necessarily unusual. In arid regions, a surface saltation layer is frequently visible in strong winds, and it has been demonstrated that such aeolian particles can be carried into the upper troposphere by deep convection and transported horizontally for considerable distances. The generation of particles as a result of chemical reactions occurring within vegetated canopies has been postulated as a cause for the blue haze phenomenon associated with forests in many parts of the world. Ocean spray is another well-known example of surface generation of particles. Resuspended particles constitute another form of atmosphere-surface interaction, thus sharing many of the features normally associated with dry deposition.

There is considerable scientific disagreement about the mechanisms involved in dry deposition. Models (such as the resistance models mentioned above) that combine knowledge of individual processes to simulate natural phenomena occasionally omit processes that are sometimes considered to be important. However, all such models enable a test to be made of scientists' ability to simulate nature on the basis of their understanding of its component parts. For some circumstances and for some chemical species, the most important factors affecting dry deposition have been formulated well enough to permit fairly accurate modeling. The summary of the dry deposition of certain chemical species that follows is based on a contribution to the Critical Assessment Review Papers on acid deposition, soon to be released by the Environmental Protection Agency.

SO_2. Uptake by plants is largely via stomates during daytime, but about 25 percent is apparently via the epidermis of leaves. At night, stomatal resistance increases substantially. When moisture condenses on the surface, resistances to transfer should decrease substantially. Deposition to masonry and other mineral surfaces is strongly influenced by the chemical composition of the surface material. To water, snow, or ice surfaces, deposition rates are influenced by the pH of the surface water and by the presence of liquid films.

O_3. Dry deposition to plants is much like SO_2, but with a significant cuticular uptake at night and with the presence of surface moisture minimizing deposition rates. Deposition to water surfaces is generally very slow.

NO_2. Similar to O_3 for deposition to plants, but with a somewhat greater resistance to transfer. Even though NO_2 is insoluble in water at low concentrations, deposition to water surfaces might be quite efficient.

NH_3. No direct measurements are yet available, but a similarity to SO_2 appears likely.

Submicrometer particles. Deposition to smooth surfaces is a minimum for particles of about 0.5-μm diameter. Deposition velocities increase as particle size increases, until the terminal settling velocity predicted by the Stokes-Cunningham formulation is reached. Very small particles are deposited at rates that are controlled by Brownian diffusivity across a limiting quasi-laminar layer in contact with the surface. For rougher surfaces, deposition velocities tend to increase.

Supermicrometer particles. Turbulence can cause particles to be deposited by inertial impaction and interception, with deposition velocities greater than the Stokes-Cunningham prediction. Particle shape is an important factor.

Sulfate particles. A value of 0.1 cm/s is often used for the deposition velocity for sulfate particles. However, recent experiments have demonstrated that deposition velocities for sulfate aerosol vary with the roughness of the surface. Values less than 0.1 cm/s seem appropriate for snow and ice, and about 0.2 to 0.3 cm/s for growing pasture and grassland. There is considerable disagreement concerning forests. Some workers use large deposition velocities (approaching 0.7 cm/s), while others prefer to continue to use the value 0.1 cm/s used in early transport and dispersion models. Phenomenological differences appear likely.

Dry deposition to the oceans remains a major unknown. Data obtained in laboratory experiments on trace gas exchange between the atmosphere and water surfaces indicate that exchange rates are limited by factors associated with the liquid phase, especially with the Henry's law constant. The deposition of hygroscopic particles is known to be influenced by their growth upon entering the region of very high relative humidity near the water surface. However, the practical significance of the effect is still being debated. Of major importance is the fact that exceedingly little information is available for dry deposition under typical open ocean conditions. The average wind speed at sea is about 8 m/s, with a highly disturbed surface and much spray. In such conditions, the relevance of experimental data obtained in laboratory experiments seems open to question. In some areas of the world ocean, such as the "Roaring Forties," the surface is sufficiently agitated that the concept of a distinct, identifiable surface between the air and the ocean becomes difficult to defend. Rather, there is an interfacial layer with properties somewhat like a gas-liquid suspension. In such conditions, exchange of trace gases and aerosol particles between the atmosphere and the ocean may be quite rapid but bidirectional. Limiting processes cannot yet be identified with confidence.

Although detailed knowledge of many of the processes involved is lacking, the ability exists to measure dry deposition fluxes in some circumstances, for some substances. Dry deposition to some surfaces can be measured directly, e.g., in the cases of accumulation on snowpacks or ice, or on some mineral and vegetative surfaces. For very large particles, deposition can be measured by exposing artificial collection surfaces or vessels since the detailed nature of the surface plays a less important role. However, until recently there has been little information on the rate of deposition of small particles and trace gases to natural surfaces exposed in natural surroundings. In the last decade, methods developed for measuring the meteorological fluxes of heat, moisture, and momentum have been extended to O_3, CO_2, SO_2, nitric acid vapor, nitrogen oxides, and various particulate pollutants, with varying degrees of success. Some of these experiments have been intensive case studies, using instrumented meteorological towers, and were intended to identify and quantify factors controlling the deposition. Other studies have used instrumented aircraft to measure spatial averages of deposition fluxes over terrain of special interest. None have yet demonstrated a capability for routine monitoring.

There are essentially two schools of thought on monitoring dry deposition. The first advocates the use of collecting surfaces and subsequent careful chemical analysis of material deposited on them. The second infers deposition rates from routine measurements of air concentration of the trace gases and aerosol particles of concern and of relevant atmospheric and surface quantities. Collecting vessels have been used for generations in studies of dustfall and gained considerable popularity following their successful use in studies of radioactive fallout during the 1950s and 1960s. The inferential methods assume the eventual availability of accurate deposition velocities suitable for interpreting concentration measurements.

In the era of concern about radioactive fallout, dustfall buckets were used to obtain estimates of radioactive deposition, especially of so-called local fallout immediately downwind of nuclear explosions. It was recognized that the collection vessels failed to reproduce the microscale roughness features of natural surfaces, but this was not viewed as a major problem because the emphasis was on large "hot" particles and the need was to determine upper limits on their deposition so that possible hazards could be assessed.

Much further downwind, so-called global fallout was found to be associated with submicrometer particles similar to those likely to be of major interest in studies of global tropospheric chemistry. However, most of the distant radioactive fallout was transported in the upper troposphere and lower stratosphere, and its deposition was mainly by rainfall. The acknowledged inadequacies of collection buckets for dry deposition collection of global fallout were of relatively little concern because dry fallout was a small fraction of the total surface flux.

The acknowledged limitations of surrogate-surface and collection vessel methods for evaluating dry deposition have caused an active search for alternative monitoring methods. In general, these alternative methods have been applied to studies of specific pollutants for which especially accurate and/or rapid response sensors are available. The philosophy of these experiments has not been to measure the long-term deposition flux, but instead to develop formulations suitable for deriving average deposition rates from other, more easily obtained information such as ambient concentrations, wind speed, and vegetation characteristics. Nevertheless, several initiatives are under way to develop micrometeorological methods for monitoring the surface fluxes of particular pollutants. Surrogate surface methods are also being improved. Although these devices share many of the conceptual problems normally associated with collection vessels, they appear to have considerable utility in some circumstances. It has been shown that deposition of small particles to surrogate surfaces is sometimes similar to that of foliage elements. However, none of the surrogate-surface or micrometeorological methods that have been identified to date has been successfully demonstrated to monitor the dry deposition of a pollutant being slowly deposited.

BIBLIOGRAPHY

Beille, S., and A. J. Alshout (1983). *Acid Deposition*. D. Reidel, Dordrecht, Holland, 250 pp.

Engelmann, R. J. (1968). The calculation of precipitation scavenging, in *Meteorology and Atomic Energy*, D. H. Slade, ed. U.S. Atomic Energy Commission.

Galloway, J. N., and D. M. Whelpdale (1980). An atmospheric sulfur budget for eastern North America. *Atmos. Environ. 14*:409-417.

Galloway, J. N., J. D. Thornton, S. A. Norton, H. L. Volchok, and R. A. N. McLean (1982). Trace metals in atmospheric deposition: a review and assessment. *Atmos. Environ. 16*:1677-1700.

Greenfield, S. M. (1957). Rain scavenging of radioactive particulate matter from the atmosphere. *J. Meteorol. 14*:115-123.

Hales, J. M. (1972). Fundamentals of the theory of gas scavenging by rain. *Atmos. Environ. 6*:635-659.

Hardy, Jr., E. P., and J. H. Harley, eds. (1958). *Environmental Contamination from Weapons Tests*. Health and Safety Laboratory Report HASL-42A. U.S. Atomic Energy Commission.

Hicks, B. B., M. L. Wesely, and J. L. Durham (1980). *Critique of Methods to Measure Dry Deposition: Workshop Summary*. EPA-600/9-80-050. U.S. Environmental Protection Agency, 69 pp. (NTIS PB81-126443.)

Lindberg, S. E., R. C. Harriss, and R. R. Turner (1982). Atmospheric deposition of metals to forest vegetation. *Science 215*: 1609-1611.

Liss, P. S., and W. G. N. Slinn, eds. (1983). *Air-Sea Exchange of Gases and Particles*, NATO ASI Series, Series C. Mathematical and Physical Sciences No. 108. D. Reidel, Dordrecht, Holland, 561 pp.

Owens, J. S. (1918). The measurement of atmospheric pollution. *Quart. J. Roy. Meteorol. Soc. 44*:149-170.

Pruppacher, H. R., R. G. Semonin, and W. G. N. Slinn, eds. (1983). *Precipitation Scavenging, Dry Deposition, and Resuspension*, Vol. 1, *Precipitation Scavenging*. Elsevier, New York, 729 pp.

Pruppacher, H. R., R. G. Semonin, and W. G. N. Slinn, eds. (1983). *Precipitation Scavenging, Dry Deposition, and Resuspension*, Vol. 2, *Dry Deposition and Resuspension*. Elsevier, New York, 731 pp.

Sehmel, G. A. (1980). Particle and gas dry deposition: a review. *Atmos. Environ. 14*:983-1012.

Shannon, J. D. (1981). A model of regional long-term average sulfur atmospheric pollution, surface removal, and net horizontal flux. *Atmos. Environ. 13*:1155-1163.

6 The Role of Modeling in Understanding Tropospheric Chemical Processes

BY R. DICKINSON AND S. LIU

PRINCIPLES OF MODELING

We have just discussed the various processes entering into tropospheric chemical cycles. These processes are studied individually to better understand them and to improve our descriptions of the quantitative relationships between the different variables entering into a process. These relationships are never exact representations of reality. They are subject to continuing improvement using better measurements and new insights. These mathematical relationships between various components of a process are referred to as "process models."

It is useful to distinguish between those variables of a process system that are internal to that system, that is, calculable from the process model, and those that are external, that is, prescribed from observation. In particular, the linkages to other processes are external variables in the formulation of an individual process.

If all the linked processes of the tropospheric chemical system are considered together with their linkages included as internal variables, we have a "system model." A system model still contains some external parameters, but if the model is sufficiently comprehensive, such parameters can be reliably prescribed for current conditions.

The system model has two basic functions. First, because it has maximized the number of variables it calculates, it provides a good opportunity to carry out extensive comparisons between the model calculated versus observed variables of the system; such comparisons help identify weaknesses in the individual process models. Second, it allows projections as to the future state of the tropospheric chemical system as various external parameters change with time. Of special interest in this context are external changes imposed by human activity, but also of interest is long-period natural variability in external conditions.

Because of the complexity of system models, they are generally integrated by means of computer programs. One important aspect of such integration is the development and use of numerically accurate procedures for solving the differential equations that are used to define the process models and, hence, system models. Tropospheric chemistry shares with meteorology a concern for a wide range of interacting scales, beginning on the scale of individual microeddies, e.g., within a smoke plume from a power plant, and ending in the global scale. Satisfactory parameterizations of the role of smaller scale processes should be one of the objectives in developing and improving system models of global tropospheric chemistry.

EXISTING MODELS

The generality and detail possible in a complete system model are limited by difficulties of interpretation as a result of its extreme inherent complexity and the large demands placed on computer and programming resources. Up to now, the necessary staff and institutional support to pursue a complete system model of tropospheric chemistry have not been developed. Furthermore, there have been considerable uncertainties in the individual processes. Consequently, the research tools that have been used in tropospheric chemistry studies for synthesis and interpretation lie between the concept of a process model and a complete system model. We shall refer to these tools simply as tropospheric chemistry models. There has been developed a wide range of these models, with their content depending on the interests and objectives of their developers as well as their access to computing resources. Basically, what distinguishes the models now used from a complete system model is that they attempt to model accurately only some of the processes of the system, the others being included only in a simplified or ad hoc fashion.

Existing models can be best classified by the issues they address. Usually these models consist of two major parts: the chemistry and the transport. Depending on the characteristics of the subject being studied, each model employs different degrees of sophistication in the treatments of chemistry and transport. At one extreme are meteorologically oriented models that obtain the motions of the atmosphere and temperature structure as three-dimensional time-varying fields by solution of the continuum equations of hydrodynamics and thermodynamics in response to realistic boundary conditions; however, these models have until recently approximated tropospheric chemistry by ignoring all species except water. Some studies are now under way using the winds generated by some regional and global meteorological models to provide transport for simplified chemical models. At the other extreme are the one-dimensional or box chemical models. As a consequence of their extremely complex processes of chemical species transformations, their transport consists essentially of empirically derived time scales for movement of species from one box or level of the model to another.

At the current state of development, one of the prominently distinguishing features of chemical models is their dimensionality. Thus there are zero-, one-, two-, and three-dimensional models. The zero-dimensional box models simulate laboratory chemical reaction measurements such as in smog chambers. Reactants in the reaction chamber are assumed to be completely mixed so that transport can be neglected. Chemistry is treated in detail by including all relevant elementary reactions. Usually Gear's code with small integration time steps is used to study the time-dependent behaviors of all reactants. Multidimensional models can be viewed as a large number of zero-dimensional models, coupled together by transport and radiation submodels, and each different because of different transport source-sinks and differing environmental conditions. In the modeling of the global tropospheric photochemistry, emphasis has been on the problems of O_3, CO, CO_2 (or carbon cycle), nitrogen cycle, and sulfur cycle. Box models and one-, two-, and three-dimensional models have been developed to study the natural chemistry and possible effects of anthropogenic activities on these species. Because of the computer resource requirements for three-dimensional modeling, full-scale chemistry has not been included. On the other hand, simpler models with full-scale chemistry usually do not successfully parameterize the important transport processes and hydrological cycle of the atmosphere.

Besides dimensionality, treatment of model time structure is also notable. For example, is the model steady state or capable of following transient changes? How does it treat diurnal and seasonal variations? Some models only calculate fast chemical transformations but prescribe as external, slowly changing species. Such models avoid the need for transport submodels because transport is primarily important for determining the distribution of the slowly changing species. Other models prescribe the species with fast chemistry, in particular OH, as external, and concentrate on the interaction among source, sinks, and transport of slow species.

One important distinction with regard to model objective is the difference between climatological and event models. This distinction arises because of the large day-to-day variability of meteorological processes, including transport. Thus a detailed case study of the processes of tropospheric chemistry over several days or less must recognize the actual transport occurring over that interval, either by explicit measurement of it over the interval, or by measuring enough initial meteorological data to allow integration of a weather forecast model for the time and space domains of interest. On the other hand, if a study is more concerned with the average behavior of the atmosphere as described by means and higher statistical moments, then there is less demand on temporal accuracy in providing the meteorological transport terms. The most effective tools in this instance are the general circulation models that obtain from first principles the statistical properties of the atmosphere by direct numerical simulation. That is, they calculate day-to-day weather variations over a long period of time that do not correspond to any particular time period but are supposed to have the same statistics as actual weather systems. In other words, they model the climate of the atmosphere system.

Much of the work on three-dimensional modeling of tropospheric chemistry up to now has been on the urban and, more recently, regional scale. The chemical models developed for these studies should be considered in developing models for global tropospheric chemistry.

Dispersion models have been developed to study the dispersion of air pollutants from power plants as well as cities. These models are concerned with urban-scale transport but usually ignore chemical transformations of air pollutants. Models that do include air pollution chemistry but employ simpler transport parameterizations are called air quality models. Most of these models are developed for metropolitan and suburban areas; occasionally regions as large as the eastern United States have been included.

Because air quality models emphasize the oxidant problem, the chemistry usually includes that of O_3, NO_x, and hydrocarbons. Studies using smog chambers have led to the development of detailed mechanisms for specific hydrocarbons. Unfortunately, the chemistry in these mechanisms is far too extensive to be incorporated into air quality computer models. In order to circumvent this problem, at least two approaches have been utilized. These involved "lumping" the hydrocarbons by classes and using a generalized reaction mechanism for these classes, or by using a carbon-bond approach, which partitions the chemical species on the basis of the similarity of their chemical bonding. These chemical models have been tested against and tuned to a variety of smog chamber data. Usually good agreement is achieved between measured and predicted concentration-time profiles for all measured species. When these reaction mechanisms are incorporated into air quality models and compared to field measurements, the agreement becomes much poorer. Discrepancies could be due to poor transport parameterizations, but there is little doubt that lack of understanding of the chemistry in the real atmosphere also contributes to the discrepancies. In particular, the chemistry of aged and diluted air pollutants may be poorly understood because it cannot be effectively tested in smog chambers. Furthermore, heterogeneous reactions are either not included or treated by oversimplified parameterizations.

Acid deposition models are used to study wet and dry deposition of acid material such as sulfur and nitrogen compounds. Acid deposition models have been developed for Europe and the eastern half of North America. The major objective of these models is to establish the source-receptor relationship of acid deposition. So far, very little chemistry is included in the acid deposition models. Constant, linear conversion rates of SO_2 to $SO_4^=$ and NO_2 to NO_3^- have been used. Wet scavenging and dry deposition are assumed to be independent of cloud types or topography. Most of the modeling effort is focused on the development of the meteorological aspect of the model. There is a clear need to incorporate into these models the full-scale fundamental chemistry and kinetics involved in the transformation of sulfur and nitrogen compounds.

In conclusion, as the field of tropospheric chemistry matures, the various kinds of models should tend to converge more toward ideal system models. This occurs, on the one hand, as modelers learn to treat more elaborate model systems and, on the other hand, as they are able to understand the error implied by various convenient approximations and hence justify these approximations when their implied error is acceptable. Global modelers and regional modelers should collaborate on those aspects of their models that are of common interest.

MODELING IN SUPPORT OF THE PROPOSED RESEARCH PROGRAMS

We discuss here the modeling programs required to provide guidance to and help synthesize the results of the research programs proposed in Part I of this report. The programs in biological sources, photochemical processes, and removal processes require the development of submodels describing the individual processes involved. These would serve three purposes: (1) help to understand better the individual process, (2) help to extrapolate from individual observational sites to regional and global averages, and (3) provide submodels to be used in a comprehensive three-dimensional meteorological model coupled to global tropospheric chemistry. By contrast, the global distributions and long-range transport program would be used to help validate the overall performance of the chemical aspects of comprehensive three-dimensional models of tropospheric chemical processes. As submodels are developed for the various subprograms, they will be incorporated into the comprehensive models.

Biological and Surface Source Models

The models required to support the biological and surface source subprogram fall into three categories: (1) global empirical models, (2) mechanistic models of biological processes, and (3) micrometeorological and oceanic models of surface transport processes.

The observational efforts in the biological source subprogram will provide measurements at individual field sites. Initial exploratory efforts will identify the ecological communities that provide significant emissions, but as a second stage, it will be necessary to obtain sufficient observations to determine annual average emissions at various sites. Variability with environmental parameters such as temperature, solar radiation, and moisture will also be obtained. However, due to the great variety and small-scale structure of biological systems, it will always be very difficult to collect sufficient data to permit straightforward numerical averaging to establish regional and global average emissions. Rather, more sophisticated approaches will be required to interpolate and extrapolate the available observations to all the non-sampled areas. Exactly the same problem occurs in summarizing other types of data obtained from ecological communities. Ecologists, in particular, have been forced to resort to empirical procedures for obtaining such parameters as net primary productivity and biomass carbon (e.g., Table 5.1) from a rather limited number of field sites (less than 100). One systematic procedure has been to correlate ecological data with readily available climatic parameters pertaining to the sampled sites, in particular rainfall and temperature. The correlations so obtained are used to transform global maps of climatic parameters into global maps of ecological parameters. Such a procedure will be used to develop empirical models (i.e., maps) for the global distributions of the various measured biological emissions.

Complementary to the development of global distributions of biological emissions will be the development of models of the detailed biological mechanisms and processes responsible for the measured emissions. These will range from models of soil or oceanic biochemical processes to models of whole leaf physiology. Their development will require intensive collaboration with experts in other biological and chemical areas outside the atmospheric chemistry community. These efforts will, however, differ from current and past modeling in these other disciplines in the following aspects. First, they will be focused on the processes responsible for providing atmospheric emissions. Because these emissions have for the most part been recognized only recently, or in some cases not yet, the other disciplines have only begun to consider how such emissions could be provided from their existing submodels. Second, this modeling effort will be focused on the whole biological system (plant-soil-microorganisms, etc.) as it interacts with the atmospheric environment. Because of the great complexity of the processes involved, a model of the whole biological system will undoubtedly require simplifications in the descriptions of biochemical processes and the treatments of differences between species of organisms.

Modeling the effects of soil microorganisms would require modeling the environment where the processes occur. For example, the question of methane production requires a model of the diffusion of CO_2, H_2, and CH_4 from the production site to the atmosphere to address the question as to whether increasing levels of CO_2 could increase methane production.

Boundary layer and surface transport models are required to describe the movement between ocean or land surfaces and the atmosphere. In the case of oceanic processes, such models require consideration of oceanic as well as atmospheric boundary layers and the effects at the ocean interface of wave breaking, i.e., the movement of air bubbles on the ocean side and spray droplets on the atmospheric side.

The transfer of gases from and to a surface generally involves near-stationary diffusion-like transport processes that are represented in terms of effective resistances or conductances. That is, if c_a represents the concentration of a gas in the atmospheric mixed layer, and this gas is maintained at some concentration c_s at some surface, then the rate of flux of the species to the mixed layer from the surface is modeled as given by $(c_s - c_a)/r_t$, where r_t is the total resistance of the diffusion processes between the atmospheric mixed layer and the surface.

The most thoroughly studied gas transfer process, for example, is that of water vapor from soil and foliage. The water vapor concentration next to the mesophyllic cells inside a leaf is that of saturation at the temperature of the leaf. To reach the mixed layer, the water must pass through leaf stomata, the leaf boundary layer, the leaf canopy, and a roughness sublayer above the canopy before reaching the atmospheric mixed layer. Each of these barriers is modeled by one or more resistance in series or parallel. This description in terms of resistance, although somewhat simplistic, provides the maximum level of detail that can be practicably matched to models of atmospheric transport above the mixed layer and validated by micrometeorological observations.

In most cases, the surface boundary conditions are not as easily modeled as that of water vapor. Surface boundary conditions for species of interest would be one of the practical outputs of the modeling in the biological source subprogram. For example, rather detailed models are now available for leaf photosynthesis that provide the concentrations of CO_2 within the leaf cavity. Plants exert physiological control over water losses through stomatal closure; the stomatal resistance is significant not only for leaf exchange of water and CO_2 but also for SO_2, NO_2, O_3, and NH_3, and is modeled in terms of soil moisture and root resistances. Detailed boundary conditions for other gases, in particular SO_2, NO_2, O_3, and NH_3, and biologically emitted sources need to be established.

Field programs, together with continuation of laboratory (i.e., wind tunnel) studies should provide the data needed to develop and refine models of oceanic gas transfer processes, in particular for those species where surface boundary layers within the ocean provide additional resistance to their flux between ocean and atmosphere.

The micrometeorological processes of gaseous and particulate exchange within complex vegetated canopies are still poorly understood in detail, and need to be further studied through combined observational and theoretical approaches to generate improved models to be used in modeling gas and particle exchange.

A special effort should also be made to develop parameterizations for the suspension into the atmosphere of continental soil aerosols, so that generation of these aerosols can be simulated in meteorological models.

The Monin-Obukhov similarity theory gives an adequate basis for modeling fluxes through the surface mixed layer over horizontally homogeneous terrain. Its adequacy over more complex terrain, for example, with only sparse trees, is still an open question. This inadequate understanding of the micrometeorology above complex terrain introduces uncertainty into the modeling of gaseous and particulate fluxes between the atmosphere and these surfaces, and so into the average regional and global fluxes.

Modeling Noncyclic Transformation and Removal Processes in Clouds

Clouds and precipitation play important roles in the removal, transport, and transformation of species in element cycles. For instance, wet removal is probably one of the most effective sinks for nitrogen and sulfur compounds. Important species such as SO_2, N_2O_5, and perhaps NO_3 may go through fast aqueous transformation in cloud droplets. Furthermore, cloud convection may be an efficient vertical transport mechanism for trace gases and aerosol particles.

In order to evaluate these processes quantitatively, a cloud-removal model should include detailed treatments of the physical and chemical mechanisms involved. The cloud model should be a submodel of a meteorological model that includes self-consistent and realistic treatments of heat, moisture, and momentum transports. Physical aspects of the model include the parameterization of radiation, condensation and evaporation, and stochastic coalescence and breakup. Chemical aspects of the model include both homogeneous gas-phase and liquid-phase reactions as well as heterogeneous reactions. In the clean atmosphere, chemical species treated within the cloud model should include at least O_3, odd-nitrogen species, hydrogen radicals, H_2O_2, sulfur species, CO, and CH_4 and its oxidation products. In the polluted atmosphere, nonmethane hydrocarbons and their oxidation products, metals such as Mn and Fe, and graphitic carbon should also be considered.

Many fundamental parameters of gas-to-particle reactions and kinetics need to be studied in the laboratory and by modeling. For instance, sticking coefficients for above-mentioned gases on various types of aerosols under atmospheric conditions need to be measured and investigated theoretically.

Currently, cloud models exist that include various degrees of sophistication in treating the physical processes mentioned above. Their incorporation into a self-consistent meteorological mesoscale model is in progress. On the other hand, photochemical cloud-removal modeling is at a rudimental stage. Some advances have been made recently, due primarily to the worldwide interest in the acid deposition problem. Because these processes are subscale for global models and sometimes for regional models, proper parameterization of their effects will be a crucial step in modeling these processes.

A program that consists of well-coordinated field and laboratory measurements and model studies is needed to gain an adequate understanding of the cloud-removal processes. Each important mechanism involved, both physical and chemical, needs to be tested through iterative intercomparisons among modeling, field, and laboratory studies. Only then can a realistic cloud-removal model be developed based on the parameterization of these mechanisms.

Modeling Fast-Photochemical Cycles and Transformations

Some of the outstanding problems confronting the modeling of fast-photochemical transformations are (1) the prediction of the concentrations of OH and HO_2 in the ambient atmosphere and their dependence on the concentrations of NO_x and hydrocarbons; (2) the budget of O_3; (3) the mechanisms and rates of the oxidation of NO_x and SO_2 to nitrate and sulfate, involving both homogeneous and heterogeneous reactions; and (4) the nature and rates of fast transformations involving atmospheric hydrocarbons and their oxidation products. Unlike the modeling of cloud-removal processes, mathematical techniques of modeling fast gas-phase photochemical transformations are well developed. Computer models dealing with fast-photochemical transformations exist for polluted urban air as well as for clean background troposphere. The major deficiency in the understanding of fast-photochemical transformations lies in the lack of laboratory and field data. For instance, concentrations of key species such as OH, HO_2, and H_2O_2 have not been reliably measured, and kinetic data of many key reaction rate constants have not been determined under tropospheric conditions.

Recent developments in field and laboratory measurement techniques have made it possible to measure some of the above-mentioned key species and reaction rate constants. Further major advances in the understanding of fast-photochemical transformations can be made by coordinating modeling with the field and laboratory studies.

Heterogeneous processes may play some important roles in the fast photochemistry. As discussed in the previous section, major advances in measurement technology and theoretical treatment are needed in this area.

Modeling Global Distributions and Long-Range Transport with a Three-Dimensional Meteorological Model

A three-dimensional model of tropospheric chemical processes linked to meteorological and climatic processes or, in brief, a Tropospheric Chemistry System Model (TCSM) is an important tool for the overall synthesis and theoretical guidance of the GTCP. A recommended institutional framework for the development and operation of such a TCSM is given in the next subsection. Here we outline some of the research studies that one or more such TCSMs would carry out in support of the global distributions and long-range transport subprogram and, more generally, for modeling exploration of the tropospheric chemical system. Two classes of studies would be carried out with TCSMs. First would be studies intended to validate and possibly develop the ability of models to simulate long-range transport and global distributions and variability of long-lived chemical species. These studies would compare model results with the data sets obtained through the global distribution and long-range transport subprogram of GTCP and, if neccessary, develop the model improvements required for satisfactory validation. Most of the modeling studies would be carried out in a climatological framework, but one or more detailed event studies would be performed in conjunction with intensive periods of field data collection. Such studies could also be carried out with more simple chemistry as needed to simulate the sources and sinks of the long-lived species.

The second class of studies would emphasize the simulation of medium-lived species. At present, the sources and sinks of these species are not sufficiently well known for such studies to be used to test model transports. Rather, these studies, assuming adequate model transport submodels, would explore the role of meteorological processes in determining the spatial distributions and temporal variability of the medium-lived species. Such modeling studies, for example, could address the question as to how important are continental pollution sources of sulfur and odd nitrogen for atmospheric distributions at remote sites.

The exploration of such questions would help improve interpretation of the data on many of the species to be monitored in the global distribution and long-range transport subprogram. Medium-lived species of special current interest include NO-NO_2-HNO_3, NH_3, SO_2, sulfate-nitrate aerosol, and continental soil aerosol. The soil aerosol is of interest not only because of its optical-radiative effects, but also as a source of Ca, which can raise the pH of droplet aerosols.

Institutional Framework for Development and Application of a Three-Dimensional Tropospheric Chemistry System Model (TCSM)

The study of tropospheric chemistry, as it has developed over the past decade, has largely been in an exploratory phase of study with emphasis primarily on the development of new instruments and concomitant pioneering measurements of previously unseen species. It is the thesis of our report that tropospheric chemistry as a discipline is ripe to become a more mature science with large-scale field programs devoted to the systematic collection of required data. Essential to the successful application of these data to advance scientific understanding are not only submodels of the various processes studied but also system models of the overall tropospheric chemical system. Such models necessarily include the meteorological processes that transport and in other ways interact with the chemical species. The meteorological processes are best provided through versions of atmospheric general circulation models (GCMs) that have been especially designed to satisfactorily provide not only large-scale tracer transports in the free atmosphere but also transport through the planetary boundary layer across the tropopause and through cloud processes. These models also require physically based cloud submodels, a good description of land surfaces, and adequate treatment of the solar radiation driving tropospheric photochemistry.

The three-dimensional distribution of chemical species should be represented with spatial and temporal resolution comparable to that of the meteorological variables. The distribution of these species is determined by meteorological transport and source and removal processes and also by wet and dry chemical transformations.

It is recommended that one or more research groups be established, building on current modeling strengths, to develop such models in the time frame of the Global Tropospheric Chemistry Program. These groups should be strongly committed to carrying out the systems modeling studies required by the Global Tropospheric Chemistry Program, as well as efforts in developing the critical submodels required for successful application of the TCSM. They should contain expertise in both chemical and meteorological modeling and maintain close contacts with the observational subprograms of the Global Tropospheric Chemistry Program. The computer and programming resources necessary for successfully carrying out the task should be made available.

BIBLIOGRAPHY

Anthes, R. A. (1983). Regional models of the atmosphere in middle latitudes (a review). *Mon. Weather Rev. 111*:1306–1335.

Bass, A. (1980). Modeling long-range transport and diffusion, in *Conference Papers, Second Joint Conference on Applications of Air Pollution Meteorology, New Orleans, La.*, pp. 193–215.

Demerjian, K. L. (1976). *Photochemical Diffusion Models for Air Quality Simulation: Current Status. Assessing Transportation-Related Impacts*. Special Report 167. Transportation Research Board, National Research Council, Washington, D.C., pp. 21–33.

Lerman, A. (1979). *Geochemical Processes. Water and Sediment Environments*. Wiley, New York, 481 pp.

Logan, J. A., M. J. Prather, S. C. Wofsy, and M. B. McElroy (1981). Tropospheric chemistry and a global perspective. *J. Geophys. Res. 86*:7210–7254.

Mahlman, J. D., and W. J. Moxim (1978). Tracer simulation using a global general circulation model: results from a midlatitude instantaneous source experiment. *J. Atmos. Sci. 35*:1340–1374.

Turner, D. B. (1979). Atmospheric dispersion modeling, a critical review. *J. Air Pollut. Control. Assoc. 29*:502–519.

7 Tropospheric Chemical Cycles

TROPOSPHERIC CHEMISTRY AND BIOGEOCHEMICAL CYCLES

BY C. C. DELWICHE

The composition of the troposphere is determined to a large extent by reactions in the biosphere that maintain a quasi-stable tropospheric composition that would not persist in the absence of biological activity. The existence of molecular oxygen and nitrogen in the atmosphere is the most obvious consequence of this activity. These substances are present as a result of slow reaction kinetics and are not at thermodynamic equilibrium.

The concept of geochemical "cycles" of elements is not new, but only recently has the significance of biological activity in these cycling processes been appreciated. Elemental cycles developed by the atmospheric chemist, the biologist, the geochemist, and others all have different features of importance, depending on the interests of the reporting scientist. In the sections that follow we will present only a brief overview of some of these cycles to place them in perspective from the standpoint of the atmospheric chemist.

Most of the elements considered here have at least one volatile component of biological origin. Most of the cycles reflect the alternate oxidation and reduction of compounds in the energy metabolism of one or another life form. The process most commonly recognized is that of the photosynthetic-respiration sequence involving carbon and oxygen. Other reactions, such as those of nitrogen fixation and denitrification and the reactions of sulfur oxidation and reduction, are ancillary expressions of the primary processes involving carbon. They are generally dependent upon the carbon cycle for their operation, although the compounds of sulfur are themselves grist for a photosynthetic energy input.

Most of these major cycles have been altered in some of their features on the global scale by a factor of 2 or more as the result of human activity. Fossil fuel burning, although only 10 percent of respiration as a source of atmospheric CO_2, gives an annual increment of about 0.3 percent. Industrial nitrogen fixation and the use of legumes have about equaled "natural" nitrogen fixation, and sulfur from fossil fuel combustion and mining activities has about equaled the natural sources of atmospheric sulfur. Other processes, such as erosion of soil and the injection of some heavy metals like lead into the atmosphere, probably have altered natural cycles even more.

The closeness with which these cycles are coupled frequently is not appreciated in attempting to predict the consequence of their perturbation by human or other influences. Fundamentally, this coupling has its source in the energy demands of living organisms. The total system is drained of all the energy extractable from any reaction that can yield energy in significant amounts, and so tends to move toward a median energy level, expressed otherwise by Lovelock and Margulis in their treatment of the Gaia hypothesis.

Many of the compartments involved in biogeochemical cycles are shown in Figure 7.1. For our purposes, we

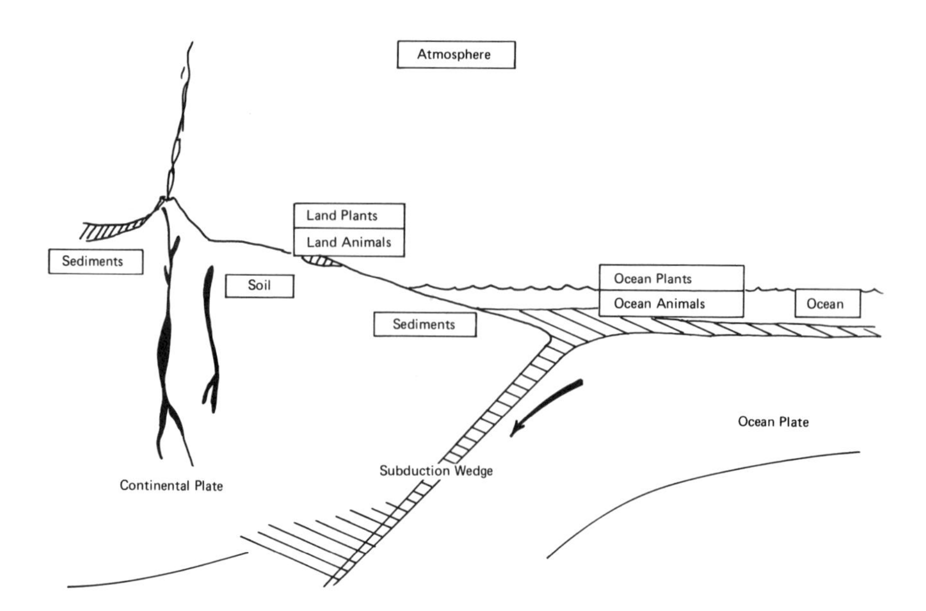

FIGURE 7.1 Diagrammatic representation of major compartments of biogeochemical cycles as discussed in the text. Although movement of oceanic plates, volcanic activity, and many other processes involved in these cycles are discontinuous, they are treated as steady-state processes for the purpose of developing mean estimates of their significance.

will not consider long-term (hundreds of thousands of years) cycle features, such as the sedimentary cycle or processes of subduction and subsequent volatilization through volcanism, except as the results of these processes contribute to the annual flux of a given element. For comparison, Figure 7.2 and Tables 7.1 and 7.2 give the distribution of four of the primary elements of interest (carbon, nitrogen, oxygen, and sulfur) between various "compartments" or "pools" in the environment and the estimated rates of transfer between them. It is important to remember that models of this type are intended as thinking tools, giving only the best estimates of the magnitude of the fluxes and burdens. Uncertainties of a factor of 2 or more are not unusual, and only in a secondary way is this uncertainty important to the analysis of problems or to planning.

A number of features have been omitted from Figure 7.2 for the sake of clarity. For example, the large pool of volatiles in magma is not considered except for an indication of volcanic sources where appropriate. Most of these volcanic sources are assumed to be the return of volatiles subducted with sediments, but some probably are truly "juvenile," representing an out-gassing of the magma that has been taking place (at a diminishing rate) since the earth was formed. The magnitude of this juvenile source relative to the recycling of subducted materials is controversial and not pertinent to the arguments we explore here.

Several points are evident from an examination of this table:

1. The major pools of the various elements are a function of their chemistry. Most of the nitrogen is in a partially "reduced" form in the atmosphere; most of the carbon is in carbonate rocks, bicarbonate ion in the ocean, or more reduced materials in soil sand sediments, with only a small (but important) fraction in the atmosphere. Sulfur is divided between the sulfate of the oceans and evaporites or in sediments, the atmosphere containing only a small amount in transit between these pools.

2. Biological processes are major factors in the movement of elements between the various pools, but, in general, the biosphere constitutes only a small fraction of the total.

3. Oxidation-reduction reactions in biological systems are responsible for most of the transfer taking place. The separation of charge of biological processes probably has created a broader range of oxidation potentials than existed before life developed on the planet. Thus there probably are both more oxidizing and more reducing conditions than existed before the appearance of life. The former is part of scientific lore, but the latter frequently is overlooked.

4. The concentration of oxygen in the atmosphere is determined not by the rate of photosynthesis, but by the degree to which reduced compounds (particularly those of carbon, nitrogen, sulfur, and iron) can be kept buried.

5. The partition of compounds between various compartments is a function of the energy balances involved. Thus, for example, the concentration of CH_4 in

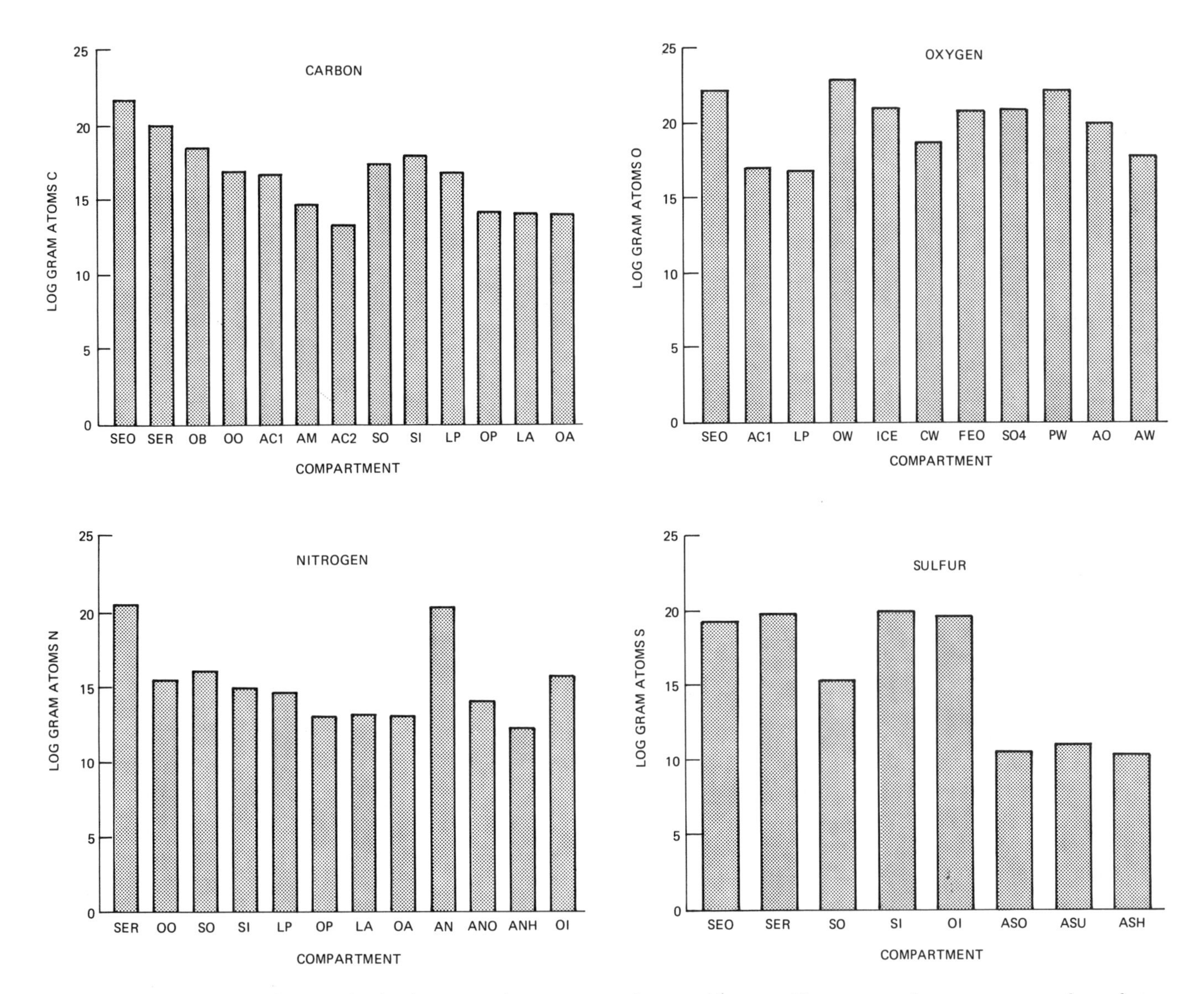

FIGURE 7.2 Pool sizes of interest for the elements carbon, oxygen, nitrogen, and sulfur. Ordinates give the log of the pool size in gram-atoms of the element. Logarithmic presentation is necessary because of the wide range in pool size. Unit increment on the scale represents a factor of 10 in pool size. Although some pool sizes are known with reasonable accuracy, others are accurate only to a factor of 2 or more. Numeric values are also presented in Table 7.1. Values compiled from various sources including Delwiche and Likens (1977); Garrels, Mackenzie, and Hunt (1975); and Söderlund and Svennson (1976).

TABLE 7.1 Pool Sizes of Interest for Carbon, Oxygen, Nitrogen, and Sulfur

	Element			
Compartment	Carbon	Oxygen	Nitrogen	Sulfur
SEO Oxidized sediments	50.8	152.		0.21
SER Reduced sediments	1.04		0.1	0.60
OB Ocean bicarbonate	0.032			
OO Ocean organic	0.00083		6.2 E-5	
AC1 Atmospheric CO_2	0.00054	0.0011		
AM Atmospheric CH_4	5.2 E-6			
AC2 Atmospheric CO	2.0 E-7			
SO Soil organic	0.0025		0.00021	2.2 E-5
SI Soil inorganic	0.010		1.1 E-5	0.81
LP Land plants	0.00069	0.00065	5.7 E-5	
OP Ocean plants	1.5 E-6		1.2 E-7	
LA Land animals	1.2 E-6		1.4 E-7	
OA Ocean animals	1.1 E-6		1.2 E-7	
OW Ocean water		761		
ICE Ice		9.16		
CW Continental water		0.055		
FEO In iron oxice		6.		
SO_4 In sulfates		8.5		
PW Sediment pore water		177		
AO Atmospheric O_2		0.76		
AW Atmospheric water		0.0058		
AN Atmospheric N_2			2.8	
ANO Atmospheric N_2O			1.3 E-6	
ANH Atmospheric NH_3			2.0 E-8	
OI Ocean inorganic			7.1 E-5	0.41
ASO Atmospheric SO_2				3.4 E-10
ASU Atmospheric sulfate ion				1.0 E-9
ASH Atmospheric reduced S				1.9 E-10

NOTES: Values are in units of 10^{20} gram-atoms of the element. Elements of igneous rock and magma are not included in this compilation. Where no values are given, the pool is not applicable, insignificantly small, or unknown. The code letters used correspond with those of Figure 7.2.

the atmosphere probably is a direct reflection of the energy relationships of microbial processes.

6. The consequences of human alteration of these cycles are best interpreted in terms of rates. Although the total system probably could accommodate large perturbations if sufficient time were allowed, the rate constants for many of the processes considered here are of the order of tens of thousands of years or more, and human activities on time scales of decades or centuries are not accommodated.

This short overview of biogeochemical cycles is intended to serve as a backdrop against which to examine atmospheric cycles of more immediate concern to this report. Details of these chemical cycles are available elsewhere (see bibliography at the end of each cycle section).

BIBLIOGRAPHY

Ahrens, L. H. (1979). *Origin and Distribution of the Elements*. Pergamon, New York, 537 pp.

Bremner, J. M., and A. M. Blackmer (1978). Nitrous oxide: emission from soils during nitrification of fertilizer nitrogen. *Science 199*:295–296.

Broda, E. (1975). The history of inorganic nitrogen in the biosphere. *J. Mol. Evol. 7*:87–100.

Broda, E. (1975). *The Evolution of the Bioenergetic Process*. Pergamon, Oxford, 211 pp.

Broecker, W. S., T. Takahashi, H. M. Simpson, and T.-H. Peng (1979). Fate of fossil fuel carbon dioxide and the global carbon budget. *Science 206*:409–418.

Delwiche, C. C. (1970). The nitrogen cycle. *Sci. Amer. 223*:137–146.

Delwiche, C. C., and B. A. Bryan (1976). Denitrification. *Ann. Rev. Microbiol. 30*:241–262.

Delwiche, C. C., and G. E. Likens (1977). Biological Response to Fossil Fuel Combustion Products, in *Global Chemical Cycles and Their Alterations by Man*, Werner Stumm, ed. Dahlen Konferenzen, Berlin, pp. 73–88.

Garrels, R. M., F. T. Mackenzie, and C. Hunt (1975). Chemical Cycles in the Global Environment. William Kaufmann, Los Altos, California.

Garrels, R. M., A. Lerman, and F. T. Mackenzie (1976). Controls of atmospheric O_2 and CO_2: past, present and future. *Amer. Sci. 64*:306–315.

Holland, H. D. (1978). *The Chemistry of the Atmosphere and Oceans*. Wiley, New York.

TABLE 7.2 Selected Transfer Rates Between Compartments

I Process	II From, To	III Quantity	IV Source Ratio	V Sink Ratio
Carbon				
Photosynthesis				
land	AC1, LP	4036	0.0747	0.0585
ocean	AC1, OP	2080	6.4 E-4	13.7
Fossil fuel combustion	SER, AC1	388	3.7 E-6	7.2 E-3
Biological CH_4 production	SO(?), AM	26	1.1 E-4	0.69
Atmosphere-ocean (CO_2) exchange	AC1, OB	8190	0.074	2.6 E-3
Wildfire	LP, AC1	126	1.8 E-3	2.3 E-3
Oxygen				
Photosynthesis				
land	CW, AO	8072	1.5 E-3	1.1 E-4
Fossil fuel combustion	AO, OW	1160	1.5 E-5	1.5 E-8
Nitrogen				
N fixation				
land	AN, LP	6.9	2.5 E-8	0.12
ocean	AN, OP	0.724	2.6 E-9	0.060
industrial	AN, SI	2.83	1.0 E-8	2.6 E-4
Denitrification				
land	SI,AN	8.5	7.7 E-3	3.0 E-8
ocean	OI, AN	2.86	4.0 E-4	1.1 E-8
Sulfur				
Fossil fuel combustion	SER, ASO	2.0	3.3 E-8	58.
Wildfire	LP, ASO	0.82	1.0 E-3	24.1
Biological reduction				
land	SI, ASH	0.12	1.5 E-9	6.3
ocean	OI, ASH	0.085	2.1 E-9	4.47
Volcanic return	SER, ASO	0.12	2.0 E-9	3.5

NOTES: The symbols used in Column II correspond with those of Table 7.1. Rates are in units of teragram (1 E-12 grams) atoms per year. Column IV gives the ratio of the quantity transferred to the source quantity; Column V gives the ratio of the quantity transferred to the sink quantity.

Holser, W. T. (1977). Catastrophic chemical events in the history of the ocean. *Nature 267*:403–408.

Junge, C. E. (1972). The cycles of atmospheric gases—natural and man-made. *Quart. J. Roy. Meteorol. Soc. 98*:711–729.

Kellogg, W. W., R. D. Cadle, E. R. Allen, A. L. Lazrus, and E. A. Martell (1972). The sulfur cycle: man's contributions are compared to natural sources of sulfur compounds in the atmosphere and oceans. *Science 175*:587–596.

Kvenvolden, K. A., ed. (1974). *Geochemistry and the Origin of Life.* Dowden, Hutchinson and Ross, 422 pp.

Li, Yuan-Hui (1972). Geochemical mass balance among lithosphere, hydrosphere, and atmosphere: the Gaia hypothesis. *Tellus 26*:1–10.

Margulis, L., and J. E. Lovelock (1978). The biota as ancient and modern modulator of the earth's atmosphere. *Pure Appl. Geophys. 116*:239–243.

Ponnamperuma, C. (1977). *Chemical Evolution of the Early Precambrian.* Academic, New York, 221 pp.

Söderlund, R., and B. H. Svensson (1976). The global nitrogen cycle, in *Nitrogen, Phosphorus and Sulphur—Global Cycles*, B. H. Svensson and R. Söderlund, eds. SCOPE Report 7. *Ecol. Bull. Stockholm 22*:23–72.

Sokolova, G. A., and G. I. Karavaiko (1964). *Physiology and Geochemical Activity of Thiobacilli.* Translated from Russian, 1968. 283 pp.

WATER (HYDROLOGICAL CYCLE)

BY R. DICKINSON

Water is such an important component of the environment that it is not surprising to realize that it is also one of the more important atmospheric species from the viewpoint of chemistry. Furthermore, the general framework that is used in this section to consider the cycles of other atmospheric trace constituents is also appropriate for water. Its distribution in the atmosphere is determined by the balances between sources, sinks, and transport, as illustrated in Figure 7.3. Water occurs in the atmosphere in three phases—vapor, liquid, and solid—and all three phases interact strongly with the other chemical cycles. The transformations between phases need special emphasis in viewing water as an atmospheric chemical.

SOURCES

With the exception of a small source by CH_4 oxidation in the stratosphere, and minuscule amounts yielded by some tropospheric reactions, the sources for water are entirely at the earth's surface. Water is removed from the earth's surface because of higher vapor pressures maintained at surface interfaces than within the atmosphere. On the global average, 1.0 m of water per year moves from the surface to the atmosphere and falls again as precipitation. The water vapor at wet interfaces is maintained at the saturation vapor pressure by equilibrium between the wet surface and its immediately adjacent molecular boundary layers. However, transport and removal processes in the atmosphere act to reduce water vapor pressure over much of the atmosphere to values below saturation. Furthermore, surface materials are often warmer than the overlying atmosphere because they absorb solar radiation. Some combination of lower temperature and lower relative humidity for the overlying air makes its water vapor partial pressure and mixing ratio lower than that of the surface. The consequent gradient in free energy drives water from the surface. Meteorologists often approximate the upward flux of water from the surface, F_w, by an expression of the form

$$F_w = C_w \rho_a V (q_s - q_a),$$

where C_w is a bulk transfer coefficient (under some conditions deductible from micrometeorological theory); ρ_a = density of the air; q_s = water vapor mixing ratio at the surface (e.g., the saturation mixing ratio evaluated at the temperature of the surface); q_a = water vapor mixing ratio in the air, evaluated at some reference level, usually 2 m above the ground or 10 m above the ocean; and V = magnitude of the wind at the reference level. Over oceans, $C_w = 1.4 \times 10^{-3}$ with some dependence on wind speed and wave height.

About 70 percent of the earth is covered by water and about 75 percent of the water entering the atmosphere comes from the ocean surface. The remaining 25 percent undergoes the interesting and complex physics of hydrological processes on land. At the simplest level, we can distinguish between evaporation from nonphotosynthesizing surfaces and transpiration. Evaporation

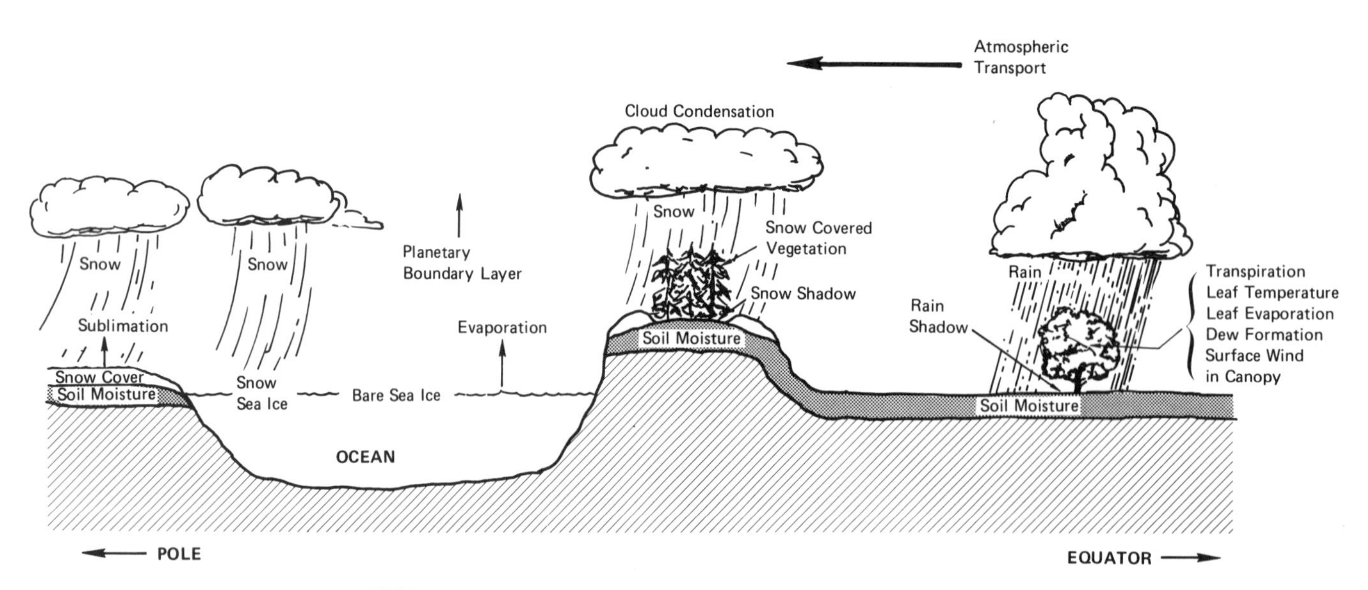

FIGURE 7.3 Features of the hydrological cycle in the atmosphere.

occurs as described above, with two additional complications: (1) the remaining water on relatively dry surfaces will be bound by surface tension and other stronger forces that will lower the water vapor pressure, and (2) the rate of water removal may be limited by the maximum rate at which water can diffuse from the interior of the soil or other object to the surface. Water transpiring from plants passes through the stomata of leaves, which generate enough diffusional resistance to lower significantly the water vapor on the outside of a leaf from its saturation value. The stomatal resistance changes with various environmental factors, including inability of roots to supply enough water because of soil dryness. Water loss by vegetation through stomata is believed to be primarily an accidental consequence of the need for plants to move CO_2 into their leaves to supply their photosynthetic cycles. Most other gas exchange between higher plants and the atmosphere also occurs through the stomata. Interception is another process involving vegetation that is of concern to hydrologists. The leaves and other plant parts become coated with water that can reevaporate, without the water progressing farther into the ground.

The saturation vapor pressure of water varies strongly with temperature according to the Clausius-Clapyron relationship. Thus saturation water vapor pressures near the surface and hence evaporation are much larger in the tropics than in high latitudes.

TRANSPORT AND DISTRIBUTION

Water vapor moves from the surface, through the planetary boundary layer, and then into the free atmosphere, where it is redistributed horizontally and vertically by atmospheric wind systems until it undergoes gas-to-droplet conversion. On a global average, a column of air holds about 27 kg/m^2 of water. Water in vapor form has an average lifetime of about 10 days and can move large distances (thousands of kilometers or more) before conversion to droplets. Liquid and ice particles generally have lifetimes of several hours or less and so are carried distances of 100 km or less before reconversion to the gas phase or removal by precipitation.

Because precipitation rates do not have as strong a latitudinal dependence as evaporation rates, large-scale atmospheric transport moves a significant fraction of the water evaporated in tropical latitudes into middle and high latitudes. This transport is one of the major processes for maintaining temperatures at high latitudes warmer than implied by radiative-convective equilibrium within a vertical column.

Motion processes on various scales are intimately connected to the gas-to-droplet conversions and droplet removal processes of precipitation systems described below.

The mixing ratio of water vapor in the troposphere varies over 4 orders of magnitude, from a few parts per hundred in the tropics near the surface to less than one part per thousand over the poles at the surface and to a few parts per million near the tropopause. This variability is explained to zeroth order by assuming a fixed relative humidity and noting that the mixing ratio varies with its saturated value. The reason relative humidity is not too variable, with sufficient averaging, is understood in terms of the role of atmospheric motion systems. By continuity, at any one time about half of the atmosphere is moving upward and is constrained to relative humidity near 100 percent by precipitation processes. The rest of the atmosphere is moving downward and drying the air to values much lower than saturation (e.g., near 10 percent). Combining the upward and downward streams gives an average relative humidity near 50 percent. As suggested by this discussion, instantaneous water concentrations at a given atmospheric level in the free atmosphere and given location have about a factor of 10 variability depending on the instantaneous vertical motion patterns.

TRANSFORMATION AND SINKS

In terms of chemical reactions of the water molecule itself, the most important role of water in the atmosphere is as a source for OH through the reaction,

$$H_2O + O(^1D) \rightarrow 2OH.$$

The production of the OH radical is fundamental to all the elemental cycles and is discussed in more detail in each of the other cycles sections. In the form of droplets, water provides the medium for numerous heterogeneous and homogeneous aqueous-phase reactions that are also fundamental to all element cycles in the troposphere.

On the microscopic scale, atmospheric water vapor is converted to droplets or snowflakes by migration to condensation centers, initially submicrometer cloud condensation nuclei. Growth of the droplets or flakes continues by further water vapor diffusion. When sizes of several micrometers or so are reached, further droplet growth occurs by collisional coalescence until the drops reach sufficiently large size (~100 μm) that their fall velocity exceeds the velocity of upward air motion. Their fall velocity is determined by the balance between downward gravitational acceleration and viscous (Stokes) drag, and so increases with increasing radius.

From a macroscopic viewpoint, water vapor condenses because atmospheric motions have produced water mixing ratios near their saturation values. The saturation mixing ratio decreases strongly with altitude because of its temperature dependence. Thus water condensation is driven primarily by upward transport via atmospheric motions. Conversely, sinking air tends to be cloudfree and of low relative humidity. The latent heat released by condensation processes can be of major importance in maintaining or enhancing atmospheric vertical motions. Two kinds of precipitation systems are distinguished, according to whether the latent heat is their primary drive or merely a positive feedback. Convective precipitation systems are driven by the latent heat they release. These generally occur on a horizontal scale with a fine structure of the order of 1 km and a large-scale organization of the order of 10 to 100 km. Layered precipitation systems are driven by upward motions in large-scale atmospheric wind systems forced by other modes of atmospheric instability. Convective precipitation can occur within layered systems.

BIBLIOGRAPHY

Baumgartner, A., and E. Reichel (1975). *The World Water Balance*. Elsevier, Amsterdam.

L'vovich, M. I. (1979). *World Water Resources and Their Future*. American Geophysical Union, Washington, D.C., 415 pp.

OZONE

BY H. LEVY II

Ozone (O_3) is both an important oxidant in its own right and a prerequisite for the production of hydroperoxyl and hydroxyl radicals. These radicals play key roles in most of the elemental cycles and control the atmospheric lifetimes of many of the short- and medium-lived trace gases. Besides its chemical role, O_3 is a significant absorber of long-wave radiation. Changes in the concentration of tropospheric O_3 will not only change the chemical lifetimes of many trace gases, but may also affect the climate.

SOURCES

The major sources of tropospheric O_3 are stratospheric injection and in situ photochemical production. There is also a small indirect contribution from the combustion source of NO_2.

The stratospheric injection of O_3 has been observed directly in the region of "tropopause folds," inferred from radioactivity measurements, and calculated from general circulation/transport models. These different approaches all arrive at a cross-tropopause flux in the range 3-12 $\times$ 10^{10} molecules per square centimeter per second.

In situ photochemical production occurs both in the polluted boundary layer and in the free troposphere as a whole. Significant production of oxidant, in particular O_3, has been clearly demonstrated in polluted urban environments. Not only has the production been simulated in smog chambers, but highly elevated O_3 concentrations are frequently observed in areas with high concentrations of hydrocarbons and NO_x. What is not known at this time is the importance of this smog source to the global troposphere. Summertime measurements of O_3 at 500 mb would suggest that this production extends up into the middle troposphere over regions of surface pollution. Positive correlations between fluctuations in simultaneous CO and O_3 vertical profiles have also been observed in the free troposphere, particularly over land at midlatitude in the northern hemisphere. This has been interpreted as demonstrating that O_3 has the same source region, the polluted boundary layer, as CO. A realistic estimate of the contribution from the polluted boundary layer is not yet available.

These same smog reactions,

$$HO_2 + NO \rightarrow NO_2 + OH,$$

and

$$RO_2 + NO \rightarrow NO_2 + OH,$$

followed by

$$NO_2 + h\nu \overset{O_2}{\rightarrow} NO + O_3,$$

should occur throughout the troposphere. Numerous theoretical calculations have predicted column production rates in the background troposphere of the range 1-10 $\times$ 10^{11} molecules per square centimeter per second or more. These calculations are, however, completely dependent on theoretical predictions of the peroxy radical concentrations and on predicted or measured concentrations of NO. At this time, there are many uncertainties in both the calculations and the measurements. Therefore, while the calculated production rates are much higher than the stratospheric injection rates, they are also less certain. On the other hand, they are so much larger that they suggest an important role for photochemical production in the troposphere.

SINKS

The two demonstrated removal paths for O_3 are destruction at the earth's surface and in situ photochemical destruction. A third, the fast reaction of O_3 with biologically emitted organics in the surface layer, is very difficult to separate from surface deposition and may end up being included in many measurements of surface destruction rates.

The surface destruction rate, frequently expressed as a surface deposition velocity, is highly variable depending not only on the nature of the surface, but, in the case of vegetation, on the type of vegetation, time of year, and even time of day. Various methods have been used to measure deposition velocity over a number of surfaces. These methods include: a direct measure of loss inside a box that covers a particular surface; indirect measurements based on inferring a flux from a measured vertical gradient; and an indirect measurement using the eddy correlation technique that calculates an eddy flux. Surface deposition velocities, while highly variable, do appear to separate into two main categories: (1) Land either bare or covered with vegetation has values of deposition velocity (W_0) that range from 2.0 cm/s for daytime forests and cultivated crops to 0.2 cm/s over nighttime grassland. Bare land falls in the low end of this range. (2) Water, snow, and ice surfaces have values in the range 0.1 to 0.02 cm/s. Estimates of average global fluxes to the surface have been made with different values for deposition velocity as a function of surface type and different values for O_3 concentration in the surface layer. There is considerable uncertainty in the distribution of surface types over the globe, deposition velocities for particular surfaces, and the global distribution of O_3 in the surface layer. Nonetheless, these calculations predict deposition fluxes in the same range as predictions of stratospheric injection.

The other half of O_3 photochemistry is photochemical destruction. At this time, the major removal path is thought to be:

$$O_3 + h\nu \rightarrow O(^1D) + O_2,$$

followed by

$$O(^1D) + H_2O \rightarrow 2\,OH.$$

A number of other mechanisms have been suggested: the destruction of O_3 by HO_x radicals; the oxidation of NO_x to nitrate and HNO_3 and their resulting deposition; the reactions involving the oxidation of halogens, particularly I, in the maritime boundary layer. Recent analyses of some regional boundary layer data in the equatorial Pacific strongly supports the existence of a photochemical removal process with an effective column removal rate of the order of 1-2 $\times$ 10^{11} molecules per square centimeter per second. This removal rate is much larger than the estimated surface deposition flux and is needed to explain the extremely low O_3 mixing ratios (5 to 10 ppb) that were observed at that time. Again, as for photochemical production, a realistic estimate of the global importance of this process requires accurate calculations of radical concentrations and detailed knowledge of other trace gas concentrations.

DISTRIBUTION/CLIMATOLOGY

A global data base, sufficient to produce a coarse resolution O_3 climatology, is urgently needed for the field of tropospheric chemistry. Not only is it needed to produce a global distribution of OH, the principal oxidizing species in the various elemental cycles, but it is needed to provide a framework for tropospheric photochemistry as a whole. Due to its high variability in the troposphere, relative standard deviations in the range of 25 to 100 percent, a realistic global data base will require relatively high spatial and temporal resolution.

Both the Dobson network and satellite observations provide a global field of total O_3. Unfortunately, approximately 90 percent of the total O_3 resides in the stratosphere, and existing techniques are not able to accurately extract the small fraction of the signal that represents the troposphere. Therefore these global fields are, at best, of qualitative use.

The best existing tropospheric data set is provided by individual ozonesonde stations that are now measuring or have in the past measured vertical profiles of O_3 from the ground to the middle stratosphere on a more or less regular basis. Data from stations still operating are being archived by the Canadian Department of the Environment. Unfortunately, there are a number of very serious problems with this data set:

1. A number of different types of sensors have been used, many of which were never accurately intercalibrated. Serious doubts have recently been raised about the absolute accuracy of the ozonesonde measurements in the troposphere, particularly for the older types of sondes that are no longer in operation or available for intercomparisons with current devices. Previous intercomparisons of operational devices alone have raised serious doubts about combining measurements from different research groups with different devices into a single data set.
2. Even if all the available data were useful, the global coverage is completely inadequate. Almost all the stations are in the northern hemisphere, and most of them are at midlatitude. There are a few in the high latitudes of the northern hemisphere, one in the tropics, and one operating (we hope) in the southern hemisphere at Aspendale, Australia. There are no stations in operation in any of the oceans, even at midlatitudes in the northern hemisphere. If all the sites that are no longer operating are included, there is minimal improvement in the global coverage.

While a global data set does not exist, careful analysis of either individual station data or individual networks using a common sensor and measurement protocol has produced many useful insights:

1. In all cases the mean profiles of O_3 increase with height.
2. Where it has been analyzed, the variance profile has a maximum in the upper troposphere and in the boundary layer with a minimum in the middle troposphere. A more detailed analysis of variance in the Aspendale, Australia, data finds in the troposphere that it is dominated by synoptic and shorter time scales.
3. The profiles show a spring maximum with, in the case of the midlatitude northern hemisphere continental stations, a continuation of this maximum into the summer. When analyzed, the variance also appears to be higher in the spring.
4. Both mean values and variability are greater at midlatitudes in both hemispheres than in the tropics, but the tropical data base is very limited. When data from a common instrument are considered, there is still some evidence for a midlatitude maximum, particularly in the lower troposphere.
5. An analysis of the North American network finds a significant east-west asymmetry on even the regional

scale. However, there is not enough longitudinal resolution in the data to determine the extent of zonal asymmetry in the global O_3 field.

A number of north-south transects through the middle and upper troposphere are available. The one data set that provides more than a single snapshot has severe troubles with absolute calibration of the sensor and was from the upper troposphere with the likelihood of aircraft incursions into the lower stratosphere at midlatitude and high latitude. A few single snapshot transects with relatively accurate O_3 sensors in the middle troposphere are also available. For this very sparse data set, O_3 is a minimum in low latitudes, the values either level off or decrease from midlatitude to high latitude, and the maxima occur at midlatitude in the hemispheric spring. Although these few data sets are *not* time mean latitudinal fields, they may have captured certain latitudinal features of O_3. Given the 25 percent relative standard deviations observed in profile data in the middle troposphere, it is also possible that these few profiles are atypical. A repeated series of flights over the same path with an accurate and validated sensor is certainly needed.

By far, the best time series data are available from continuous measurements of surface O_3. Unfortunately, the boundary layer is frequently very unrepresentative of the troposphere as a whole. The mean values may be strongly affected by local meteorology and surface removal, as well as local photochemical production and destruction. It is not clear what, if anything, can be inferred about the global troposphere from these excellent time series. In remote clean regions, the data appear to have many of the features observed in O_3 profiles with the addition of significant and currently unexplained diurnal fluctuations. They do appear to have significantly lower variability than is observed in the boundary layer of the profile data. In regions with pollution sources of NO_x, they show concentration maxima extending from spring into summer.

The final sources of distribution data are field generated by general circulation/transport models. A recent calculation of tropospheric O_3 with only stratospheric injection and surface removal, photochemical production and destruction having been excluded, has produced a tropospheric climatology of O_3 representative of the *model* meteorology. To the extent that the model meteorology is representative of atmospheric meteorology, the model field should represent the transport contribution to the real O_3 field. Outside of the boundary layer in general and continental regions with significant anthropogenic pollution in particular, the real O_3 field may be dominated by real atmospheric transport. This should be particularly true of the variance in the O_3 field. Therefore, the model variance fields may be quite useful in designing observational networks for O_3.

ISSUES

The three key issues involving tropospheric O_3 are as follows:

1. Its climatology (i.e., tropospheric distribution of mean values and higher moments);
2. The process or combination of processes that exert dominant control on its climatology;
3. The possible existence of long-term changes in the mean concentration and the causes of such trends if they do exist.

It is obvious that these three issues are intertwined. Furthermore, it is obvious that the first requirement is the development of a reliable data set. A few stations making very accurate long-term measurements for the detection of trends are needed, along with a significantly larger number making accurate measurements for a few years to establish at least a coarse global climatology. Coupled with this is the continued development and refinement of both the theory and numerical modeling of tropospheric transport and fast photochemistry.

The earliest view of tropospheric O_3 had it being transported down from the stratosphere and being destroyed at the earth's surface. Other than boundary layer variability, which would result from the large inhomogeneity of the surface destruction process, the distribution and variability would be controlled by meteorological processes on all scales. This view is still supported by much of the observational data.

In the early 1970s, an active photochemistry was proposed for the clean troposphere, which led to the prediction of large photochemical production and destruction rates. These predictions depended on many reactions that have not been quantitatively confirmed in the real troposphere and have as inputs species concentrations that were not well known. Nonetheless, the calculated photochemical production and destruction rates were much larger than measurements and estimates of stratospheric injection and surface removal. Furthermore, there were observations, particularly at midlatitude, that supported a strong role for photochemistry in the summer.

Recently, a unification of the transport and photochemical theory was proposed in which photochemical production occurred primarily in the upper troposphere with the precursor NO_x being injected from the stratosphere. Ozone destruction would then dominate in the lower troposphere where NO_x was very low. This O_3 destruction has been observed in one set of data taken in the tropical Pacific boundary layer. This theory depends critically on a tropospheric NO_x distribution, which increases strongly with height.

A recent general circulation/transport model study

reexamines the classical transport theory. It finds considerable agreement between observations and model results. At this time, there are not enough data to establish any theory, and the gathering of such data has the highest priority in O_3 research.

BIBLIOGRAPHY

Chameides, W. L., and J. C. G. Walker (1973). A photochemical theory of tropospheric ozone. *J. Geophys. Res. 78*:8751–8760.

Chatfield, R., and H. Harrison (1977). Tropospheric ozone, 2: variations along a meridional band. *J. Geophys. Res. 82*:5969–5976.

Crutzen, P. J. (1974). Photochemical reactions initiated by and influencing ozone in unpolluted tropospheric air. *Tellus 26*:47–57.

Fabian, P., and P. G. Pruchniewicz (1977). Meridional distribution of ozone in the troposphere and its seasonal variation. *J. Geophys. Res. 82*:2063–2073.

Fishman, J., and W. Seiler (1983). Correlative nature of ozone and carbon monoxide in the troposphere: implications for the tropospheric ozone budget. *J. Geophys. Res. 88*:3662–3670.

Galbally, I. R., and C. R. Roy (1980). Destruction of ozone at the earth's surface. *Quart. J. Roy. Meteorol. Soc. 106*:599–620.

Junge, C. E. (1962). Global ozone budget and exchange between stratosphere and troposphere. *Tellus 14*:364–377.

Lenschow, D. H., R. Pearson, Jr., and B. B. Stankov (1981). Estimating the ozone budget in the boundary layer by use of aircraft measurements of ozone eddy flux and mean concentration. *J. Geophys. Res. 86*:7291–7298.

Liu, S. C., D. Kley, M. McFarland, J. D. Mahlman, and H. Levy, II (1980). On the origin of tropospheric ozone. *J. Geophys. Res. 85*:7546–7552.

Mahlman, J. D., H. Levy, II, and W. J. Moxim (1980). Three-dimensional tracer structure and behavior as simulated in two ozone percursor experiments. *J. Atmos. Sci. 37*:655–685.

Oltmans, S. J. (1981). Surface ozone measurements in clean air. *J. Geophys. Res. 86*:1174–1180.

Pittock, A. B. (1977). Climatology of the vertical distribution of ozone over Aspendale (38S, 145E). *Quart. J. Roy. Meteorol. Soc. 103*:575–584.

FIXED NITROGEN CYCLE

BY S. LIU AND R. CICERONE

CURRENT ISSUES

The global nitrogen cycle is similar to the cycle of carbon and sulfur in that it has important atmospheric, marine, and soil components. In the atmosphere, as in the biosphere, fixed (or odd) nitrogen compounds are especially important. Further, fixed nitrogen is a limiting nutrient in many soils and water bodies. Its scarcity has caused man to intervene by producing artificially fixed nitrogen. This industrial production, when added to the nitrogen oxides fixed inadvertently in high-temperature combustion processes, has brought about the present situation in which it can be demonstrated that man is strongly perturbing the nitrogen cycle regionally and globally.

We will focus the discussion of the nitrogen cycle on individual fixed nitrogen compounds of atmospheric importance. Species of interest include odd-nitrogen compounds ($NO + NO_2 + NO_3 + N_2O_5 + HNO_2 + HNO_3 + HNO_4$ + peroxyacetyl nitrate (PAN) + other organic nitrates), NH_3, HCN, and N_2O, and the aqueous-phase species NO_3^-, NO_2^-, and NH_4^+. Nitrous oxide is included because of its significance as the principal source of stratospheric odd nitrogen, which flows downward and can affect the upper troposphere.

Odd-nitrogen species play a central role in tropospheric O_3 and H_xO_y chemistry. It is well known that odd-nitrogen species together with nonmethane hydrocarbons are the major precursors of high urban O_3. There is also evidence that odd-nitrogen species emitted by anthropogenic activities may contribute to the formation of high rural O_3 that is frequently observed in the summer over rural North America and Europe. Ozone, NO_2, PAN, and HNO_3 are major air pollutants that may be health hazards and/or damage plants under some conditions. There are also suggestions that the O_3 throughout the northern hemisphere might have been increased by pollution-derived nitrogen oxides and CO. This is of interest because infrared absorption by tropospheric O_3 is a significant factor in the energy balance of the atmosphere.

Nitrate is a major anion in regions of high acid deposition in Europe and North America. There are indications that long-range transport deposition of NO_3^- along with $SO_4^=$ takes place over distances of hundreds to thousands of kilometers. In addition, through the effect of odd-nitrogen species on H_xO_y chemistry, odd-nitrogen species may also influence the rate of oxidation of SO_2 to $SO_4^=$. It is also possible that odd-nitrogen species play key roles in the oxidation of hydrocarbons, CO, and other reduced gases in the troposphere.

Ammonia is the primary basic gas in the atmosphere, and it partially neutralizes sulfuric and nitric acid in precipitation. On the other hand, NH_3 can probably be oxidized to odd-nitrogen species in the troposphere, and the products contribute to the acidity of precipitation. The photochemistry of atmospheric NH_3 is not well understood, and a quantitative evaluation of the production rate of odd nitrogen from NH_3 is not yet available. Ammonium is an important component of continental aerosol particles and may help stabilize other compounds in the aerosol particles. Deposition of atmospheric NH_3, NH_4^+, and nitrate represents an important nutrient input to the biosphere in some regions.

The importance of N_2O lies mainly in its control of stratospheric O_3. Changes in stratospheric O_3 may have significant effects on tropospheric chemistry and on biospheric processes. Furthermore, NO_x produced from N_2O in the stratosphere is transported down to the troposphere and may affect tropospheric NO distributions, especially in unpolluted areas. The observed global increases of N_2O indicate that the global nitrogen cycle is not in a perfectly balanced steady state.

SOURCES

Emissions from fossil fuel burning (primarily high-temperature combustion), stratospheric intrusion, subsonic aircraft, biomass burning, lightning, soil exhalation, oxidation of atmospheric NH_3, and photolysis of marine NO_2^- are the likely major sources of odd-nitrogen species. Annual fluxes from the first three sources are probably known within a factor of 2. Uncertainties for fluxes from the other sources are much greater.

The estimated global annual source strength for each of these sources is presented in Table 7.3. Despite the large uncertainties in these estimates, it can be seen that anthropogenic odd-nitrogen emissions (i.e., fossil fuel burning and most of the biomass burning) are significant if not dominant sources of odd-nitrogen species. In view of the importance of atmospheric odd-nitrogen species, there is an urgent need to evaluate the effect of these anthropogenic emissions on the distribution of tropospheric O_3, H_xO_y, CO, hydrocarbons, and so on. Because the removal of odd-nitrogen species is primarily through wet and dry deposition, which is more efficient in the lower troposphere, upper tropospheric sources such as lightning, stratospheric intrusion, and aircraft may exert important control on the global odd-nitrogen distribution.

Significant sources of NH_3 may be animal waste, ammonification of humus followed by emission from

TABLE 7.3 Global Odd-Nitrogen Sources

Sources	Source Strength (Tg N/yr)
Fossil fuel burning	15–25
Subsonic aircraft	0.15–0.3
Stratospheric intrusion	0.5–1.5
Biomass burning	1–10
Lightning	2–20
Soil exhalation	1–10
Oxidation of NH_3	≤5
Photolysis of marine NO_2^-	≤5

soils, losses of NH_3-based fertilizers from soils, and industrial emissions. Estimates for natural NH_3 emissions are available, but are not reliable. An upper limit of about 50 Tg N/yr has been estimated for global emissions of NH_3 from undisturbed land. The other NH_3 sources (primarily from animal waste) are due to anthropogenic activities. Yearly anthropogenic emissions from the United States are estimated to be about 3 Tg N/yr. Apparently, there is not enough NH_3 emitted to neutralize the production of H_2SO_4 over North America.

Because NH_3 is readily absorbed by surfaces such as water, plants, soil, and atmospheric aerosol particles, its residence time should be very short in the lower troposphere. In fact, NH_3 emitted from vegetation-covered soil may have little chance of escaping to the atmosphere. Thus considerable caution must be used in estimating the emission rates of NH_3.

Nitrification and denitrification of natural soluble fixed nitrogen, industrial fertilizers, and fixed marine nitrogen, and power plant emissions are the primary sources of N_2O. All these sources are poorly quantified.

Hydrogen cyanide is emitted from cyanogenic plants, steel industry processes, and smoldering combustion of vegetation. No quantitative estimates are available.

DISTRIBUTIONS

Modeling the distribution of odd-nitrogen species has been severely hampered by the poor understanding of heterogeneous removal processes and the present inability to model realistically hydrological processes in the atmosphere. Current knowledge of the distribution of odd-nitrogen species is primarily from some infrequent and sparsely distributed measurements of uneven quality.

The distributions of NO, NO_2, HNO_3, and particulate NO_3^- are relatively better known, at least in continental regions, because there are established measurement techniques available. There have been a few long path optical absorption measurements of NO_3 in the surface air. PAN, HNO_2, and other organic nitrates have been observed but not adequately quantified in the troposphere. The distribution of odd-nitrogen species in the upper troposphere is probably more uniform than in the lower troposphere since this region is relatively far from large surface sources and sinks, as well as surfaces on which heterogeneous reactions can take place. Observed upper tropospheric NO_x (NO + NO_2) mixing ratios range from 0.1 to 0.4 ppbv. For HNO_3 the range is about 0.3 to 1.5 ppbv. The major sources of upper tropospheric odd-nitrogen species are probably stratospheric intrusion and lightning. It has been suggested that O_3 produced in the upper troposphere may be the major natural source for tropospheric O_3. According to model calculations, there are probably significant amounts of HNO_4 and PAN in the upper troposphere. Measurements of these species are needed. In the lower troposphere, observed mixing ratios of NO_x range from 0.01 ppbv in clean oceanic air to 500 ppb for highly polluted urban air. Similar mixing ratio ranges can be found for HNO_3 and NO_3^-.

There has been no attempt to model the atmospheric distribution of NH_3 or particulate NH_4^+. As in the case of odd-nitrogen species, observed data are too few. Most of the measurements to date are for particulate NH_4^+; gas-phase measurements of NH_3 are practically nonexistent. Sensitive and specific analytical techniques are needed for NH_3 and should be feasible with some effort.

The distribution of N_2O in the troposphere is well known. There is slightly more in northern latitudes, and the global concentration is rising slowly. About 160 pptv of HCN has been observed at midlatitudes, and little, if any, vertical or latitudinal variation is expected.

TRANSFORMATIONS

Gas-phase reactions of odd-nitrogen species have been studied extensively. Additional laboratory kinetic studies of the reactions of NO_3, N_2O_5, HNO_3, PAN, and other organic nitrates are still needed, especially under tropospheric conditions. Reactions of odd-nitrogen species with nonmethane hydrocarbons and their products also need to be quantified. Field studies of NO_3, the ratio of NO to NO_2, and their relation to O_3, HNO_3, and NO_3^- have revealed many interesting, perplexing results that may have important implications to tropospheric photochemistry.

Heterogeneous processes, including atmospheric heterogeneous transformations and surface deposition, play crucial roles as sinks of odd-nitrogen species and in interconversions among odd-nitrogen species. Lack of reliable laboratory determinations of the key reaction paths and rates is probably the most serious difficulty in attempting to understand heterogeneous processes involving odd-nitrogen species. In particular, sticking coefficients, aqueous reaction rates, and equilibrium constants need more attention.

Most of the natural and anthropogenic odd-nitrogen species emissions are in the form of NO. Usually NO reaches photochemical equilibrium quickly with NO_2 through reactions

$$NO + O_3 \rightarrow NO_2 + O_2 \quad (7.1)$$

$$NO + HO_2(RO_2) \rightarrow NO_2 + OH(RO) \quad (7.2)$$

$$NO_2 + h\nu \rightarrow NO + O(^3P), \quad (7.3)$$

where R denotes a hydrocarbon radical. Reactions (7.1) and (7.3) are fast but do not result in net gain or loss of O_3 because O quickly associates with O_2 to form O_3. Under most circumstances (when NO > 10 pptv) in the troposphere, reactions (7.2) and (7.3) result in net production of O_3 over urban as well as rural areas. These reactions may also play an important role in the budget of O_3 in the entire troposphere.

Oxidation of NO_2 by OH will form HNO_3, which is readily removed from the troposphere by heterogeneous processes such as precipitation scavenging and surface deposition. This is one of the major sinks of odd-nitrogen species and a major source of acidity in precipitation. PAN and HNO_2 formed through reactions (7.4) and (7.5) and nighttime reactions play important roles as temporary reservoirs for NO_x and hydrogen radicals in the photochemistry of the polluted atmosphere.

$$CH_3COO_2 + NO_2 \rightleftarrows CH_3COO_2NO_2\,(PAN) \quad (7.4)$$

$$OH + NO + M \rightarrow HNO_2. \quad (7.5)$$

Unlike HNO_3, PAN is apparently not scavenged by precipitation. The major sink for PAN is thermodissociation followed by reaction of CH_3COO_2 with NO. Because of the low temperature in the upper troposphere, PAN may have a tropospheric lifetime of a few months or longer. Therefore, if PAN is transported to or manufactured in the upper troposphere, it may serve as an important odd-nitrogen reservoir and thus a carrier for long-range transport of odd-nitrogen species. Other organic nitrates and HNO_3 may play the same role as PAN in the upper troposphere. Reaction of HNO_4 with OH may be an important sink for OH in the upper troposphere. This reaction is probably not important in the lower troposphere because of the fast thermal dissociation rate for HNO_4.

Formation of NO_3 and N_2O_5 by reactions

$$NO_2 + O_3 \rightarrow NO_3 + O_2 \quad (7.6)$$

$$NO_3 + NO_2 + M \rightleftarrows N_2O_5 + M \quad (7.7)$$

are important nighttime reactions for odd nitrogen. If N_2O_5 or NO_3 is scavenged by aerosol particles as indicated by recent field measurements of NO_3, they would constitute a significant sink for odd nitrogen and a major source for HNO_3 or particulate NO_3^-.

The photochemical transformations among odd-nitrogen species discussed above are summarized in Figure 7.4. The more speculative transformations are indicated by dashed arrows. Minor reactions have been omitted.

The most important gas-phase reaction for NH_3 is probably oxidation by OH. Little is known about subsequent reactions that involve NH_2 and its products. Reaction of NH_3 with HNO_3 to form NH_4NO_3 may be a significant sink for both NH_3 and HNO_3. However, the reaction rate constant has not been measured. Most of atmospheric NH_3 is believed to be removed by wet and dry deposition. Like odd-nitrogen species, quantitative information on heterogeneous processes is not available.

Quantitative aspects of cycling N_2O through the atmosphere are known within a factor of 2. Nitrous oxide exhibits no known tendency to react in the troposphere. Accordingly, its chemical lifetime is long (about 150 years), and its accumulation in the atmosphere can lead to significant infrared greenhouse effects. Discovered only recently, HCN could be the dominant odd-nitrogen compound in the background troposphere. Currently at about 160 ppt, its oxidation comprises an in situ source of NO_x. Its apparent slow reactions probably imply a small source (0.5 Tg N/yr) of tropospheric NO_x.

REMOVAL

Odd-nitrogen species, NH_3, and their counterparts in the particulate form are removed by cloud and precipitation scavenging, as well as dry deposition. These processes are poorly understood. Important constraints on removal rates and/or sources can be obtained through measurements of deposition rates of NH_4^+ and NO_3^-. Because of recent emphasis on acid rain research, wet deposition rates for NH_4^+ and NO_3^- are routinely measured over Europe and North America. Dry deposition may account for more than half of the total deposition of these species, but a routine measurement program does not exist. Deposition data are extremely sparse over the oceans. Atmospheric NH_4^+ and NO_3^- may be important sources of fixed nitrogen to the oceans in some regions, along with NO_3^- runoff in rivers.

Photolysis and oxidation by $O(^1D)$ in the stratosphere are the only known sinks for N_2O. This removal rate (about 10 Tg N/yr) is known within a factor of 2; accordingly, one can calculate from the background concentration of N_2O of 300 ppb an atmospheric chemical lifetime (about 150 years) to within a factor of 2.

Oxidation by atmospheric OH radicals is probably the major sink for HCN. Mechanistic understanding of the reactions involved is not yet complete.

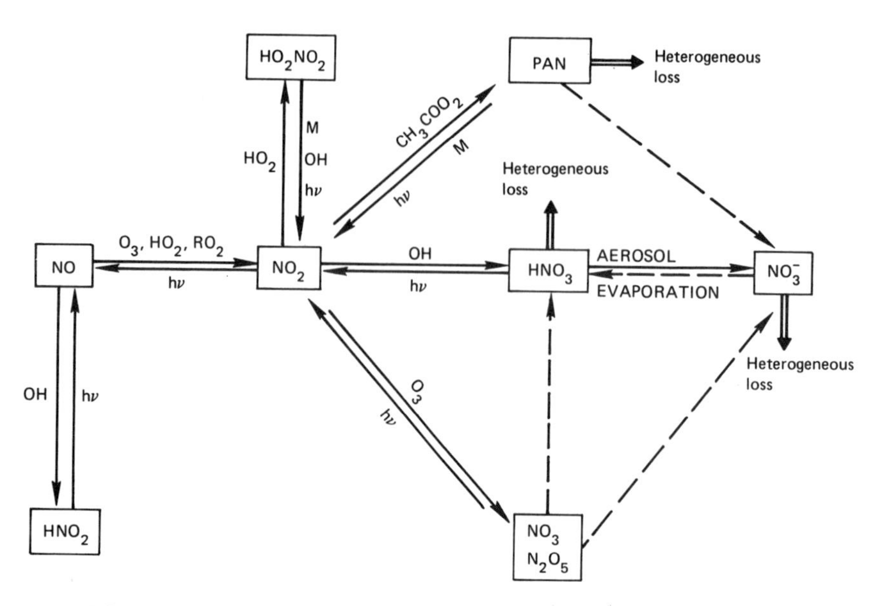

FIGURE 7.4 Photochemical transformations among odd-nitrogen species.

BIBLIOGRAPHY

Bauer, E. (1982). *Natural and Anthropogenic Sources of Oxides of Nitrogen (NO_x) for the Troposphere*. P-1619. Institute for Defense Analysis, Arlington, Va.

Calvert, J. G., and W. R. Stockwell (1982). The mechanism and rates of the gas phase oxidation of sulfur dioxide and the nitrogen oxides in the atmosphere, in *Acid Precipitation: SO_2, NO, NO_2 Oxidation Mechanisms: Atmospheric Considerations*, Ann Arbor Science, Ann Arbor, Mich.

Crutzen, P. J. (1979). The role of NO and NO_2 in the chemistry of the troposphere and stratosphere. *Annu. Rev. Earth Planet. Sci. 7*:443–472.

Dawson, G. A. (1977). Atmospheric ammonia from undisturbed land. *J. Geophys. Res. 82*:3125–3133.

Demerjian, K. L., A. J. Kerr, and J. G. Calvert (1974). The mechanism of photochemical smog formation. *Adv. Environ. Sci. Technol. 4*:1–262.

Ehhalt, D. H., and J. W. Drummond (1982). The tropospheric cycle of NO^x, in *The Proceedings of NATO Advanced Study Institute on Chemistry of the Unpolluted and Polluted Troposphere*. Reidel, Hingham, Mass.

Galbally, J. E., and C. R. Roy (1978). Loss of fixed nitrogen from soils by nitric oxide exhalation. *Nature 275*:734–735.

Huebert, B. J., and A. L. Lazrus (1980). Tropospheric gas-phase and particulate nitrate measurements. *J. Geophys. Res. 85*:7322–7328.

Kley, D., J. W. Drummond, M. McFarland, and S. C. Liu (1981). Tropospheric profiles of NO_x. *J. Geophys. Res. 86*:3153–3161.

Logan, J. (1983). Nitrogen oxides in the troposphere: global and regional budgets. *J. Geophys. Res. 88*:10785–10808.

Platt, U., D. Perner, J. Schroder, C. Kessler, and A. Toennissen (1981). The diurnal variations of NO_3. *J. Geophys. Res. 86*:11965–11980.

Singh, H. B., and L. J. Salas (1983). Peroxyacetyl nitrate in the free troposphere. *Nature 302*:326–328.

Söderlund, R., and B. H. Svensson (1976). The global nitrogen cycle, in *Nitrogen, Phosphorus, and Sulfur: Global Cycles*. SCOPE Report 7. *Ecol. Bull. (Stockholm) 22*:23–74.

SULFUR CYCLE

BY R. HARRISS AND H. NIKI

CURRENT ISSUES

Human activities strongly influence the tropospheric sulfur cycle in certain regions of the world, particularly in and downwind of populated areas. The literature on the reaction and transformation of SO_2 and its distribution and transport in eastern North America and western Europe is voluminous. A reading of even selected portions of the literature on sulfur cycling illustrates that measurements and models of sulfur on the regional scale are providing a relatively consistent understanding of the sources, transport, and fate of anthropogenic emissions of SO_2. However, with the possible exception of reasonably well-understood processes related to local and mesoscale impacts of anthropogenic SO_2, global sulfur cycle studies are in the infancy stage.

Among the general categories of tropospheric sulfur sources, anthropogenic sources have been quantified the most accurately, particularly for the OECD countries. Research on fluxes of sulfur compounds from volcanic sources is now in progress. However, very few generally accepted measurements are available for either concentrations or fluxes of SO_2, H_2S, DMS, DMDS, CS_2, COS, and other sulfur species derived from natural biogenic sources. Measurement techniques have been inadequate until recently; serious questions still remain concerning flux determinations. Tables 7.4, 7.5, and 7.6 summarize most of the data available in the open literature.

What do the existing data indicate in terms of interesting hypotheses and the design of future global studies? First, natural sources of reduced sulfur compounds are highly variable in both space and time. Variables, such as soil temperature, hydrology (tidal and water table), and organic flux into the soil, all interact to determine microbial production and subsequent emissions of reduced sulfur compounds from anaerobic soils and sediments. For example, fluxes of H_2S, COS, CS_2, $(CH_3)_2S$, $(CH_3)_2S_2$, and CH_3SH can vary by several orders of magnitude on time scales of hours and space scales of meters in a coastal environment. A second interesting aspect of existing data on biogenic sources of reduced sulfur relates to the origin of relatively high SO_2 values measured in the mid-troposphere over the tropics and in the southern hemisphere during GAMETAG.

SOURCES AND DISTRIBUTIONS

Current estimates of global sources of atmospheric sulfur are based on very few data and will not be discussed in detail here. Several recent comprehensive reviews are cited at the end of this section for the reader unfamiliar with previous attempts to estimate sulfur sources. We briefly summarize available information on sources of COS, CS_2, DMS, and H_2S to the troposphere in the following paragraphs; these are the major biogenic sulfur species with a clearly identified role in tropospheric chemistry.

Carbonyl Sulfide (COS)

Carbonyl sulfide is the most abundant gaseous sulfur species in the troposphere. Concentrations of COS are approximately 500 + 50 pptv, with no detectable systematic variations vertically or latitudinally. This constant concentration with altitude and latitude suggests a relatively long atmospheric lifetime, estimated to be around 2 years.

Current estimates of global sources and sinks of COS are summarized in Table 7.6. It is important to note that these source and sink estimates are derived by extrapolation of a very limited data base and are subject to large uncertainties. Recent data suggest that oceanic regions of high biological productivity and organic content, particularly coastal waters and upwelling areas, are a major global source of COS. Experiments in coastal waters

TABLE 7.4 Approximate Tropospheric Concentration Range of Selected Sulfur Compounds in Unpolluted Air

Location	Atmospheric Concentrations[a] (ng/m^3)				
	H_2S	DMS	CS_2	COS	SO_2
Ocean boundary layer	<5–150	<2–200	1200–1550	50–70	<15–300
Temperate continental boundary layer	20–200	PD	1200–1550	PD	<15–300
Tropical coastal boundary layer	100–9000	PD	1200–1550	PD	PD
Free troposphere	PD	<2(PD)	1200–1550	<10	30–300

[a]PD indicates poorly determined at this review.

TABLE 7.5 Biogenic Emissions of Sulfur Compounds (emission rate in g $S/m^2/yr$)

Location	H_2S		DMS		CS_2		COS	
	Avg.	Max.	Avg.	Max.	Avg.	Max.	Avg.	Max.
Salt marsh	0.55	41.5		3.84				
			0.006					
	0.5	100	0.66	2.5		0.2		0.03
	72	381	0.093			1.13		6.36
Freshwater marsh	0.6	1.27						
Inland soils (U.S.)	0.001				0.001		0.002	
Swamps and tidal flats	0.044							
Sediments of shallow coastal area	~19	~2000						
Soils of humid equatorial forests	0.07	2.6						
Soils of temperate regions	0.044	0.24						
Open ocean			0.106					

indicate that COS is produced by photooxidation of dissolved organic matter independent of salinity, plant metabolism, or bacterial activity.

Some authors have suggested that the oceans may be a net sink for COS from the atmosphere. An intensive research program concerned with the production, distribution, and emissions of COS from coastal and oceanic environments will be required to quantify the role of the marine environment as a source of this compound.

Soils can also be a source of COS to the troposphere. Coastal salt marsh soils appear to be a "hot spot" for COS emissions, but the small area of these soils limits the role of marshes as a major global source. Measurements from a variety of soils in the United States were used to calculate the global soil source of COS shown in Table 7.6. The total absence of data on COS emissions from tropical soils introduces significant uncertainties into estimates of the global soil source. Efforts to quantify COS emissions from soils will probably be complicated by large variations in both space and time. Microbial processes that produce COS are influenced by soil moisture, nutrients, soil organic content, and other physiochemical variables.

Combustion processes are also thought to be a significant global source of COS. These processes include biomass burning, fossil fuel burning, and high-temperature industrial processes involving sulfur compounds. Again, it must be emphasized that these estimates are based on very limited data and may change significantly as new data become available.

During periods of low volcanic activity, COS may be a major source of sulfur to the stratosphere, resulting in the formation of the stratospheric aerosol layer that influences the earth's climate. The anthropogenic sources of COS identified in Table 7.6 represent approximately 25 percent of the total source strength, supporting speculations of possible effects on climate within the next century. Because of the importance of COS in the global sulfur cycle, its sources, atmospheric chemistry, and sinks are a critical scientific issue.

Carbon Disulfide (CS_2)

The abundance and distribution of CS_2 in the troposphere are not well known. Available measurements in the literature at this date show a typical range from approximately 15 to 30 pptv in surface nonurban air to 100 to 200 pptv in surface polluted air. The concentra-

TABLE 7.6 Global Sources and Sinks of Carbonyl Sulfide

	Estimate	Range
Sources (Tg/yr)		
Oceans	0.60	0.3–0.9
Soils	0.40	0.2–0.6
Volcanoes	0.02	0.01–0.05
Marshes	0.02	0.01–0.06
Biomass burning	0.20	0.1–0.5
Coal-fired power plants	0.08	0.04–0.15
Automobiles, chemical industry, and sulfur recovery processes	0.06	0.01–0.3
Subtotal	1.4	≤3
CS_2-COS: CS_2-photochemistry and OH reactions	0.60	0–2
Total	2	≤5
Global burdens (Tg)	4.7 (500 pptv)	3.8–5.2
Lifetime (yr)	2–2.5	≥1
Sinks (Tg/yr)		
OH reaction	0.8	0.1–1.5
Stratospheric photolysis	0.1	≤0.2
O reaction	0.03	—
Other	1.1	≤3.3

NOTES: The estimated emissions are consistent with observed distributions of COS and CS_2 according to a global mass balance. All combinations of emissions within the ranges given above may not be consistent.

SOURCE: From Khalil and Rasmussen, 1984.

tion of CS_2 appears to decrease rapidly with altitude, indicating ground sources and a relatively short atmospheric lifetime. The primary removal mechanism for CS_2 in the troposphere is thought to be reaction with OH, producing COS and SO_2. The reaction rate constants for oxidation of CS_2 are poorly known, and the relative importance of CS_2 as a precursor for atmospheric COS and SO_2 is an unresolved issue.

The primary natural sources of COS and CS_2 are thought to be similar. The one available set of measurements of CS_2 in seawater indicates that concentrations are highest in coastal waters.

Dimethylsulfide $(CH_3)_2S$

Dimethylsulfide (DMS) is the most abundant volatile sulfur compound in seawater with an average concentration of $\sim 100 \times 10^{-9}$ g/l. This compound is produced by both algae and bacteria. The evidence for a biogenic origin for DMS has come from laboratory measurements of emissions produced in pure, axenic cultures of marine planktonic algae and field measurements of emissions from soils, benthic macroalgae, decaying algae, and corals. Extensive oceanographic studies have shown direct correlations between DMS concentrations in seawater and indicators of phytoplankton activity. The vertical distribution, local patchiness, and distribution of DMS in oceanic ecozones exhibit a pattern very similar to primary productivity. Selected groups of marine organisms such as coccolithophorids (i.e., a type of marine planktonic algae) and stressed corrals are particularly prolific producers of DMS. The calculated global sea-to-air flux of sulfur as DMS is ~ 0.1 g S/m^2/yr, which totals to approximately 39×10^{12} g S/yr. A more limited set of measurements has been made in coastal salt marshes with DMS emissions commonly in the range of 0.006 to 0.66 g S/m^2/yr.

Hydrogen Sulfide (H_2S)

Knowledge of natural sources of H_2S to the troposphere is still rudimentary. Preliminary studies have shown that anaerobic, sulfur-rich soils (e.g., coastal soils and sediments) emit H_2S to the atmosphere, albeit with strong temporal and spatial variatons. Hydrogen sulfide fluxes at a single location can vary by a factor of up to 10^4 depending on variables such as light, temperature, Eh, pH, O_2, and rate of microbial sulfate reduction in the sediment. The presence of active photosynthetic organisms or a layer of oxygenated water at the sediment surface can reduce or stop emissions due to rapid oxidation of H_2S.

Agricultural and forest soils can also be a source of H_2S to the atmosphere. Measurements by several investigators suggest that maximum emissions from nonmarine soils are associated with wet tropical forest soils. It is likely that many soils that appear to be aerated contain anaerobic microhabitats suitable for microbial sulfate reduction; the magnitude of H_2S emissions will depend on the net effects of many processes that influence production, transport in the soil, oxidation rates, and exchange at the soil-air interface.

Photochemical sources for atmospheric H_2S have been proposed to occur through a combination of the following reactions:

$$OH + COS \rightarrow SH + CO_2,$$

$$OH + CS_2 \rightarrow COS + SH,$$

$$SH + HO_2 \rightarrow H_2S + O_2,$$

$$SH + CH_2O \rightarrow H_2S + HCO,$$

$$SH + H_2O_2 \rightarrow H_2S + HO_2,$$

$$SH + CH_3OOH \rightarrow H_2S + CH_3O_2,$$

and

$$SH + SH \rightarrow H_2S + S.$$

Removal of H_2S is thought to be accomplished by

$$OH + H_2S \rightarrow H_2O + SH,$$

resulting in a lifetime of approximately 1 day. In situ photochemical production from COS and CS_2 precursors is the most likely source of H_2S measured in remote ocean air. Atmospheric concentrations of H_2S in continental air are highly variable, resulting from a complex interaction of factors determining ground emissions, in situ photochemical production, and atmospheric lifetime.

TRANSFORMATIONS AND SINKS

The oxidation of SO_2 to H_2SO_4 can often have a significant impact on the acidity of precipitation, currently an issue of national and international concern. A schematic representation of some important transformations and sinks for selected sulfur species is illustrated in Figures 5.11, 5.13, and 7.5. The oxidation of reduced sulfur compounds, such as H_2S, CS_2, $(CH_3)_2S$, and others, leads to the production of acids or acid precursors such as SO_2, $SO_4^=$, and CH_3SO_3H. Subsequent oxidation steps involving a combination of homogeneous and heterogeneous reactions lead to the production of H_2SO_4, which is removed from the atmosphere by wet and dry deposition processes (see Chapter 5). Carbonyl sulfide appears to be relatively inert in the troposphere and is primarily destroyed in the stratosphere.

In terms of experiments to elucidate the fast photochemistry of this system, measurement schemes will be

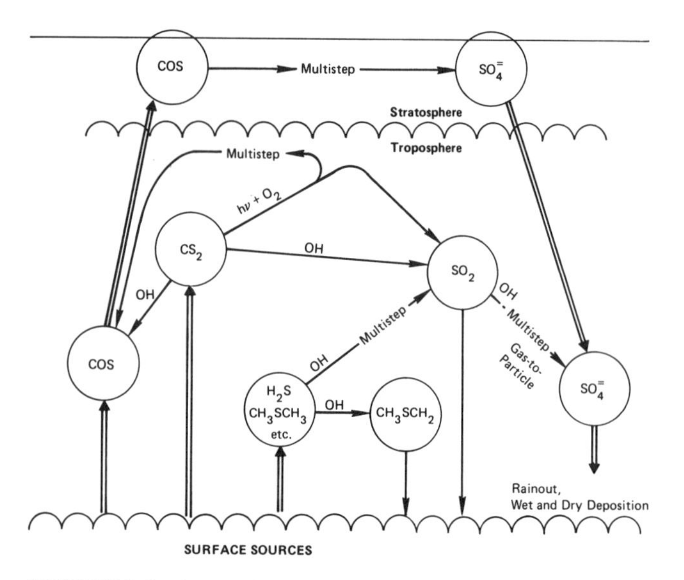

FIGURE 7.5 A tentative scheme for the oxidation and removal of atmospheric sulfur species.

needed to verify the chemical pathways by which reduced sulfur species are oxidized to SO_2 and $SO_4^=$. It is probable that it will be useful to carry out these experiments in a variety of different environments, including areas of intense sulfur emissions (e.g., swamps, tidal flats, and marshes) as well as remote marine areas. Unfortunately, present understanding of the distributions of atmospheric sulfur species and the elementary chemical reactions involved in the previously described oxidation chains is quite poor. In addition, the instrumentation necessary to measure many of the key atmospheric constituents has yet to be developed. Once this task is completed, it will be possible to design specific fast-photochemistry experiments to selectively study various facets of the atmospheric sulfur system.

In the case of H_2S oxidation, for instance, it is believed that oxidation is initiated by reaction with OH, i.e.,

$$H_2S + OH \rightarrow H_2O + SH,$$

and is followed by an as yet unconfirmed reaction sequence that produces SO_2 as an end product. The lifetime of H_2S in the atmosphere seems highly variable based on limited field measurements. In situ studies of H_2S oxidation kinetics in a variety of environments (e.g., swamps, salt marshes, and mangroves) would be extremely useful to improved understanding of the sulfur cycle.

Recent studies of DMS photooxidation provide important data on reaction mechanisms and products. The major gas-phase sulfur product produced in outdoor smog chamber experiments was SO_2. Substantial formation of light-scattering aerosol particles was observed, with inorganic sulfate and methane sulfonic acid as major components of the aerosol. Fourier transform infrared methods have been used to quantify products of the reaction of $HO + CH_3SCH_3$ in the presence of C_2H_5ONO and NO. Methyl thionitrite (CH_3SNO) was observed as an intermediate product, with SO_2 and CH_3SO_3H as major products. These studies serve as models of important photooxidation sinks for reduced sulfur species.

ROLE OF CLOUDS AND AQUEOUS-PHASE CHEMISTRY

As indicated in Chapter 5 of this report, aqueous-phase chemistry (i.e., in cloud and raindrops) plays a major role in the oxidation of SO_2 to H_2SO_4. Current thinking also suggests that clouds may be the dominant transport conduit for movement of SO_2 and other relatively short-lived reduced sulfur species to mid-tropospheric altitudes. Sulfur dioxide produced below cloud base may be injected directly into the free troposphere by updrafts associated with clouds or may dissolve or react with cloud droplets, depending on a variety of poorly quantified physical and chemical variables. Evaporation of cloud droplets may produce small sulfate-rich aerosol particles that subsequently act as cloud condensation nuclei. If the transport and reaction mechanisms mentioned in this paragraph are active over large areas of the nonurban troposphere, they contribute to explanations for acid rain in remote oceanic regions and higher SO_2 in the free troposphere than in underlying ocean boundary layer air. Once in the middle to upper troposphere, SO_2 may have a much longer lifetime with potential for long distance transport beyond the synoptic scale.

Future field experiments will need to measure a variety of species including SO_2, $(CH_3)_2S$, H_2O_2, and methane sulfonic acid in gas, liquid, and solid phases where appropriate. Combined ground and aircraft measurements focused on the role of cloud and aqueous-phase processes are a high priority.

BIBLIOGRAPHY

Andreae, M. O. (1980). Dimethylsulfoxide in marine and freshwaters. *Limnol. Oceanogr. 25*:1054-1063.

Andreae, M. O., and H. Raemdonck (1983). Dimethyl sulfide in the surface ocean and the marine atmosphere: a global view. *Science 221*:744-747.

Aneja, V. P., A. P. Aneja, and D. F. Adams (1982). Biogenic sulfur compounds and the global sulfur cycle. *J. Air Pollut. Control Assoc. 32*:803-807.

Barnard, W. R., M. W. Andreae, W. E. Watkins, H. Bingemer, and H. W. Georgi (1982). The flux of dimethyl sulfide from the oceans to the atmosphere. *J. Geophys. Res. 87*:8787-8793.

Brown, K. A. (1982). Sulfur in the environment: a review. *Environ. Pollut. 3*:47-80.

Chatfield, R. B., and P. J. Crutzen (1984). Sulfur dioxide in remote oceanic air: cloud transport of reactive precursors. *J. Geophys. Res.* (in press).

Delmas, R., J. Baudet, J. Servant, and Y. Baziard (1980). Emissions and concentrations of hydrogen sulfide in the air of the tropical forest of the Ivory Coast and of temperate regions in France. *J. Geophys. Res. 85*:4468-4474.

Ferek, R. J., and M. O. Andreae (1984). Photochemical production of carbonyl sulfide in marine surface waters. *Nature 307*:148-150.

Graedel, T. E. (1977). The homogeneous chemistry of atmospheric sulfur. *Rev. Geophys. Space Phys. 15*:421-428.

Graedel, T. E. (1979). Reduced sulfur emission from the open oceans. *Geophys. Res. Lett. 6*:329-331.

Herrmann, J., and W. Jaeschke (1984). Measurements of H_2S and SO_2 over the Atlantic Ocean. *J. Atm. Chem. 1*:111-123.

Husar, R. B., J. P. Lodge, and D. J. Moore (1978). *Sulfur in the Atmosphere*. Pergamon, New York.

Ingvorsen, K., and B. B. Jorgensen (1982). Seasonal variation in H_2S emission to the atmosphere from intertidal sediments in Denmark. *Atm. Environ. 16*:855-864.

Jones, B. M. R., R. A. Cox, and S. A. Penkett (1984). Atmospheric chemistry of carbon disulphide. *J. Atm. Chem. 1*:65-86.

Jorgensen, B. B. (1982). Ecology of the bacteria of the sulfur cycle with special reference to anoxic-oxide interface environments. *Phil. Trans. Roy. Soc. London B298*:543-561.

Khalil, M. A. K., and R. A. Rasmussen (1984). Global sources, lifetimes, and mass balances of carbonyl sulfide (COS) and carbon disulfide (CS_2) in the earth's atmosphere. *Atm. Environ.* (in press).

Kritz, M. A. (1982). Exchange of sulfur between the free troposphere, marine boundary layer, and the sea surface. *J. Geophys. Res. 87*:8795-8803.

Lawson, D. R., and J. W. Winchester (1979). Atmospheric sulfur aerosol concentrations and characteristics from the South American continent. *Science 205*:1267-1269.

Logan, J. A., M. B. McElroy, S. C. Wofsy, and M. J. Prather (1979). Oxidation of CS_2 and OCS: source for atmospheric SO_2. *Nature 281*:185-188.

Maroulis, P. J., A. L. Torres, A. B. Goldberg, and A. R. Bandy (1980). Atmospheric SO_2 measurements on Project GAMETAG. *J. Geophys. Res. 85*:7345-7349.

McElroy, M. B., S. C. Wofsy, and N. Dak Sze (1980). Photochemical sources for atmospheric H_2S. *Atm. Environ. 14*:159-163.

Moller, D. (1984). On the global natural sulphur emission. *Atm. Environ. 18*:29-39.

National Research Council (1978). *Sulfur Oxides*. Committee on Sulfur Oxides, Assembly of Life Sciences, National Academy of Sciences, Washington, D.C.

National Research Council (1981). *Atmosphere-Biosphere Interactions: Toward a Better Understanding of the Ecological Consequences of Fossil Fuel Combustion*. Commission on Natural Resources, National Academy of Sciences, Washington, D.C.

Nguyen, B. C., B. Bonsang, and A. Gaudry (1983). The role of the ocean in the global atmospheric sulfur cycle. *J. Geophys. Res. 88*:10903-10914.

Niki, H., P. D. Maker, C. M. Savage, and L. Breitenbach (1980). Fourier transform study of the OH radical initiated oxidation of SO_2. *J. Phys. Chem. 84*:14-16.

Rasmussen, R. A., M. A. K. Khalil, and S. D. Hoyt (1982). The oceanic source of carbonyl sulfide. *Atm. Environ. 16*:1591-1594.

Servant, J., and M. Delapart (1982). Daily variations of the H_2S content in atmospheric air at ground level in France. *Atm. Environ. 16*:1047-1052.

Shriner, D. S., C. R. Richmond, and S. E. Lindberg (1980). *Atmospheric Sulfur Deposition*. Ann Arbor Science, Ann Arbor, Mich.

Slatt, B. J., D. Natusch, J. M. Prospero, and D. L. Savoie (1978). Hydrogen sulfide in the atmosphere of the northern equatorial Atlantic Ocean and its relation to the global sulfur cycle. *Atm. Environ. 12*:981-991.

CARBON CYCLE

BY H. NIKI, R. DUCE, AND R. DICKINSON

CURRENT ISSUES

Although CO_2 is the primary carbonaceous trace gas in the biogeochemical cycle of carbon, it is relatively unreactive in the troposphere. For this reason, we will separate the discussion of the carbon cycle into two parts—reactive carbon compounds currently recognized to have potentially important atmospheric and biospheric components, and CO_2. The two most abundant reactive carbon compounds in the global atmosphere are CH_4 and CO. Both compounds play central roles in O_3 and H_xO_y cycles through chemical interaction involving common reactive intermediates (free radicals, e.g., OH, HO_2, and CH_3OO). It is highly significant that the photochemical oxidation of CH_4 can lead to the formation of a large variety of oxygenated products (e.g., CH_2O, CH_3OOH, HCOOH, and CH_3OOOH). Atmospheric in situ formation rates of these individual compounds are critically controlled by H_xO_y and NO_x chemistry. Moreover, some of these compounds can be removed from the atmosphere heterogeneously prior to the subsequent oxidation to CO and, eventually, to CO_2. Thus the atmospheric transformation of CH_4 involves complex chemical feedback mechanisms. Anthropogenic perturbation of CO and CH_4 cycles and its overall atmospheric impact are of great current interest. In addition, numerous nonmethane hydrocarbons (NMHCs) (e.g., saturated and unsaturated C_2-C_9 compounds, isoprene, and terpenes), particulate organic carbon (POC) (including C_9-C_{28} alkanes and C_{12}-C_{30} fatty acids), and elemental or soot carbon are present in the global troposphere. Interest in NMHCs in the global troposphere has been increasing in recent years because of their potential importance as a source of CO and as reactants interacting with O_3, H_xO_y, and NO_x chemistry. However, at present, the global sources, distribution, fluxes, and transformations of NMHCs represent perhaps the greatest deficiency in understanding the active carbon cycle in the troposphere.

Elemental carbon is primarily of concern because of its radiative properties and its surface characteristics. It is the only important aerosol component that can absorb a significant amount of visible light. Changes in the global distribution of elemental carbon may thus be related to changes in the retention of solar radiative energy in the troposphere. Elemental carbon particles also have a very active surface that effectively adsorbs many trace gases. The role of elemental carbon in the global troposphere as a sink for trace gases and in heterogeneous reactions in general has yet to be evaluated. Of particular importance may be its activity as a catalyst for reactions in clouds and precipitation.

Estimates of budgets of several active carbon species are summarized in Table 7.7. Key issues and existing uncertainties are indicated below for the global cycles of these individual or classes of compounds.

THE CYCLES OF REACTIVE CARBON COMPOUNDS

Sources

Major direct sources of CO appear to be primarily anthropogenic, e.g., fossil fuel use and tropical biomass burning. The latter source is still poorly quantified in terms of spatial and seasonal variations. Photochemical production from CH_4 is well recognized as a major indirect source of CO. This source is probably known within a factor of 3, reflecting uncertainties in existing estimates of the global yearly average OH concentration and of the transformation mechanism. Oxidation of natural NMHCs, e.g., isoprene and terpenes, leading to CO production is potentially very important, but the magnitude of this source is highly uncertain, as mentioned earlier. Other minor global sources thus far identified include oxidation of anthropogenic hydrocarbons, emissions by plants, wood used as fuel, forest wildfires, and the ocean.

Several sources may be contributing comparable quantities of CH_4 to the global troposphere, e.g., rice paddy fields, natural wetlands (swamps and marshes), enteric fermentation (ruminants and termites), and biomass burning. However, available data on all these sources are severely limited, and more extensive flux measurements are needed for reasonably quantitative estimates. Clearly, many of these sources have direct or indirect anthropogenic components and are subjected to significant perturbation, a probable cause for the recent increase in the atmospheric CH_4 level.

The major volatile NMHCs in the global atmosphere are probably isoprene (C_5H_8) and various types of monoterpenes emitted from vegetation. Although there exist relevant baseline data on emission rates of vegetation classes and effects of meteorological conditions (sunlight, temperature, and humidity), more extensive flux measurements, especially in tropical regions, are needed for sufficiently accurate estimates of global annual emission rates. Note that ongoing deforestation in tropical jungles may already be having a significant

TABLE 7.7 Budgets of Carbon Species

Gas	Direct Source per Year and Source Identification	Secondary Source per Year and Source Identification	Removal by	Atmospheric Lifetimes	Transport Distances Δx, Δy, Δz (km); Volume Mixing Ratios in the Unpolluted Troposphere
CO	$4\text{-}16 \times 10^{14}$ g CO Biomass burning 6.4×10^{14} g CO Industry $0.2\text{-}2 \times 10^{14}$ g CO Vegetation	$3.7\text{-}9.3 \times 10^{14}$ g CO Methane oxidation $4\text{-}13 \times 10^{14}$ g CO C_5H_8, $C_{10}H_{16}$ Oxidation	30×10^{14} g CO OH 4.5×10^{14} g CO Uptake by soils	2 months	4000, 2500, 10 50-200 ppbv
CH_4	$0.3\text{-}0.6 \times 10^{14}$ g CH_4 Rice paddy fields $0.3\text{-}2.2 \times 10^{14}$ g CH_4 Natural wetlands 0.6×10^{14} g CH_4 Ruminants $<1.5 \times 10^{14}$ g CH_4 Termites $0.3\text{-}1.1 \times 10^{14}$ g CH_4 Biomass burning 0.2×10^{14} g CH_4 Gas leakage		4×10^{14} g CH_4 OH	7 years	Global 1.5-2.0 ppmv
C_5H_8, $C_{10}H_{16}$	8.3×10^{14} g C Trees		8.3×10^{14} g C OH	10 hours	400, 200, 1 0-10 ppbv

NOTES: Diffusion distances in E-W, S-N, and vertical directions (in km) over which concentrations are reduced to 30 percent by chemical reactions. Lifetimes and removal rates calculated with [OH] = 6×10^5 molecules per cubic centimeter.
SOURCE: Crutzen, 1983.

impact on these sources. In addition, numerous light hydrocarbons (C_2-C_5) are emitted in potentially significant amounts from a variety of sources including the ocean, plants, natural gas leakage, biomass burning, and fossil fuel combustion. However, for the majority of these compounds, the major contributors have not been identified. For compounds such as C_2H_2 and C_6H_6, there are no known significant biogenic sources. Existing estimates of the global emission rates of these NMHCs are largely based on their observed atmospheric concentrations and removal rates by OH radicals.

Particulate organic carbon (POC) consists of a complex mixture of hydrocarbons, alcohols, acids, esters, organic bases, and other polar organic compounds. Gas-to-particle conversion appears to be the main source of POC in the global atmosphere, since the total POC mass is generally in the small particle range with radii <1 μm. Photochemically produced species, including free radicals, may participate in this process. Several potentially significant sources for the gaseous precursors of POC are land (vegetation), aquatic systems (microbiota of marine and lacustrine environments), petroleum seepages, and pollution (fossil fuel combustion and biomass burning). Existing speculations on the total source strength for POC are based on rough estimates of the tropospheric POC distribution and the mean tropospheric residence time of aerosol particles (4 to 7 days). Thus the range of uncertainty is believed to be greater than one order of magnitude for the natural sources. To unravel complex interactions of numerous, as yet poorly characterized, factors controlling POC formation, concerted efforts in several research areas are needed. In particular, many solubilities and rate parameters need to be determined. The transfer processes across the interface at the aerosol particle surface require better quantitation.

The primary source of elemental carbon is combustion processes. While industrial combustion processes are significant in many urban areas, biomass burning—both natural and as a result of human activities—is also very important globally. Elemental carbon is proving to be an excellent tracer of the long-range transport of combustion aerosol particles.

Distribution

The tropospheric distribution of CO has been studied in considerable detail and is known to be rather variable. It exhibits an interhemispheric gradient of about a factor of 3, with the highest (150 to 200 ppbv) at the middle and higher latitudes of the northern hemisphere corresponding to the large anthropogenic fluxes there. Seasonal variations with summer maxima and winter minima have been recently identified. Further extensive measurements of the seasonal and latitudinal distribution in both the boundary layer and the free troposphere, particularly in the tropics where the major sources are suspected to exist, are critical for obtaining a better understanding of its sources, transformation, and sinks.

It is reasonably well established that CH_4 is distributed fairly uniformly throughout the global troposphere (excepting urban atmospheres) at approximately 1.6 ppmv. However, as much as 0.1 ppmv interhemispheric gradient with higher values in the northern hemisphere has been observed. There is also recent evidence for approximately a 1 to 2 percent annual global increase in CH_4 in the late 1970s and early 1980s. Thus accurate long-term measurements of this radiatively active gas are warranted. Carbon isotope ratio measurements may provide a useful clue to this increase. In addition, it is crucial to obtain information on atmospheric concentrations of oxygenated products, e.g., CH_2O and CH_3OOH, derived from CH_4. Very limited measurements are currently available on the diurnal variation of CH_2O. There are no field data on CH_3OOH because of the lack of adequate monitoring instrumentation.

Isoprene and terpenes (e.g., α-pinene, β-pinene, δ-3-carene, camphene, and d-limonene) are very short-lived in the atmosphere, and their distribution is confined well within the boundary layer. Up to a few parts per billion by volume of these compounds have been observed in warm and humid forested areas. Thus far, there have been no field flux determinations of these compounds because of the lack of fast-response detection instrumentation to utilize eddy correlation or profiling techniques. Concomitant product measurements are also required, but at present, little is known about the ensuing oxygenated species (e.g., carbonyls and organic peroxides). On the other hand, some baseline data are becoming available for the vertical and latitudinal distribution of C_2-C_5 NMHCs. Since chemical lifetimes of these compounds are short or at least comparable to the time scale of tropospheric transport, their distributions are highly variable, with higher values in the middle and high latitudes of the northern hemisphere. Light alkanes such as ethane and propane reach concentrations as large as 2 to 3 ppbv. Relatively high concentrations of such very reactive alkenes as propene ($\sim$0.2 ppb) are found rather uniformly distributed over the world ocean suggesting an important marine source for these hydrocarbons. More extensive, simultaneous monitoring of these NMHCs and oxygenated products (e.g., carbonyls, peroxides, and organic nitrates) is needed to determine their atmospheric roles.

Carboxylic acids, particularly formic and acetic acids, are a common constituent of both aerosol particles and rain and have also been observed in the vapor phase. There is evidence that these weak organic acids can contribute a significant fraction of the free acidity of precipitation at locations far from urban areas. Considerable additional research is needed to evaluate accurately the role of organic acids in global-scale precipitation chemistry.

The relatively few data available for POC indicate concentrations of 0.1 to 0.5 $\mu g/m^3$ STP in marine air and about 1 $\mu g/m^3$ STP in nonurban continental air. The composition of POC in terms of neutral compounds (60 percent), acids (30 percent), and bases (10 percent) appears to remain remarkably constant in ground-level air. For n-alkanes (C_{10}-C_{28}), the relative portions in the particulate phase are often much less than expected from the saturation vapor pressures, possibly due to heterogeneous reactions. Elemental carbon concentrations are often over 1 $\mu g/m^3$ in continental regions, particularly near urban areas. There are relatively few data available in remote areas. Concentrations over the North and South Atlantic range from about 0.05 to 0.2 $\mu g/m^3$, while concentrations during the winter at Barrow, Alaska, averaged $\sim$0.3 $\mu g/m^3$. Considerably more data are required on the spatial and temporal distribution of both individual and classes of vapor phase and particulate organic compounds in remote areas over a wide variety of potentially significant ground-level sources. Anthropogenic contributions should be better characterized by comparing emission estimates with measured ambient concentrations.

Transformations and Sinks

The main removal process for CO is in situ oxidation to CO_2 by OH radicals. Inherent to existing estimates of the global rate of this process are uncertainties associated with OH concentration and with the rate constant. Estimates of the global annual OH production and distribution are commonly assumed to be accurate within a factor of 2 to 3. The rate constant for this reaction may be uncertain by as much as 50 percent. Although it has been measured extensively with great accuracy in the presence of inert diluent gases at low pressures, it is known to exhibit peculiar dependence on O_2 pressure. To date, there have been no direct measurements of this rate constant under tropospheric conditions. Further study of the combined pressure and temperature effects is needed. Soil bacterial uptake is an additional sink for CO. The removal rate by this sink appears to be rather minor on the global scale, but may be substantial in the boundary layer over land areas.

The OH-radical-initiated oxidation of CH_4 is primarily responsible for the removal and transformation of this compound, although recent studies have shown that CH_4 is consumed in the soils of both temperate and tropical forests. In the presence of sufficient NO, the

OH oxidation of CH_4 leads to the formation of CO via photooxidation of the oxygenated intermediate CH_2O. With little NO present, CH_3OOH becomes a major intermediate in place of CH_2O, but its subsequent fate is uncertain. The CH_3OOH does not photodissociate readily, but may undergo sufficiently fast in situ removal by OH. The rate constant and mechanism for the HO-CH_3OOH reaction, as well as heterogeneous removal of CH_3OOH, must be established. In addition, because of its relatively long tropospheric lifetime (~7 years), CH_4 escapes into the stratosphere in significant amounts and plays an important photochemical role in the O_3 cycle. Further quantitation of the stratospheric removal rate is needed.

Nonmethane hydrocarbons are generally far more reactive with OH than is CH_4. In addition, some of the double-bonded NMHCs, particularly terpenes, can be removed by O_3 at a rate comparable to or even greater than those of the corresponding OH reactions. For light alkanes, there is a fair body of laboratory evidence that their oxidation mechanisms are somewhat analogous to that of CH_4, although they give rise to a large variety of classes of oxygenated products including carbonyls (RCHO and RCR′O; R and R′ = organic group), peroxides (ROOH), peroxynitrates ($ROONO_2$), peroxyacids (RC(O)OOH), and acids (RC(O)OH). Again, large uncertainties exist in their oxidation mechanisms at low NO concentrations and in the subsequent gas phase and heterogeneous removal processes for the oxygenated products. Not much is known about the mechanisms and key intermediate products for the OH- and O_3-initiated oxidation of unsaturated compounds, e.g., isoprene and terpenes, either under laboratory conditions or in the atmosphere, primarily because of the lack of adequate methodologies and detection systems.

Cloud and precipitation processes appear to be most important for the removal of POC. Aqueous photochemistry involving OH and O_3 may play a particularly important role in transforming POC, as indicated by the presence of a variety of oxygenated compounds in POC. Dry deposition processes involving the sea, soil, and vegetation are also potentially significant sinks. More quantitative knowledge is required on both gaseous and particulate fluxes associated with all of these processes. Since some of the needed analytical techniques and methodologies are available, further field and laboratory studies in this area should be encouraged.

THE CYCLE OF CARBON DIOXIDE

The species discussed up to this point represent the minute fraction of the total environmental carbon that is chemically active in the atmosphere. However, it should be recognized that the cycling of these active species is embedded in and ultimately controlled by the slower biological, oceanographic, and geologic reservoirs of carbon. These slower processes are briefly summarized here.

Carbon dioxide is the dominant form of carbon storage in the atmosphere. Vegetation and soils store organic carbon, whereas the oceans store some organic carbon but mostly inorganic carbon in the form of carbonate ions. Ocean sediments and terrestrial rocks store vast amounts of carbonate and organic carbon, mostly accumulated over millions of years from oceanic biological debris.

Carbon is mostly of atomic weight 12, but about 1 percent is of weight 13 and a minute amount of weight 14. This ^{14}C is unstable with an approximately 5000-year lifetime and is constantly being created in the atmosphere by galactic cosmic rays transforming nitrogen. It serves as a useful clock and thus a tracer of transfer from the atmosphere to other reservoirs. The ^{13}C serves as a tracer for sources of carbon since its relative concentration differs between organic and inorganic carbon.

Over land, carbonate and silicate rocks are dissolved by terrestrial waters, which in doing so act to convert atmospheric CO_2 into bicarbonate ions. This carbon is converted back to CO_2, in part by the conversion of bicarbonate into the carbonate of sea creatures. Further CO_2 is released by volcanoes, presumably after having been squeezed from carbonate rocks in their conversion to silicate rocks as a result of subduction of oceanic plates under continental plates. Such processes are of interest in explaining how there could have been 5 to 10 times more CO_2 in the atmosphere in past geological times than there is now. Such an abundance of CO_2 is currently the most favored hypothesis for explaining the warm Cretaceous climates of 100 million years ago. On intermediate time scales of thousands of years, deep oceanic processes involve the concentrations of Ca^{++} ions and possibly were related to biological activity dominant in controlling atmospheric CO_2. Such processes may have reduced atmospheric concentrations of CO_2 near the end of the last ice age to values not much more than half of current concentrations.

On the time scales of seasons to centuries, large carbon exchanges occur between the atmosphere, oceanic near-surface waters, and living and dead vegetation, the latter mostly in soils. All these reservoirs are of roughly the same magnitude. The atmosphere loses and gains annually about 10 percent of its carbon content by exchange with the other reservoirs. It is these relatively short-term exchanges and the physical and biological processes accompanying them that are of most relevance to questions of global tropospheric chemistry.

First, the transfers of CO_2 between the atmosphere

and the land biosphere or oceans occur by means of the same physical processes whereby other trace gases are transferred. To the extent that more accurate measurements can be made of CO_2 transfer processes than of those of other gases, study of CO_2 fluxes improves the understanding of the fluxes of other gases. For example, terrestrial vegetation removes O_3, SO_2, and NO_2 by absorption through leaf stomata, the primary function of the stomata being to capture CO_2 from the atmosphere as a major plant nutrient. As another example, atmospheric CH_4 comes from soils, where it is produced by anaerobic bacterial decomposition of organic carbon. However, most of the soil bacterial decomposition is aerobic, so that more than 99 percent of the soil organic carbon released to the atmosphere is in the form of CO_2. The CH_4 is oxidized in the atmosphere to CO and eventually CO_2 as discussed earlier, thus completing the cycle between biosphere and atmosphere that was initiated by plant stomatal uptake.

Second, the seasonal and latitudinal variations of CO_2, although relatively small, are measured with considerable accuracy so that CO_2 would provide a useful tracer of atmospheric motions if its sources and sinks could be adequately modeled. Seasonal variations are largely driven by northern hemisphere seasonal growth of plants whose rapid removal of CO_2 from the atmosphere from May to August or so depletes atmospheric concentrations by about 1 ppm. Decay processes, which restore CO_2 to the atmosphere, vary more smoothly over the annual cycle though operating faster during the warmer seasons. Latitudinal atmospheric variations are driven in part by the excess of fossil fuel combustion in the northern hemisphere and in part by transfers from warm tropical oceans which are supersaturated with respect to CO_2. The addition of CO_2 to the atmosphere from fossil fuel combustion has increased atmospheric CO_2 concentrations from 315 to 340 ppmv in the last 25 years since measurements began and has led to a national program to study the possible climate change due to the warming from those increases, as well as details of the exchange of CO_2 among atmosphere, oceans, and biosphere.

INTERACTION WITH OTHER CYCLES

Carbon monoxide interacts directly with the H_xO_y cycle via free radical reactions, thereby profoundly affecting the cycles of virtually all elements of atmospheric interest. For example, OH radicals are removed primarily by CO in the global troposphere. This process serves as the principal source of HO_2 radicals, which, in turn, interact with species such as NO_x and O_3. The OH radicals are regenerated to some extent by some of these secondary reactions (e.g., $HO_2 + NO \rightarrow OH + NO_2$ and $H_2O_2 + h\nu \rightarrow 2OH$). Thus, because of complexities in the chemical feedback mechanism, the overall atmospheric impact of CO requires thorough numerical evaluation.

Methane interacts with cycles of other elements primarily through CO formation. The major oxygenated product CH_2O can provide an additional significant source of HO_2 radicals. It may also be possible that the short-lived NO_2-containing product CH_3OONO_2 serves as an intermediate for accelerating the gas-to-particle conversion of NO_x. Similarly to H_2O_2, the peroxy product CH_3OOH may play a catalytic role in the aqueous chemistry of other elements.

In general, the OH-initiated oxidation of NMHCs involves oxygenated free radicals and intermediates of the form R_xO_y (R = organic group), which can interact directly or indirectly with NO_x, H_xO_y, and O_3. Conspicuous formation of O_3, PAN, and organic aerosols by the photochemical reactions of NMHC with NO_x is a well-known phenomenon in urban atmospheres, i.e., photochemical smog, and is probably occurring in the tropical regions during biomass burning. These relevant reactions can take place to a significant extent in the global atmosphere, e.g., PAN formation. A large variety of R_xO_y species are also produced by the reactions of O_3 with double-bonded NMHCs. Some of these R_xO_y compounds appear to interact with SO_2 and with H_2O as well. However, the identity and chemical behavior of the majority of these species remain largely unknown.

As in the gaseous phase, organic compounds in both particulate and aqueous phases may undergo significant photo- and dark-reactions with H_xO_y, O_3, NO_x, SO_x, and trace metals. A detailed assessment of chemical interactions in atmospheric aerosol particles is hampered by a scarcity of relevant rate data.

BIBLIOGRAPHY

Andreae, M. O. (1983). Soot carbon and excess fine particle potassium: Long-range transport of combustion-derived aerosols. *Science 220*:1148–1151.

Atkinson, R., K. R. Darnal, A. Lloyd, A. M. Winer, and J. N. Pitts, Jr. (1979). Kinetics and mechanism of the reaction of the hydroxyl radical with organic compounds in the gas phase. *Adv. Photochem. 11*:375–488.

Berner, R. A., A. S. Lasaga, and R. M. Garrels (1983). The carbonate-silicate geochemical cycle and its effect on atmospheric carbon dioxide over the past 100 million years. *Amer. J. Sci. 283*:641–684.

Brewer, P. G., G. M. Woodwell, L. Machta, and R. Revelle (1983). Past and future atmospheric concentrations of carbon dioxide, Chapter 3 in *Changing Climate*. National Academy Press, Washington, D.C., pp. 186–265.

Broecker, W. S. (1982). Ocean chemistry during glacial time. *Geochim. Cosmochim. Acta 46*:1689–1705.

Bufalini, J. J., and R. R. Arnts, eds. (1981). *Atmospheric Biogenic Hydrocarbons*, Vol. 1, *Emissions*. Ann Arbor Science, Ann Arbor, Mich.

Clark, W. C., ed. (1982). *Carbon Dioxide Review: 1982*. Oxford University Press, New York.

Crutzen, P. J. (1983). Atmospheric interactions-homogeneous gas reactions of C, N, and S containing compounds, Chapter 3 in *The Major Biogeochemical Cycles and Their Interactions*. SCOPE 21. B. Bolin and R. Cook, eds., Wiley, New York, pp. 67–112.

Duce, R. A., V. A. Mohnen, P. R. Zimmerman, D. Grosjean, W. Cautreels, R. Chatfield, R. Jaenicke, J. A. Ogren, E. D. Pellizzari, and G. T. Wallace (1983). Organic material in the global troposphere. *Rev. Geophys. Space Phys. 21*:921–952.

Ehhalt, D. H., R. J. Zander, and R. A. Lamontagne (1983). On the temporal increase of tropospheric CH_4. *J. Geophys. Res. 88*:8442–8446.

Graedel, T. E. (1979). Terpenoids in the atmosphere. *Rev. Geophys. Space Phys. 17*:937–947.

Keller, M., T. J. Goreau, S. C. Wofsy, W. A. Kaplan, and M. B. McElroy (1983). Production of nitrous oxide and consumption of methane by forest soils. *Geophys. Res. Lett. 10*:1156–1159.

Khalil, M. A. K., and R. A. Rasmussen (1983). Sources, sinks, and seasonal cycles of atmospheric methane. *J. Geophys. Res. 88*:5131–5144.

Logan, J. A., M. J. Prather, S. C. Wofsy, and M. B. McElroy (1981). Tropospheric chemistry: a global perspective. *J. Geophys. Res. 86*:7210–7254.

National Research Council (1983). *Changing Climate*. National Academy Press, Washington, D.C., 496 pp.

Ogren, J. A., and R. J. Charlson (1983). Elemental carbon in the atmosphere: cycle and lifetime. *Tellus 35B*:241–254.

Peterson, E., and D. T. Tingey (1980). An estimate of the possible contribution of biogenic sources to airborne hydrocarbon concentrations. *Atmos. Environ. 14*:79–81.

Rudolph, J., and D. H. Ehhalt (1981). Measurements of C_2 and C_5 hydrocarbons over the northern Atlantic. *J. Geophys. Res. 86*:11959–11964.

Seiler, W., and J. Fishman (1981). The distribution of carbon monoxide and ozone in the free troposphere. *J. Geophys. Res. 86*:7255–7266.

Wolff, G. T., and R. L. Klimiseh, eds. (1982). *Particulate Carbon: Atmospheric Life Cycle*, Plenum, New York.

Zimmerman, P. R., R. B. Chatfield, J. Fishman, P. J. Crutzen, and P. L. Hanst (1978). Estimates on the production of CO and H_2 from the oxidation of hydrocarbon emissions from vegetation. *Geophys. Res. Lett. 5*:679–682.

HALOGENS

BY R. CICERONE

CURRENT ISSUES

The family of halogen elements (fluorine, chlorine, bromine, and iodine) display many similarities in their chemical behavior, and they occur widely in the natural geochemical environment. Although regular progressions and order appear in the properties of gases and solutions of one halogen versus another (i.e., inter-halogen differences), there are also many accessible forms for each halogen in the atmosphere. For example, each halogen exists in a number of volatile forms; thus there are halogen-containing gases in the earth's atmosphere. Also, for each halogen there are volatile inorganic and organic species. In the liquid phase, there are dissolved halides and halide ions, and there are halide condensates on aerosol particles.

This spectrum of halogen compounds and phases and chemical reactivities gives halogens a correspondingly wide range of atmospheric behavior and of atmospheric residence times. For example, large sea-salt particles rich in halogens are airborne only for hours; some perhaloalkanes have residence times over 100 years.

Both natural and man-made halogen-containing substances are now involved in atmospheric chemistry. The outstanding questions and concerns (outlined below) in atmospheric halogen chemistry are thus a mixture of basic scientific issues and much more practical pollution problems. Thirty years ago, there was little basis for practical concern over halogenated compounds in the atmosphere. Today, there is a growing list of problems with halogen compounds. Although some important localized pollution problems exist, we are more concerned with global or semiglobal issues here. Examples include the surface sources and tropospheric sinks of CH_3Cl (an important source of stratospheric chlorine, active in stratospheric O_3 chemistry), and the potential climatic effects of chlorofluorocarbons and 1,1,1-trichloroethane, also known as methyl chloroform.

Other current questions are more qualitative because too few field data are available to permit completely quantitative hypotheses. It can also be said that these questions are profound. For example, what is the source of tropospheric HCl? In the marine atmosphere, it is plausible that sea-salt aerosol chloride can be liberated as volatile HCl as the aerosol particle becomes acidified by evaporation or by accumulation of NO_2 and SO_2 from dirtier air, but this is far from proven. What is the origin of gaseous CH_3Cl, CH_3I, and CH_3Br? While supersaturated surface waters have been observed and while there are industrial sources of CH_3Cl and CH_3Br, what other sources are there? Is biomass burning or direct biological emission important? What is the cause of the recently observed springtime Arctic atmosphere pulse of gaseous and particulate bromine? Are tropospheric photochemical reactions of iodine, chlorine, or bromine potentially important to tropospheric O_3 levels? What physical and chemical factors control the gas/particle partitioning of tropospheric halogens? What are the dominant gaseous inorganic species?

In the remainder of this discussion, we review the available information on the distributions of atmospheric halogen gases and particles, their sources, atmospheric transformations and removal processes, and their interactions with other atmospheric cycles. Many important topics of stratospheric interest are omitted here; our focus is on the troposphere.

DISTRIBUTION OF HALOGENS IN THE ATMOSPHERE

In the paragraphs that follow, the discussion proceeds through chlorine, fluorine, bromine, and iodine, roughly in descending order of existing knowledge. For each halogen, there is a further subgrouping: organic gases, inorganic gases, and aqueous or particulate halide.

Of all the halogen-bearing gaseous species in the atmosphere, the best measured are the organic chlorine gases, Cl^{o}_{g}. These data are largely from electron-capture detection gas chromatography. Global concentration patterns are well known for CCl_3F and CCl_2F_2, at least in the boundary layer. Less detailed data are available for CH_3CCl_3, CCl_4, CH_3Cl, CH_2Cl_2, $CHCl_3$, and various chloroethanes and chloroethylenes. Interhemispheric gradients exist, especially for the man-made compounds with shorter residence times. A wide range of residence times characterizes the organic chlorine gases—from approximately 100 years for $CClF_3$ and CCl_2F_2 to perhaps a month for $CHCl_3$. Temporal trends (increases) have been observed and are well documented for CCl_2F_2, CCl_3F, CH_3CCl_3, CCl_4, and several other synthetic species. Numerical data on concentrations, gradients, and trends are available in articles listed in the bibliography. The most abundant chlorocarbon is CH_3Cl (approximately 700 pptv in both hemispheres). In 1983 the concentrations of CCl_2F_2 and CCl_3F were about 320 and 200 pptv, respectively. Other organic chlorine species are present in lesser concentrations. Although very few quantitative data are available, certain relatively stable aromatic compounds like chlorobenzenes are nearly ubiquitous in the global troposphere.

Inorganic chlorine gases are known to exist in the troposphere, apparently in higher quantities in marine air and in polluted regions. Early evidence came from trapping air in chemically basic traps and analyzing the elemental contents of the trapped samples. Acids such as HCl are efficiently trapped this way, but other unidentified inorganic gases can easily yield signals that appear as HCl. Chemically specific measurement methods have been employed rarely, if at all. Spectroscopic methods, quite specific to HCl, have been employed to detect and measure tropospheric HCl, but only two brief reports of such work have ever been published. Indeed, HCl is the only inorganic chlorine species ever to have been detected with chemically or spectroscopically specific instrumentation in the troposphere. Scanty evidence suggests that there is more HCl in the marine atmosphere than in clean continental air.

Chloride in precipitation and aerosol particles were studied extensively in the 1960s, but not much since then. In marine aerosols, Cl^- is a major constituent by mass (1 to 10 $\mu g/m^3$ STP), and the seawater chlorine/sodium ratio is generally mirrored closely in marine aerosols, at least in the larger particles (diameter over 1 μm). Smaller particles often show a deficit of chlorine compared to Cl/Na in seawater. Generally, the Cl^- content of aerosol particles decreases with distance from oceans; isolated observations of higher chloride in continental aerosol particles have been attributed to large intrusions of marine air or to pollution sources. In rainfall, there is generally more chloride (up to 10 mg/l) in marine precipitation than over continents (where Cl^- is as low as 0.1 mg/l). The total annual precipitation of Cl^- is probably the largest sink of atmospheric halogen, but much of this represents removal of short-lived marine boundary layer chloride. Few data are available for Cl^- in snow.

Organic bromine gases, Br_g^o, are much less studied and are apparently present at much lower concentrations than Cl_g^o. Total Br_g^o concentrations of 14 to 68 ng/m^3 STP have been observed in clean tropospheric air; marine air generally contains more Br_g^o than continental air. Only five species of Br_g^o have been measured specifically—CH_3Br, CH_2Br_2, $CHBr_3$, $C_2H_4Br_2$, and CF_3Br; there are few, if any, southern hemispheric data; and some of these are from the Antarctic. CF_2Br_2 has been used as an air-trajectory tracer in regionally polluted air. Thus, although there are indications of higher Br_g^o concentrations over oceans than continents, there are no data available on interhemispheric differences of total Br_g^o or individual species. Recently, it has been shown that there is an annual springtime peak in gaseous and particulate bromine in the Arctic atmosphere. It is likely that most of these elevated gaseous levels (perhaps 20 times normal) are Br_g^o species.

Inorganic bromine gases, Br_g^i, have been measured collectively through methods similar to those for total Cl_g^i. Very little information is available on latitudinal variations, and no individual Br_g^i species have been detected with chemical or spectroscopic specificity. Available data suggest that Br_g^o concentrations exceed those of Br_g^i by at least a factor of 2. Also, diurnal variations of Br_g^i have been observed in the marine atmosphere.

As with gaseous bromine, there are fewer data available for bromide in aerosol particles than for chloride. Typical concentrations of particulate Br^- are 5 to 10 ng/m^3 STP in marine air. As with chloride, most of this bromide mass resides on large-diameter particles. Total gaseous bromine always exceeds total particulate bromine, usually by a factor of 5 to 20. There is some evidence for loss of bromine from aerosol for particles with time and for diurnal variations in particulate Br^- levels. The springtime Arctic bromine bloom mentioned above was observed first for particulate bromide. Approximately between mid-February and mid-May, particulate bromide levels are about 100 ng/m^3 STP compared to about 6 ng/m^3 STP for the rest of the year in the Arctic atmosphere. Br^- in precipitation has not been measured extensively, but several studies of marine precipitation suggest that the ratio of Br^- to Cl^- tends to exceed that for seawater.

Organic iodine gases, I_g^o, have been measured in the same way as Cl_g^o, but with much less extensive efforts. Total gaseous iodine values of 5 to 30 ng/m^3 STP have been observed with elemental analysis methods; it is thought that most of this I is organically bound. CH_3I has been measured with good latitude coverage by two groups. Concentrations generally range from 1 to 6 pptv, with values between 10 and 20 pptv in the marine boundary layer over biologically active oceans. There appears to be slightly more CH_3I in the northern hemisphere. While CH_3I is the only I_g^o species detected specifically to date, it is not safe to say that it is the dominant species—relatively little effort has been expended so far. No inorganic iodine gases, I_g^i, have been detected with specificity, but attempts have been made to measure total I_g^i. Values around 10 ng/m^3 STP have been reported, but it is likely that I_g^o species contributed to the apparent I_g^i values. Despite the fact that only one I_g^o and zero I_g^i species have been identified, it is clear that total gaseous I exceeds total particulate I (by a factor of 2 to 4). In turn, iodine in marine aerosols (both I^- and IO_3^-) is greatly enriched over the value expected from the iodine/chlorine ratio of seawater. Iodine enrichments of 100 to 1000 have been observed. Accordingly, it is a little surprising that gaseous I concentrations exceed those of particulate I. No global or extended-area studies of I in precipitation have been performed.

Organic fluorine gases, F_g^o, have been studied fairly extensively, at least the chlorofluorocarbon class of F_g^o species (because of interest in the sources of stratospheric chlorine). The dominant F_g^o species by concentration is CCl_2F_2. Total F_g^o values are about 1 $\mu g/m^3$ STP, and this total is growing with time because of increasing anthropogenic input and the long residence time of F_g^o species. There are very few known occurrences of C-F bonds in natural products, marine or terrestrial; F_g^o species are entirely anthropogenic. For F_g^i, there are almost no data in nonurban air. Tropospheric observations have been limited to SF_6 and to fluoride-based analyses (probably but not necessarily of HF) in polluted air.

Fluoride in particles is likely due to fluorine-containing contaminants released in industrial processes. This current view is based mostly on post-1977 data that have shown lower F^- levels in precipitation and particles than were seen earlier with cruder analytical methods. Although the question is not settled, present indications are that observed distributions of F^- in rain and particles are mostly from continental sources.

In the Element Cycle Matrices section of this document (Appendix C), a brief summary appears for the state of knowledge of distributions of halogens in the troposphere.

SOURCES OF ATMOSPHERIC HALOGENS

As in our discussion of the distributions of halogens, here we summarize briefly the sources of each halogen element, roughly in order of decreasing available knowledge.

Organic chlorine gases have natural and industrial sources. The most prevalent species, CH_3Cl, is calculated to be furnished to the atmosphere at a rate of about 2×10^6 metric tons annually; less than 5 percent of this source is industrial. Almost all other Cl_g^o species are anthropogenic, or mostly so. Based on knowledge from marine natural-products chemistry, it would not be surprising to find natural sources of $CHCl_3$, or even CCl_4. For CH_3Cl, it is suspected that natural sources include marine microbial processes and biomass burning. It is clear that there are no in situ atmospheric sources of Cl_g^o. R-Cl molecules, where R is an organic group such as CH_3, are not synthesized in the open air of the earth's oxidizing atmosphere.

The principal source of chlorocarbons and chlorofluorocarbons is from the escape of these substances from their usages as solvents and degreasers and in foam-blowing processes and refrigeration units, their release from aerosol spray cans, and their use in a variety of specialized processes in electronics, medicine, and manufacturing.

Very little direct and verified information is available on sources of Cl_g^i species such as HCl. Although its vertical and latitudinal distributions are not known, HCl is probably most concentrated (1 to 2 ppbv) in the marine boundary layer, where its residence time is perhaps 4 days. A global source of 10^8 metric tons of HCl per year would be required to maintain this concentration of HCl. Independently, it has been estimated that 3 to 20 percent of the annual input of sea-salt chloride is liberated from these particles as gaseous species, probably HCl. If so, 2-12 $\times 10^8$ tons/yr of HCl is so produced. Also, volcanoes and combustion are thought to emit perhaps 6×10^6 and 3×10^6 tons/yr, respectively, globally. Although small in comparison with the global input from volatilization of sea-salt chloride, these latter sources could dominate regionally. All of these sources are represented schematically in Figure 7.6, an outline of tropospheric halogen cycles. Processes such as the reactions of sea-salt aerosols with polluted continental air masses could also release NO_2Cl, NOCl, Cl_2, or even gaseous NaCl. Particulate chloride in the marine atmosphere results from sea-salt aerosol production. As sketched in Figure 7.6, these particles become airborne as a result of breaking waves, whitecap bubble-bursting, and impact of precipitation drops on the sea surface. Gas-to-particle conversion can also produce particulate chloride. Over continents, there are volcanic and combustion sources of HCl and particulate Cl^-.

Organic bromine sources are much less well understood, especially in light of the springtime Arctic bloom mentioned above. Neither the sources of the background or seasonally perturbed Br_g^o levels are clear. It is known that usage of the agricultural fumigant, CH_3Br, can inject some volatile CH_3Br into the atmosphere, but quantities are uncertain. Also, it is likely that some of the observed atmospheric $C_2H_4Br_2$ is from combustion of automobile and truck fuel additives. Also, bromoform ($CHBr_3$) has been observed in the Arctic Ocean surface waters. Many bromine-containing marine natural products have been identified, and further investigations are needed. A few other anthropogenic bromocarbons are also of interest. Sources of inorganic bromine gases have not been explored at all. Clearly, the tropospheric oxidation of Br_g^o species, largely by tropospheric OH, must produce some Br_g^i in situ. Sources of bromide in aerosol particles and precipitation are probably an incorporation of sea-salt bromide and scavenging of Br_g^i by clouds, rain, and aerosol particles.

Sources of tropospheric iodine have also been explored only crudely. For I_g^o species, only CH_3I has been studied. There are indications that biogenic CH_3I from the oceans, possibly from biological methylation of seawater I^-, is an important source. The direct emission of I_2 from seawater has been suggested from certain laboratory experiments in which O_3 was allowed to react

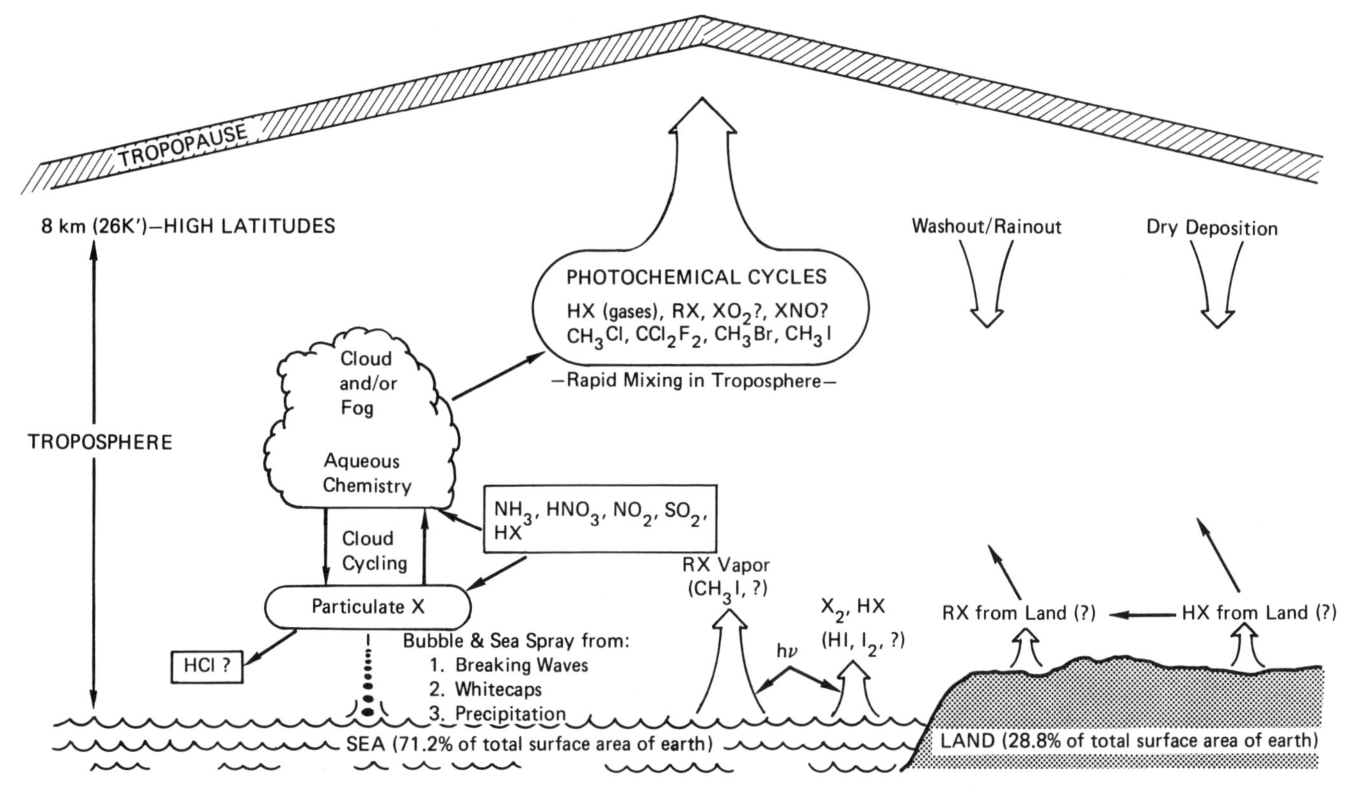

FIGURE 7.6 Schematic diagram to show processes and to exemplify key species in global tropospheric halogen cycles. X denotes F, Cl, Br, or I.

with dissolved I^-. Particulate I^- (and/or IO_3^-) is highly enriched with respect to Cl^- in marine aerosols. Compared to the seawater ratios, I^-/Cl^- is usually 100 or even 1000 times enhanced, especially on small particles. Clearly, some fractionation process is at work at the air-sea interface as particles are injected into the atmosphere. The involvement of iodine-rich organic films has been suggested. Aged aerosol particles do gather gaseous iodine to increase the I^-/Cl^- ratio further, and if so, what are the sources of I_g^i that allow this?

Fluorine sources, especially those for F_g^o, are similar to those for chlorine. In addition to the chlorofluorocarbons discussed above, a few pure fluorocarbons are also of interest. One of these, CF_4, is probably from aluminum ore processing, but possibly also from various carbon-electrode processes. Certain perfluoroethanes and perfluorocylohexanes are also entering the atmosphere now from a variety of specialized usages, often as inadvertent emissions. F_g^i species such as HF are known to be pollutants from industrial processes such as aluminum refining and cement production. Also, one can imagine that gaseous HF is released from fluoride-containing aerosol particles as these particles become drier and acidified (HF is a weak acid compared to H_2SO_4 and HNO_3). Sources of F^- in particles and precipitation include sea-salt input and industrial airborne particles.

REACTIONS AND TRANSFORMATIONS OF HALOGENS

Considering the many and complex reactions—homogeneous gas phase, heterogeneous (gas-particle) and homogeneous liquid phase—that are possible with halogens in the troposphere, research on them to date is very sparse. Consequently, very little is known about the mechanisms of halogen reactions and transformations. By contrast, stratospheric halogen reactions are limited to those in the gas phase, and these are known to be important.

Organo-halogen gases, R-X, of low molecular weight are generally volatile and not very soluble in water. Photochemical reactivity increases from fluorine to chlorine to bromine to iodine, that is, as halogens replace hydrogen atoms in compounds; C-F bonds are stronger than C-Cl, C-Br, or C-I bonds. Perfluorocarbons and perchlorocarbons are generally stable in the troposphere and are not susceptible to attack by O_3, OH, or tropospheric photons. Instead, they decompose only in the stratosphere and above when attacked by vacuum ultraviolet and electronically excited oxygen atoms. Other R-X species exhibit a wide range of photochemical reactivity. Some are photolyzed in the troposphere (e.g., CH_3I), and most are dissociated by OH attack to form inorganic halogen species.

Inorganic halogen gases are potentially important in tropospheric photochemical cycles, although no major specific role has yet been proven. Volatile species exist for each of the halogens, and a great variety of species need to be considered. For chlorine and fluorine, the stability of HCl and HF greatly slows regeneration of Cl and F atoms. Hydrogen donors, R-H, react readily with Cl and F to form HCl and HF, so the free atoms and their oxides are not thought to be very prevalent. Longer chain lengths for gas-phase catalytic processes are possible for bromine and iodine because HBr and HI are less stable once formed. Possible roles for I_g^i species in destroying O_3 and affecting other tropospheric photochemical cycles have been proposed, but great uncertainties exist. Examples include lack of information on the levels of I_g^i concentrations, unavailability of certain chemical kinetic data, and possible interferences by heterogeneous processes. Until there are some field data on specific halogen-containing inorganic species, little progress can be expected. Stratospheric investigations have provided some guidance, especially for chlorine and bromine, but direct tropospheric studies are needed.

Heterogeneous reactions and transformations need attention, but have received little. There is strong evidence that heterogeneous reactions are responsible in great part for the very existence of tropospheric HCl, yet few, if any, mechanistic studies have been performed. There is some evidence that particulate bromine concentrations increase at night and decrease by day and that gaseous bromine exhibits opposite diurnal behavior, but no studies of possible mechanisms are available. Similarly, the processes that lead to loss and/or uptake of halogens from marine aerosols of various sizes have not yet been investigated, nor have analogous processes in precipitation been studied. Such investigations are greatly hampered by a dearth of field data and of fundamental kinetic and photochemical data from the laboratory. Equilibrium-type data such as vapor pressures, even when available, are not necessarily valid when complicated multiphase, multiconstituent mixtures are to be considered. Homogeneous aqueous-phase reactions and transformations are potentially very important in clouds, rain, and water-coated aerosol particles, for example, with halogens as oxidizing agents, but virtually no research has been performed on this topic.

REMOVAL PROCESSES FOR HALOGENS

A very large spectrum of time constants exists for the residence times of various halogen-containing gases and particles, and the global atmospheric cycles of the halogens encompass both large and small reservoirs and transfer rates. For example, the annual input of sea-salt chloride into the atmosphere is around 6×10^9 tons/yr, but much of this chloride is airborne for a day or less. By contrast, the annual input of CCl_2F_2 is only about 3×10^5 tons/yr of chlorine, but it is transported to the stratosphere, and its residence time is 100 years. Accordingly, the global atmospheric cycles of halogens and the sinks for atmospheric halogens are made up of terms that are difficult or meaningless to compare.

The largest single sink for atmospheric halogens is represented by precipitation. Perhaps two-thirds of the large sea-salt particles that carry the most mass are removed by precipitation, and one-third by gravitational settling. Ninety percent of the removal occurs over oceans. Although these values have been deduced for chlorine, they are probably similar for bromine and iodine, whose sources are predominantly marine. Fluorine sinks are probably dissimilar in their distribution. As is indicated in Figure 7.6, dry deposition removes halogen gases and particles. Surface-active species like HCl, HF, or HOCl are probably most affected by dry deposition; the least-affected species are probably the organo-halogen gases. Finally, those portions of the atmospheric halogen cycles that penetrate the stratosphere, for example, perhalocarbons like CCl_2F_2, have their upward flows counterbalanced by downward return flows of HCl and HF in precipitation and dry deposition, at least in a steady state.

BIBLIOGRAPHY

Barnard, W. R., and D. K. Nordstrom (1982). Fluoride in precipitation: II. Implications for the geochemical cycling of fluorine. *Atmos. Environ. 16*:105–111.

Berg, W. W., P. D. Sperry, K. A. Rahn, and E. S. Gladney (1983). Atmospheric bromine in the Arctic. *J. Geophys. Res. 88*:6719–6736.

Chameides, W. L., and D. D. Davis (1980). Iodine: its possible role in tropospheric photochemistry. *J. Geophys. Res. 85*:7383–7398.

Cicerone, R. J. (1981). Halogens in the atmosphere. *Rev. Geophys. Space Phys. 19*:123–139.

Duce, R. A., J. W. Winchester, and R. VanNahl (1965). Iodine, bromine and chlorine in the Hawaiian marine atmosphere. *J. Geophys. Res. 70*:1775–1799.

Eriksson, E. (1959). The yearly circulation of chloride and sulfur in nature: meteorological, geochemical and pedological implications, 1. *Tellus 11*:375–403.

Eriksson, E. (1960). The yearly circulation of chloride and sulfur in nature: meteorological, geochemical and pedological implications, 2. *Tellus 12*:63–109.

Rasmussen, R. A., M. A. K. Khalil, R. Gunawardena, and S. D. Hoyt (1982). Atmospheric methyl iodide (CH_3I). *J. Geophys. Res. 87*:3086–3090.

Singh, H. B., L. J. Salas, and R. E. Stiles (1983). Methyl halides in and over the eastern Pacific. *J. Geophys. Res. 88*:3684–3690.

World Meteorological Organization (1981). *The Stratosphere 1981*. WMO Global Ozone Research and Monitoring Project Report No. 11. 503 pp.

TRACE ELEMENTS

BY R. DUCE

CURRENT ISSUES

The cycles of most trace elements in the troposphere have received relatively little attention from atmospheric chemists for several reasons. Trace elements, i.e., all elements except C, N, O, S, H, and the halogens, are present in such low concentrations that they have little impact on the overall photochemistry of the troposphere, its acid-base characteristics, or on climate. Many, if not most, trace elements are present entirely in the particulate phase and are not directly involved in gas-particle conversion processes or other aspects of gas-phase tropospheric chemistry. Being primarily present in aerosol particles, their tropospheric residence times are of the order of days to a few weeks, and there has been relatively little effort to evaluate global-scale changes in their distribution caused by human activity.

The tropospheric chemistry of many trace elements is an important part of the present-day overall biogeochemical cycles of these elements, but in most cases the tropospheric part of these cycles is poorly known. For example, the mobilization of Hg in the environment, whether it be from natural or pollution sources, is primarily through the troposphere in the gas phase, but very few data are available on the Hg concentrations in the remote troposphere, and even less is known about its chemical speciation and primary sources. Phosphorus is one of the primary nutrients in both the terrestrial and marine biosphere, and there is some evidence that tropospheric transport of P to the ocean may be significant in certain regions. However, the understanding of the spatial and temporal distribution of P in the troposphere, of its chemical forms and sources, and even of whether a long-surmised gaseous species exists is extremely primitive.

Trace elements can most conveniently be separated into two groups: Group A includes those elements that almost certainly spend their entire tropospheric lifetime on aerosol particles. Group B includes those elements for which a vapor phase, or likelihood of a vapor phase, exists. Group A includes such lithophilic elements as Al, Fe, Na, Ca, Mg, Si, V, Cr, Cu, Mn, and the rare earths. Group B includes such elements as B, Hg, Se, As, Sb, Cd, Pb, and possibly Zn and P.

SOURCES AND TRANSPORT

Trace element distribution patterns in aerosol particles are of considerable use in determining sources, transport paths, and deposition for the particles themselves. This is quite valuable since aerosol particles are an important end product for virtually all tropospheric cycles through heterogeneous and homogeneous reactions, and they play a major role in weather and climate. Through the use of interelement ratios, it is often possible to determine the sources for aerosol particles. For example, Al/Sc ratios on aerosol particles similar to that present in the earth's crust are an indication of a crustal weathering source, whereas Na/Mg ratios similar to that in the ocean suggest a marine source. The use of such "reference" elements as Al or Sc for the crust, Na or Mg for the ocean, and noncrustal V (i.e., that vanadium present on aerosol particles that is not derived from the earth's crust) for combustion of residual fuel oil or Pb for the combustion of gasoline containing tetraethyllead has proven quite useful. There are many sources that have not been so easily tagged, but efforts to determine appropriate trace element signatures are continuing. For example, B is being examined as a signature for coal burning, and As for smelter operations. This approach is potentially useful for identifying other specific sources of primary aerosols, including the terrestrial biosphere, volcanism, extraterrestrial particles, and a number of specific pollution sources. Recent efforts to identify regional source areas of aerosol particles through the use of a number of trace elements also show considerable promise. Trace elements used have included Se, Sb, As, Zn, In, noncrustal V, noncrustal Mn, and their interelemental ratios.

There is growing evidence that anthropogenic processes, followed by long-range tropospheric transport, can result in significant changes in the tropospheric and oceanic concentrations of certain toxic and essential trace elements on the near-global scale. For example, concentrations of Pb, both in the marine troposphere and in the surface waters of the Atlantic and Pacific Oceans, particularly in the northern hemisphere, are considerably elevated as a result of the burning of gasoline containing tetraethyllead on the continents and its subsequent tropospheric transport over and deposition to the oceans. The mobilization of other toxic elements such as Hg and Se by fossil fuel combustion and As by smelters and in herbicides and defoliants may be equivalent to or greater than mobilization by natural sources. In fact, as is the case for most cycles, one can make much more accurate estimates of the global source strengths from pollution sources for trace elements than from such natural sources as volcanism, the oceans, and the biosphere.

In particular, there is virtually no information on the production of vapor-phase trace elements or direct production of aerosol particles containing trace elements by the terrestrial biosphere. The apparent increased volatilization of Hg by higher plants (relative to release from unvegetated soils of comparable concentration) has been explained as a "detoxification" process, although it may simply be an expression of the ion-concentrating processes of plants combined with the reducing potentials created by the charge separation processes of metabolism. In any case, vegetation appears to have a major role in the cycling of Hg. For a number of other elements, including As, Sn, Se, Pb, and Sb, biological methylation has been observed in the laboratory, and methylated forms of many of these elements have been observed in highly polluted areas. For a few elements, ionic methyl compounds have been observed in uncontaminated regions. A further understanding of the biological production of methylated metals requires the development of specific detection capabilities for these species. The entire area of trace element release from the biosphere requires considerable effort in the future.

DISTRIBUTION

A growing data base is developing on the trace element composition of aerosol particles in remote regions. Reasonably good data are available over short time periods from the boundary layer in both polar regions and over the Atlantic and Pacific Oceans. Virtually no data are available on the vertical distribution of trace elements in these regions, however, and this information is critical to evaluate sources and fluxes of the trace elements. Information on the mass-size distribution of trace elements on aerosol particles is of considerable value in ascertaining sources and source processes for these elements. Many additional data of this type are required.

Very little is known about the chemical form of the trace elements in the vapor phase, but for elements in Group B, a vapor phase does, or is expected to, exist. Mercury apparently exists primarily as a gas in the form of elemental Hg, with evidence emerging for some organic forms as well, probably methylated. Mercury is one of the few metals whose ions can be reduced to the metallic state at reduction potentials frequently found in biological systems. The metal has an appreciable vapor pressure at 25°C ($\sim 1 \times 10^{-3}$ mm Hg); thus the presence of gaseous elemental Hg in the troposphere is not surprising.

Although the vapor phase apparently dominates tropospheric B and $B(OH)_3$ has been suggested as the primary vapor phase, no measurements have corroborated the presence of $B(OH)_3$. Although there are a number of volatile borane derivatives, their formation requires much more reducing conditions than those apparently achieved by microorganisms under anoxic conditions. Measurement of specific B species in the troposphere is clearly required before even a rudimentary understanding of the B cycle is possible.

In many respects, Se parallels S in geochemical behavior. Vapor-phase Se has been observed in remote and urban regions, where it may account for about 25 percent of the total Se present. However, except for dimethylselenide and dimethyldiselenide measured in urban and near-urban areas in Belgium, the chemical form of the vapor phase of Se is unknown.

Approximately 10 percent of the As in the marine boundary layer is apparently present in the vapor phase. The form of the vapor-phase As is also unknown, although dimethyl arsinic acid has been observed in the oxic marine environment and trimethyl arsine is known to be produced by certain fungi.

Methylated forms of Se, Sb, Pb, and other trace elements have been observed in other compartments of the environment (e.g., the ocean and plants). In the case of Sb, Hg, and Pb, it is unclear whether methylation can occur in oxic regions or only under anaerobic conditions, where microorganisms are probably of considerable importance. For all these trace elements in the vapor phase, the data base from remote regions is extremely small, in some cases being as few as 5 to 10 samples.

TRANSFORMATIONS AND SINKS

Very little information is available on transformation reactions involving trace elements in the troposphere. The residence time for the vapor phase of many trace elements in the troposphere may be very short, perhaps only minutes or hours, as is probably the case for certain As and Pb species. From mass balance considerations, the residence time of total (vapor plus particle) As in the global troposphere has been estimated as ~ 10 days. Evidence suggests that the residence time of vapor-phase As is considerably shorter, however. For some trace elements the vapor phase probably has a longer residence time than the particulate phase and is the dominant phase in the troposphere, as is likely the case for Hg and perhaps B. The residence time for elemental Hg may be as long as several months. Estimates of the residence time of vapor-phase Se, B, etc., compounds have not been made. It is likely that the removal of these trace elements is primarily governed by precipitation processes. For all these elements, woefully little is known about the specific chemical species present. Information on the chemical form of these trace elements in the vapor phase and details of their tropospheric reaction paths and rates are required before the importance of transformation reactions to their tropospheric cycles can be evaluated.

Certain trace elements, particularly transition metals, may be important as catalysts for reactions in cloud and rain droplets. For example, the oxidation of

SO_2 to sulfate in solution is enhanced by the presence of Mn ions. The importance of transition metals as catalysts depends on a number of factors, including abundance, chemical form, stable oxidation states, bonding properties, and solubility. Copper, manganese, and perhaps vanadium may be important as homogeneous catalysts in solution. For heterogeneous catalysis, solubility is not important; and the metals above, as well as Fe, Ti, and perhaps Cr, are of potential importance. The role played by trace element catalysis in aqueous atmospheric chemical reactions is largely unknown at present, but potentially very important and should receive increased attention.

BIBLIOGRAPHY

Braman, R. S., and M. A. Tompkins (1979). Separation and determination of nanogram amounts of inorganic tin and methyltin compounds in the environment. *Anal. Chem. 51*:12–19.

Brinckman, F. E., G. J. Olson, and W. P. Iverson (1982). The production and fate of volatile molecular species in the environment: metals and metalloids, in *Atmospheric Chemistry*, E. D. Goldberg, ed. Springer-Verlag, Berlin, pp. 231–249.

Cunningham, W. C., and W. H. Zoller (1981). The chemical composition of remote area aerosols. *J. Aerosol Sci. 12*:367–384.

Duce, R. A., R. Arimoto, B. J. Ray, C. K. Unni, and P. J. Harder (1983). Atmospheric trace elements at Enewetak atoll: I. Concentration, sources, and temporal variability. *J. Geophys. Res. 88*:5321–5342.

Fitzgerald, W. F., G. A. Gill, and A. D. Hewitt (1983). Air/sea exchange of mercury, in *Trace Metals in Seawater*, C. S. Wong, E. Boyle, K. W. Bruland, J. D. Burton, and E. D. Goldberg, eds. Plenum, New York, pp. 297–315.

Fogg, T. R., R. A. Duce, and J. L. Fasching (1983). Sampling and determination of boron in the atmosphere. *Anal. Chem. 55*:2179–2184.

Galloway, J. N., J. D. Thornton, S. A. Norton, H. L. Volchok, and R. A. N. McLean (1982). Trace metals in atmospheric deposition: a review and assessment. *Atmos. Environ. 16*:1677–1700.

Graham, W. F., and R. A. Duce (1979). Atmospheric pathways of the phosphorus cycle. *Geochim. Cosmochim. Acta 43*:1195–1208.

Harrison, R. M. and P. H. Laxen (1978). Natural sources of tetraalkyllead compounds in the atmosphere. *Nature 275*:738–739.

Jiang, S., H. Robberecht, and F. Adams (1983). Identification and determination of alkyl selenide compounds in environmental air. *Atmos. Environ. 17*:111–114.

Lantzy, R. L., and F. T. Mackenzie (1979). Global cycles and assessment of man's impact. *Geochim. Cosmochim. Acta 43*:511–515.

Maenhaut, W., H. Raemdonck, A. Selen, R. Van Grieken, and J. W. Winchester (1983). Characterization of the atmospheric aerosol over the eastern equatorial Pacific. *J. Geophys. Res. 88*:5353–5364.

Mosher, B. W., and R. A. Duce (1983). Vapor phase and particulate selenium in the marine atmosphere. *J. Geophys. Res. 88*:6761–6768.

Nriagu, J. O. (1979). Global inventory of natural and anthropogenic emissions of trace metals to the atmosphere. *Nature 279*:409–411.

Rahn, K. A., and D. H. Lowenthal (1984). Elemental tracers of distant pollution aerosols. *Science 223*:132–139.

Settle, D. M., and C. C. Patterson (1982). Magnitudes and sources of precipitation and dry deposition fluxes of industrial and natural leads to the North Pacific at Enewetak. *J. Geophys. Res. 87*:8857–8869.

Slemr, F., W. Seiler, and G. Schuster (1981). Latitudinal distribution of mercury over the Atlantic Ocean. *J. Geophys. Res. 86*:1159–1166.

Walsh, P. R., R. A. Duce, and J. L. Fasching (1979). Consideration of the enrichment, sources, and flux of arsenic in the troposphere. *J. Geophys. Res. 84*:1719–1726.

AEROSOL PARTICLES

BY J. M. PROSPERO

An aerosol is defined as a suspension of fine liquid and/or solid particles in a gas. In the case of interest to us, that gas is the atmosphere. Although in the strict sense of the definition the word aerosol refers to the particle and gas phases as a system, the term is often used to refer to the particle phase alone.

The subject of aerosol particles is somewhat anomalous in the general context of this section on cycles since all the other subjects focus on specific chemical species. Indeed, in most cases, the atmospheric aerosol is the end product of many of the chemical processes acting on the aforementioned species. Many of these reaction products are relatively unreactive in the aerosol phase. Because of this unreactivity and because of the relatively short residence time of aerosols in the troposphere (on the order of a week or two), the aerosol phase can be considered to be, in effect, the sink for many gas-phase species (see Figure 7.7).

Aerosol particles are treated as a separate cycle because they can have an impact on a number of important physical processes in the atmosphere. The relationship between aerosol properties and atmospheric processes is depicted in Part 1, Figure 2.2. For example, aerosols play an important role in the hydrological cycle—they can affect cloud microphysics, which, in turn, can affect the types, amounts, and distribution of rainfall. Clouds play a critical role in fixing the albedo of the earth; thus, if aerosols affect the amount, type, and distribution of clouds, then changes in aerosol concentration and properties could have the effect of changing the albedo.

Aerosol particles can affect light as it passes through the atmosphere by the mechanisms of scattering and absorption. The most obvious radiative consequence of airborne particles is the appearance of haze and degradation of visibility. Less obvious, but more important, are the possible effects of these same particles on the heat balance of the earth. Particles can cause a decrease in the amount of radiation reaching the ground, can increase or decrease the albedo, and, if the aerosol absorbs light, can cause atmospheric heating. In order to understand these optical effects, it is necessary to know the chemical composition of the aerosol and its size, because these characteristics will determine the aerosol scattering and absorption properties. Also, the composition of aerosols is often size dependent; thus a specific physically (or chemically) active species could be concentrated in a limited portion of the particle size spectrum.

Finally, there is the concern about the impact of many pollutants on health. The vector for many of these harmful species is the aerosol particle. However, the ability of aerosol particles to penetrate to the lung is dependent on

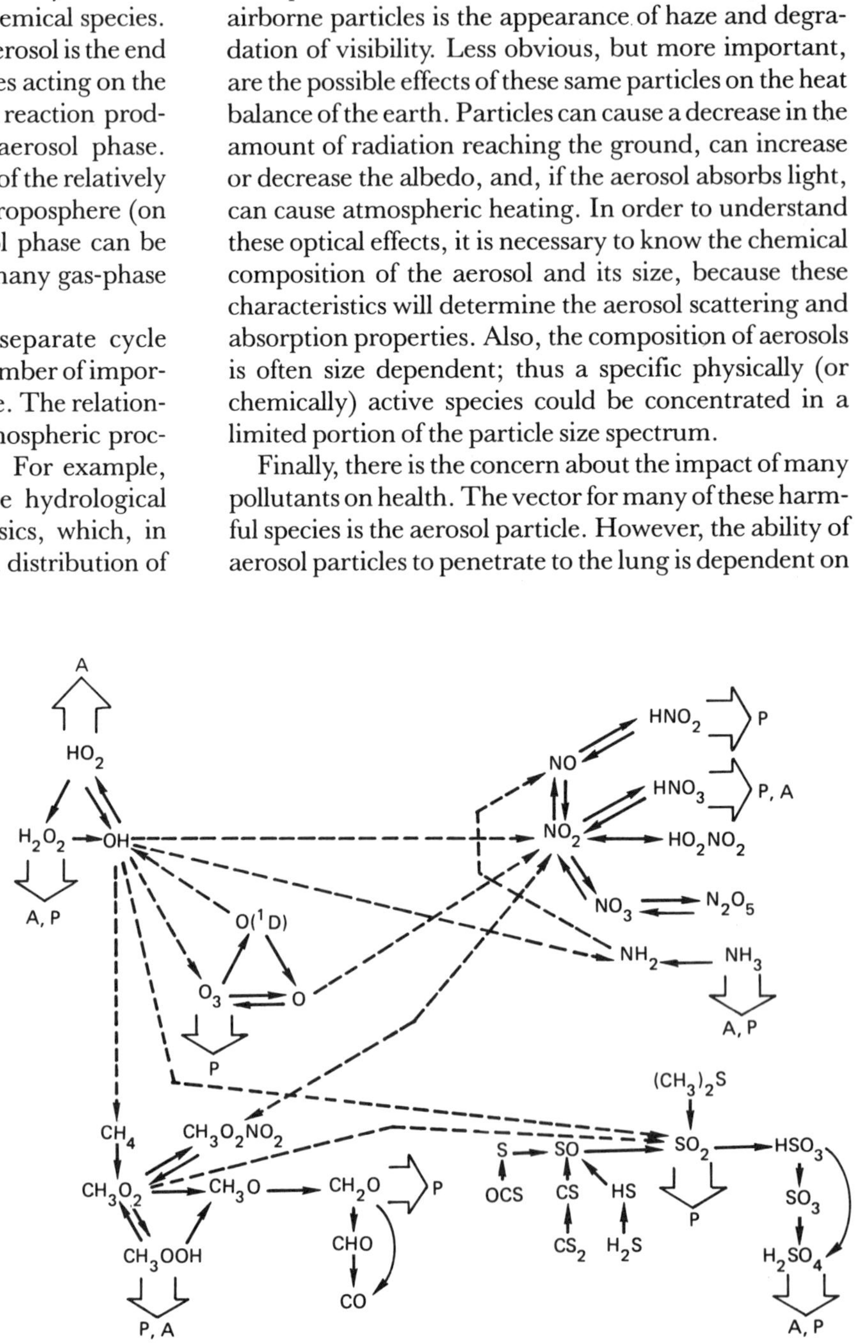

FIGURE 7.7 Aerosols as an end product of atmospheric reactions. Major reaction pathways for gas-phase constituents are depicted by solid lines. Interactions between chemical families are indicated by dashed lines. Heavy (double) arrows show key heterogeneous pathways involving aerosols (A) and precipitation (P) (Turco et al., 1982).

the aerosol size and its chemical properties as a function of size.

SOURCES

From the standpoint of production processes, aerosol particles can be categorized as being either primary or secondary. Primary aerosols are those directly emitted as a solid or a liquid, while secondary aerosols are derived from materials initially emitted as gases. Important natural primary aerosols are the salt residue from sea spray, wind-blown mineral dust, ash from volcanic eruptions, and organic particles from biota (pollen, spores, debris, etc.). A major class of anthropogenic primary aerosol is smoke particles. However, the burning of land biota is another major source of smoke; this source can be either natural or man-made, depending on how the fire was started.

There are a number of major sources of secondary aerosols. As described in preceding sections, many biological systems emit gaseous species that are eventually converted to aerosols. Finally, many of man's activities release gases that are aerosol precursors.

Estimates of annual aerosol mass production rates are summarized in Table 7.8. The quality of these data leaves much to be desired. The wide range of estimates for some species is a reflection of the poor state of knowledge of their sources. Nonetheless, it is clear that knowledge of the input rates from anthropogenic sources is much better than that of the rates from natural sources.

In the category of primary particles, natural sources (predominantly sea salt and soil dust) far outweigh anthropogenic sources. However, it must be borne in mind that the anthropogenic sources are concentrated in a relatively small area and hence will be much more significant on a local and regional scale. Obviously, the major source of sea-salt aerosol is the ocean. Although a fair amount of research has focused on the sea-salt aerosol production mechanism and especially on the rate of production as a function of various environmental parameters, there is still considerable debate on the results of such work and their interpretation. The rates of production and the physical and chemical characteristics of sea-salt aerosol are important for a number of reasons. For example, salt spray may play an important role in transporting trace metals and organic materials from the ocean into the atmosphere and in absorbing or reacting with gaseous species in the marine boundary layer. Because the oceans are so large and the mass of sea-salt aerosol is so great, these processes must be understood in order to understand global chemical cycles in general.

The major sources of soil aerosols are arid regions. Clearly, much soil dust, such as that from the Sahara, is generated by purely natural processes. However, in some cases, the distinction between natural and anthropogenic sources is not so clear. For example, recent work has shown that large quantities of soil dust are being transported out of Asia far into the North Pacific. Much of this material is believed to have been deflated from agricultural regions in China in the spring after the soils have been ploughed for planting. Also, in the United States, the primary standard for total suspended particulate materials, as defined in the Clean Air Act, is most widely violated in agricultural areas because of the mobilization of soil dust by farming activity. It is clear from Table 7.8 that there is considerable uncertainty in the rates of mobilization of soil dust. However, it is even more uncertain as to what fraction is derived from natural sources as a consequence of natural processes.

The range of estimates for production rates from volcanoes and biomass burning is extremely large. This uncertainty is a consequence of the difficulty in obtaining data on sources that are sporadic, widely dispersed, and relatively inaccessible.

TRANSPORT

Aerosols, whether primary or secondary, can be transported great distances in the atmosphere. As previously stated, large quantities of soil aerosols are routinely transported thousands of kilometers over the oceans from their sources in continental regions. Similarly, pollutant aerosols can be transported great distances. For example, of the acid species in aerosols over the northeast United States and Canada, a large fraction is derived from sources in the central United States. On a larger scale, the trace metal composition of aerosols in Arctic haze episodes suggests that the sulfate-rich particles are advected primarily from sources in Europe and Asia. These interpretations are supported by meteorological studies and trajectory computations.

Aerosol species such as sea salt and soil dust are relatively inert, and their principal physical and chemical properties remain essentially unchanged during transport in the atmosphere. Because of their relatively conservative nature, these species can serve as tracers for atmospheric transport and removal processes. Some research along these lines has already begun with some success. Given a sufficiently large data base, measurements of these species could be used to validate atmospheric transport models that are currently under development. Unreactive species are most useful for such validation because it is not necessary to make any assumptions about in situ chemistry during transport. An advantage of using aerosols as tropospheric tracers is that their lifetime is of the same order as the lifetime of a typical synoptic meteorological event, about a week.

TABLE 7.8 Estimates of Global Particle Production from Natural and Man-Made Sources (10^6 tons/yr)

Source	After Peterson and Junge (1971),[a] <5 μm	After Hidy and Brock (1971)[a]	After *Study of Man's Impact on Climate* (1971)[a] <20 μm	<6 μm[b]	Other Estimates[a]
Man-Made					
Direct particle production					
Transportation	1.8				
Stationary fuel sources	9.6				
Industrial processes	12.4				
Solid waste disposal	0.4				
Miscellaneous	5.4				
Subtotal	29.6	37-110	10-90	6-54	54
					126
Particles formed from gases					
Converted sulfates	200	110	130-200		
Converted nitrates	35	23	30-35		
Converted hydrocarbons	15	27	15-90		
Subtotal	250	160	175-325		270
Total man-made	280	269	185-415		396
Natural					
Direct particle production					
Sea salt	500	1095	300	180	1000-2000
Windblown dust	250	60-360	100-500	60-300	70
					60-360
					128 ± 64
					200 ± 100
Volcanic emissions	25	4	25-150	15-90	4.2
Meteoric debris	0	0.02-0.2			1-10
					0.02-0.2
Forest fires	5	146	3-150		
Subtotal	780	1610	428-1100		1730
Particles formed from gases					
Converted sulfates	335	37-365	130-200		
Converted nitrates	60	600-620	140-700		160
Converted hydrocarbons	75	182-1095	75-200		154
Subtotal	470	2080	345-1100		1319
Total natural	1250	3690	773-2200		3049
Grand Total	1530	3959	958-2615		3445

[a] For references see Bach (1976) or Prospero et al. (1983).
[b] Values are for particles <6 μm as recomputed by Jaenicke (1980).

SOURCE: Bach, 1976.

TRANSFORMATIONS

Gases can react in the atmosphere to produce nonvolatile products that end up in the aerosol phase. It is clear from Table 7.8 that for many species the quantities of aerosol produced in this manner equal or exceed the quantities emitted directly as particles. This is true for sulfate and nitrate species that are currently of interest because of their role in the formation of acid rain.

Clouds play a dominant role in the formation, modification, and removal of aerosols. The condensation of water vapor on particles, and the phoretic, diffusive, or inertial capture of particles by droplets lead to the incorporation of particles within the aqueous phase. Solution reactions, including those with dissolved gases, become possible, and transformations can occur. If subsequent droplet growth leads to precipitation, the aerosol is removed from the atmosphere. However, if the droplet reevaporates, as it does in over 90 percent of the cases, then the aerosol is regenerated, but its size and composition are changed. Cloud cycling is probably the major mechanism for modifying the atmospheric aerosol in the lower troposphere. In contrast to the processes of particle interactions and coagulation, which are reasonably well understood, the cloud cycling aspects of aerosol-hydrometeor-gas interaction are poorly understood.

REMOVAL PROCESSES

Precipitation is the major mechanism for the removal of aerosols from the atmosphere. For example, it has been shown that 80 to 90 percent of the radioactive fallout deposited on the earth's surface was brought down in precipitation. In turn, the composition of precipitation is determined to a considerable extent by the composition of the aerosol phase.

Gravitational removal is important only for relatively large particles (i.e., those larger than about 10-μm diameter). Sedimentation will be important for soil aerosols close to the source area and for sea-salt aerosols. Sedimentation is generally not important for anthropogenic materials. Pollution control measures have sharply reduced the rates of emission of large particles. On the other hand, the size of secondary aerosols is less than 1-μm diameter, and consequently, these aerosols have a very small settling velocity.

DISTRIBUTION

Because of the relatively short residence time of aerosols in the atmosphere, their distribution will be closely linked to the distribution and activity of sources and to the controlling meteorological phenomena. Thus, in order to characterize any trends in the concentration and distribution of aerosols, it will be necessary to sample frequently on a broad spatial scale that encompasses the suspected major source regions and the dominant meteorological systems. Some well-conceived regional sampling programs are currently in place. However, on a larger scale, with a few exceptions, current sampling efforts leave much to be desired from the standpoint of the species studied, the quality of the data, the frequency of sampling, and the location of stations.

CONCLUSIONS

We can identify a number of areas that warrant further research on aerosols:

1. The role of aerosols in geochemical transport and anthropogenic impacts. The atmosphere is an important mode of transport for many species. For example, the anthropogenic emissions of sulfur, mercury, and lead to the atmosphere already exceed the stream loads for these elements, while the emissions of copper, arsenic, zinc, tin, selenium, molybdenum, antimony, and silver are within a factor of 10 of stream fluxes.

2. Gas-particle processes. The photochemical and chemical reactions that initially transform gases into secondary reaction products are extremely complex and not yet fully understood. The particles formed by these mechanisms are mostly in the "fine particle" size range (i.e., submicrometer). It is estimated that several hundred million tons of fine particles are formed every year as a consequence of the emission and subsequent reaction of natural and anthropogenic gaseous species.

The role of organics in aerosols is poorly understood. Yet, the concentration of particulate organic carbon in the atmosphere is quite high. For example, in most ocean areas, the mean value is comparable to that of mineral aerosols and non-sea-salt sulfate and nitrate. Gas-to-particle conversion appears to be a major mechanism for the production of fine-particle carbon over the oceans and also over the continents. Unfortunately, there are very few concurrent measurements available for both the vapor and the particulate phases.

3. The role of aerosols in the hydrological cycle. It is clear that clouds play a major role in the formation and removal of aerosols. As stated earlier, any process that acts on clouds could have an impact on weather and climate. There is now sufficient evidence to conclude that anthropogenic emissions, especially of sulfur and nitrogen species, do indeed have an impact on cloud microphysics. Given the importance of sulfur in the hydrological cycle and bearing in mind that about half of the global flux of sulfur has an anthropogenic origin, there is good cause for concern that man may be altering weather on a larger scale. Of particular interest are the possible effects on the urban and regional scale where the magnitude of the anthropogenic sulfur sources is dramatically higher—for example, in the eastern United States, where it is 10 times that of natural sources.

The assessment of the impact of man on weather or climate is difficult for a number of reasons. One has to do with the fact that climate is subject to variations that are completely natural in origin. Thus, until there is a better understanding of the mechanisms that determine climate, it will be difficult to ascertain the ways in which it has been changed, or might be changed, by anthropogenic activities. Therefore, research efforts directed at elucidating impacts must be balanced by efforts directed at gaining an understanding of basic processes. In the case of aerosols, one of the basic processes of great importance is the role of aerosols in cloud physics.

Although the impact of aerosols on the hydrological cycle from the standpoint of precipitation quantity and distribution cannot be quantitatively assessed with certainty, it can be stated with certainty that there has been a very marked impact on precipitation quality, most notably in the increased acidity of rain.

4. Radiative transfer. There are a number of different aerosol types that are important and that could play

a major role in climate by altering the radiation budget of the earth:

a. Volcanic debris. Most important in this category are the sulfur species that are injected into the stratosphere, where they are oxidized to sulfuric acid droplets, which have a residence time of years.

b. Soil dust. About one-third of the surface of the continents is arid and a potential source of soil dust. Here the impact of man on soil mobilization is a major concern.

c. Elemental carbon (soot). This material is a highly efficient absorber of radiation; thus it is important that its abundance and distribution be measured. However, scientists have a very poor idea of the global budget of carbon because much carbon is produced in remote regions by slash-and-burn agricultural practices and because carbon aerosols are difficult to measure with current analytical techniques.

5. Characterization of temporal and spatial trends. A recurring conclusion in the assessments of the possible impact of aerosols on climate is that there is a serious lack of information about the composition, concentration, and physical properties of aerosols and their temporal and spatial variability.

Despite the undeniable evidence that anthropogenic materials are being transported over great distances, there is no evidence that the particles have significantly reduced atmospheric transmission in remote regions. For example, there is no evidence of any long-term decrease in transmission in the data from the Mauna Loa Observatory or from the Smithsonian Astronomical Observatories. However, we do not mean to minimize the possibility that such increases may be in the process of occurring; we merely emphasize that, with current measurement techniques and the length of the records on hand, one cannot separate any trends, if they exist, from the natural variability of the atmospheric aerosol. Indeed, it is essential that the natural processes be understood before any anthropogenic effects can be identified.

BIBLIOGRAPHY

Bach, W. (1976). Global air pollution and climatic change. *Rev. Geophys. Space Phys. 14*:429–474.

Barry, R. G., A. D. Hecht, J. E. Kutzbach, W. D. Sellers, T. Webb, and P. B. Wright (1979). Climatic change. *Rev. Geophys. Space Phys. 17*:1803–1813.

Blanchard, D. C. (1984). The production, distribution and bacterial enrichment of the sea-salt aerosol, in *The Air-Sea Exchange of Gases and Particles,* W. G. N. Slinn and P. Liss, eds. D. Reidel, Boston, Mass., pp. 407–454.

Charlson, R. J., and H. Rodhe (1982). Factors controlling the acidity of natural rainwater. *Nature 295*:683–685.

Dockery, D. W., and J. D. Spengler (1981). Indoor-outdoor relationships of respirable sulfates and particles. *Atmos. Environ. 15*:335–343.

Edmonds, R. L., ed. (1979). *Aerobiology, The Ecological Systems Approach, US/IBP Synthesis Series*, Vol. 10. Dowden, Hutchinson and Ross, Stroudsburg, Pa., 386 pp.

Friedlander, S. K. (1978). Aerosol dynamics and gas-to-particle conversion, in *Recent Developments in Aerosol Science*, D. T. Shaw, ed. Wiley, New York, pp. 1–24.

Galloway, J. N., G. E. Likens, W. C. Keene, and J. M. Miller (1982). The composition of precipitation in remote areas of the world. *J. Geophys. Res. 87*:8771–8786.

Hinds, W. C. (1982). *Aerosol Technology*. Wiley, New York, 424 pp.

Holland, W. W., A. E. Bennett, I. R. Cameron, C. du V. Florey, S. R. Leeder, R. S. F. Schilling, A. V. Swan, and R. E. Waller (1979). Health effects of particulate pollution: reappraising the evidence. *J. Epidemiol. 110*:525–659.

Jaenicke, R. (1980). Atmospheric aerosols and global climate. *J. Aerosol Sci. 11*:577–588.

Lodge, Jr., J. P., A. P. Waggoner, D. T. Klodt, and C. N. Crain (1981). Non-health effects of airborne particulate matter. *Atmos. Environ. 15*:431–482.

Pewe, T. L., ed. (1981). Desert dust: origin, characteristics and effect on man. *Geol. Soc. Amer. Spec. Pap. 186*, 303 pp.

Podzimek, J. (1980). Advances in marine aerosol research. *J. Res. Atmos. 14*:35–61.

Prospero, J. M. (1981). Aeolian transport to the world ocean, in *The Sea*, Vol. 7, *The Oceanic Lithosphere*, C. Emiliani, ed. Wiley Interscience, New York, pp. 801–874.

Prospero, J. M., V. Mohnen, R. Jaenicke, R. Charlson, A. C. Delany, J. Moyers, W. Zoller, and K. Rahn (1983). The atmospheric aerosol system: an overview. *Rev. Geophys. Space Phys. 21*:1607–1629.

Rahn, K. A. (1981). Relative importance of North America and Eurasia as sources of Arctic aerosol. *Atmos. Environ. 15*:1447–1456.

Seiler, W., and P. J. Crutzen (1980). Estimates of gross and net fluxes of carbon between the biosphere and the atmosphere from biomass burning. *Clim. Change 2*:207–247.

Subcommittee on Airborne Particles (SAP) (1979). *Airborne Particles*. National Research Council. University Park Press, Baltimore, Md., 343 pp.

Toon, O. B., and J. B. Pollack (1980). Atmospheric aerosols and climate. *Amer. Sci. 68*:268–278.

Turco, R. P., O. B. Toon, R. C. Whitten, R. G. Keesee, and P. Hamill (1982). Importance of heterogeneous processes to tropospheric chemistry: Studies with a one-dimensional model, in *Heterogeneous Atmospheric Chemistry*. Geophysical Monograph Series, Vol. 26, R. Schryer, ed. American Geophysical Union, Washington, D.C., pp. 231–240.

8 Instrumentation Development Needs for Use of Mass-Balance Technique

BY D. LENSCHOW

In Part I of this report, several of the investigations of the proposed research program require the measurement of fluxes of chemicals. Such measurements are required in the mass-balance or box-budget experiments proposed under the general heading "Noncyclic Transformation and Removal," and also for field studies of dry deposition and for advanced experiments under the heading "Sources of Atmospheric Chemicals, Biological and Abiological." In this brief section, we outline the principles of such measurements and the general requirements they place on instrumentation.

The budget for the mean concentration of a species s can be written as

$$\frac{\partial \bar{s}}{\partial t} + \bar{u}_i \frac{\partial \bar{s}}{\partial x_i} + \frac{\partial \overline{w's'}}{\partial z} = \bar{P}_s - \bar{L}_{s'} \qquad (8.1)$$

where the overbar denotes an average over a time or horizontal distance long enough to obtain a stable estimate, and the prime denotes a departure from the mean. For aircraft flux measurements, the time or distance over which a mean is calculated is typically at least 5 min or 25 km. This equation states that the time rate of change, plus the mean advection of s, plus the vertical turbulent flux divergence of s is equal to the mean internal sources $\bar{P}_s$ and sinks $\bar{L}_s$ of s. Integrating vertically and applying (8.1) to a convective boundary layer that is vertically well mixed to a depth h and capped with a well-defined inversion layer,

$$h \frac{\partial <s>}{\partial t} + <\bar{u}> h \frac{\partial <s>}{\partial x} - <\bar{u}> \Delta s \frac{\partial h}{\partial x}$$

$$- (\overline{w's'})_0 - w_e \Delta s = <\bar{P}_s - \bar{L}_s> h, \qquad (8.2)$$

where the brackets denote an average through the boundary layer, Δs is the change in s across the top of the boundary layer, and w_e is the entrainment velocity through the boundary layer top. If the boundary layer is not uniformly mixed, it may still be possible to integrate (8.1), but additional terms will result.

The terms on the left side of (8.2) can be measured in a well-designed experiment by using aircraft over a horizontally homogeneous surface. The aircraft must be equipped to measure mean horizontal air velocity and mean and turbulent fluctuations of the species s, as well as turbulent fluctuations of the vertical velocity.

As pointed out by Lilly (1968), it may be possible to estimate w_e by the relation $w_e = -(\overline{w's'})_h/\Delta s$, where $(\overline{w's'})_h$ is the vertical flux at h, if flux measurements of a conserved variable are obtained throughout the boundary layer, and extrapolated through the capping inversion, and Δs is well defined. Lenschow et al. (1982) have done this for O_3 in a marine boundary layer. Measurements of surface flux $(\overline{w's'})_0$ can be obtained by extrapolating flux measurements from an aircraft at several levels in the boundary layer down to the surface. Surface flux can also be obtained from ground-based sites, where a variety of techniques have been used in addition to direct eddy correlation measurements (Hicks et al., 1980). These techniques include (1) box methods, where an enclosure is placed over a surface of interest and the time rate of change of the species concentration is measured; and (2) profile or gradient methods, where differences in concentration at several levels close to the surface can be related to flux at the surface. These methods require only mean concentration measurements. However, the box method assumes that the enclosure does not affect the surface flux, or at least that such effects can be corrected, and the profile method requires accurate measurements of concentration differences (on the order of 1 percent of the mean concentration or better), as well as supporting micrometeorological measurements to determine the surface stress and stability. Further details on use of different platforms and sensors for measurements of fluxes and other micrometeorological variables are presented by Dobson et al. (1980) and in a special issue of *Atmospheric Technology*.[1]

Absolute accuracy for turbulence measurements is not important, as long as the sensor output does not drift significantly during the period of a flux-measuring run (typically about 10 min), and the sensor gain is known accurately. This is because a flux measurement is calculated from departures from a mean value; the mean of each quantity is not used in the calculation of a vertical turbulent flux. However, the fluctuations must be resolved. In a convective boundary layer where the fluctuations are generated by a flux at the surface, the measured resolution of a trace species should be at least about 10 percent of its surface flux divided by the convective velocity,

$$w_* = [(g/T)(\overline{w'\theta'})_0 h]^{1/3},$$

where $(\overline{w'\theta'})_0$ is the surface virtual temperature flux and g/T is the buoyancy parameter (gravitational acceleration divided by the mean temperature). Typically, w_* is of the order of 1 m/s.

In order to resolve the eddies important for vertical transport through most of the convective boundary layer, a sensor must resolve wavelengths at least as small as 30 m and preferably 5 m. Close to the surface, the requirements are even more stringent since there the eddies scale with height above the ground. At 10-m height, for example, wavelengths as small as 3 m must be resolved accurately to measure all the significant contributions to the turbulent flux (Kaimal et al., 1972). For a stably stratified boundary layer (e.g., the clear-air nocturnal boundary layer over land), somewhat smaller spatial resolution is required to resolve the turbulent fluctuations. A sample rate of 50 per second would be desirable if aircraft measurements are proposed for this situation.

For measurements above the boundary layer, in the free troposphere and lower stratosphere, a slower sampling rate seems adequate. Possibly two per second would be satisfactory. If at all possible, however, a higher rate (say, 5 to 10 per second) should be used.

Whatever rate is used, it should be remembered that the Nyquist frequency (i.e., the maximum frequency that can be resolved by a spectral decomposition of the data) is half the sampling rate. Thus a 20-per-second sampling rate will resolve 10-Hz spectral variables. At an aircraft speed of 100 m/s, this is equivalent to a 10-m wavelength.

Another factor to remember is that a first-order instrument time constant must be multiplied by $\sim 2\pi$ before taking the reciprocal in order to estimate the response of the instrument in the frequency domain. Thus, if one has a first-order time constant of 0.1 s, the amplitude of its variance is reduced to 72 percent of the variance of the input signal at a frequency of 1 Hz. For flux calculations, the phase angle between the input and the output is also important. At 1 Hz, the output signal lags the input by 32° in the above example. This lag reduces the contribution to the eddy flux at 1 Hz to 85 percent of the input signal. This phase-angle requirement means that sampling of variables must be simultaneous, or at well-defined time intervals so that corrections can be made, if necessary, to match the phase angles of variables before calculating fluxes.

For flux measurements, measurement noise that is not correlated with the vertical velocity does not contribute to the vertical flux. The noise may, however, necessitate a longer averaging time. This is important to keep in mind for measurements of trace gases that involve counting a limited number of photons. In this case, the noise is likely to have a flat "white noise" spectrum with a Poisson distribution, while the signal spectrum typically decreases with frequency with a $-5/3$ slope. Thus the noise may surpass the signal above some frequency that depends on the magnitude of both the signal and the noise.

By evaluating the left side of (8.1), it is possible to estimate the net internal production or loss of s ($\overline{P_s}$ and $\overline{L_s}$). This has been done to estimate the net photochemical production of O_3 in the boundary layer (Lenschow et al., 1981). In addition, the species flux at low levels in the boundary layer is a direct measurement of surface deposition (or emission) if chemical reactions are not significant on a time scale of a few hundred seconds or less. Even if the lowest measurement level is several tens

[1] "Instruments and Techniques for Probing the Atmospheric Boundary Layer." *Atmospheric Technology* No. 7, 1975, edited by D. H. Lenschow. Available from National Center for Atmospheric Research, P.O. Box 3000, Boulder, CO 80307-3000.

of meters above the surface, the surface flux may be estimated by extrapolating to the surface a flux profile obtained from measurements at several levels through the boundary layer, since the flux profile is linear for a conserved species in a horizontally homogeneous well-mixed boundary layer.

Because of the usefulness of resolving turbulent fluctuations of trace species in the boundary layer, it is important to develop this capability for a larger number of species. In this way, measurements of chemical and photochemical production and loss, surface sources and sinks, and transport through and across the top of the boundary layer can be obtained for direct comparison with model predictions, or as fundamental data in themselves.

In evaluating the mean concentration budget, the horizontal mean advection term is obtained by measuring the horizontal gradient of a species s. The required accuracy of this measurement can be estimated by assuming that the removal at the surface is equal to the horizontal advection term. Accuracy of surface removal can be specified in terms of a deposition velocity, which is defined as $v_d = -(\overline{w's'})_0/\bar{s}$. Equating the surface removal to the advective term in (8.2), one obtains

$$\frac{\delta s}{s} \simeq v_d \left(\frac{L}{\bar{u}h}\right), \quad (8.3)$$

where L is the horizontal distance across which the difference $\delta s \equiv s_2 - s_1$ is measured. As an example to illustrate the magnitude of the horizontal changes expected in an aircraft experiment, let $L = 10^5$ m, $\bar{u} = 5$ m/s, and $h = 10^3$ m. Thus

$$\frac{\delta s}{s} = 20\,[\mathrm{s\,m^{-1}}]v_d. \quad (8.4)$$

For many species, a reasonable accuracy goal for measuring v_d is 5×10^{-4} m/s. Thus $\delta s/s = 1$ percent. In many cases, L can be increased by as much as a factor of 10. Therefore, $\delta s/s = 1$ to 10 percent.

A potential alternative to eddy correlation flux measurements is implementation of the eddy accumulation technique (Desjardins, 1977). In this technique, mean concentrations of trace species in two gas samples are measured. The rate of flow of the sampler is controlled to be proportional to the magnitude of the vertical air velocity. One sample is obtained from upward moving air, and the other is obtained from downward moving air. The difference in concentrations between the two airstreams is proportional to the vertical flux. The main advantage of this technique is that fast-response concentration measurements are not required for flux measurements; instead, fast-response, accurate, and sensitive air flow control is necessary. However, very accurate mean concentration differences are required. These latter two requirements may preclude application of this technique for many species, particularly if their removal rate is small.

Surface-tower techniques provide an important complement to the aircraft methods described above. Whereas aircraft provide direct measurements of spatial averages of dry deposition fluxes, tower instrumentation provides a more detailed investigation of the factors that control these fluxes on a time-evolving basis. A comprehensive study would necessarily involve both techniques. Instrumentation developed to meet the requirements for aircraft eddy flux applications will also satisfy the requirements for tower operation. In some cases, however, the requirement for rapid response can be relaxed slightly, and sometimes it can be replaced by a demand for extremely accurate difference measurements. This is the case if the desire is for instruments suitable for measurement of concentration gradient instead of covariance.

One of the important applications of direct flux measurements is to provide detailed knowledge of deposition velocities for various species, and the variables that determine them. These deposition velocities can then be used in numerical models to parameterize surface fluxes. Field verification of these modeling studies requires concentration data from a network of surface observation sites. Simple, but reliable sampling methods need to be developed for this purpose. Methods analogous to high-volume filtration for airborne particles appear to offer special promise. Such methods are already in operation in some networks (e.g., in Canada and Scandinavia), and the methods need to be improved to permit routine and inexpensive operation on a global basis.

BIBLIOGRAPHY

Desjardins, R. L., 1977. Energy budget by an eddy correlation method. *J. Appl. Meteorol. 16*:248–250.

Dobson, F., L. Hasse, and R. Davis, 1980. *Air-Sea Interaction Instruments and Methods*. Plenum, New York, 801 pp.

Hicks, B. B., M. L. Wesely, and J. L. Durham, 1980. Critique of methods to measure dry deposition: Workshop summary. EPA-600/9-80-050. Environmental Protection Agency, Washington, D.C., 83 pp.

Kaimal, J. C., J. C. Wyngaard, Y. Izumi, and O. R. Cote, 1972. Spectral characteristics of surface-layer turbulence. *Quart. J. Roy. Meteorol. Soc. 98*:563–589.

Lenschow, D. H., R. Pearson, Jr., and B. B. Stankov, 1981. Estimating the ozone budget in the boundary layer by use of aircraft measurements of ozone eddy flux and mean concentration. *J. Geophys. Res. 86*:7291–7297.

Lenschow, D. H., R. Pearson, Jr., and B. B. Stankov, 1982. Measurements of ozone vertical flux to ocean and forest. *J. Geophys. Res. 87*:8833–8837.

Lilly, D. K., 1968. Models of cloud-topped mixed layers under a strong inversion. *Quart. J. Roy. Meteorol. Soc. 94*:292–309.

9 Instrument and Platform Survey

BY V. MOHNEN, F. ALLARIO, D. DAVIS, D. LENSCHOW, AND R. TANNER

Of central importance in our considerations of the requirements for technology development and platforms in any future tropospheric chemistry program were analytical techniques for the measurement of the various chemical species in the troposphere. We developed an up-to-date overview of the present analytical capabilities on the basis of a request for information that was sent to over 250 atmospheric scientists. Obviously, the survey relies heavily on the response that we obtained, and no claims for completeness are made here. A similar survey on aircraft platforms was conducted by the Working Group for Coordination of Research Aircraft and collected by NCAR's Research Aviation Facility, who kindly made the results of the survey available to us.

Information on oceanographic platforms is available from the Commander, Naval Oceanography Command, NSTL Station, Bay Street, St. Louis, and we have reviewed this information. Finally, we examined the current state-of-the-art in remote sensing technology applicable to tropospheric chemistry research. The results of the instrumentation survey and information on aircraft and oceanographic platforms are presented in this chapter, as is a very brief discussion of remote sensing technology.

INSTRUMENTATION FOR IN SITU MEASUREMENTS

To attain the long-term goals and objectives of the Global Tropospheric Chemistry Program, sensitive instrumentation will be required for both the measurement of chemical species in the remote troposphere and the elucidation of critical reaction mechanisms and rates in laboratory studies. Although currently available instrumentation is adequate to initiate some of the exploratory phases of major future field programs in the proposed Global Tropospheric Chemistry Program, this instrumentation is not adequate to carry out the detailed research program outlined in this report. Currently available sensors cover a broad range from low-technology, low-cost, in situ sensors with limited accuracy and sensitivity to high-technology, delicate, accurate, but costly bench-type instrumentation that still requires considerable development and intercomparison before field deployment. In addition, there are no instruments yet available for the measurement of certain critical species in the global troposphere such as HO_2.

While concentration measurements for most of the major species in tropospheric chemical cycles are possible today within the planetary boundary layer, many of these measurements cannot be made in the free troposphere, where concentrations are significantly lower. In addition, relatively few instruments are capable of making in situ measurements with a frequency greater than one measurement per second. The higher frequency of observation (better than 1 Hz) is necessary in order to obtain flux measurements with airborne platforms. Absolute calibrations, instrument intercomparisons, and other quality control procedures during all research efforts in the Global Tropospheric Chemistry Program are required. Attainment of the long-term goals of the

Global Tropospheric Chemistry Program depends critically on the further development and careful intercalibration of current instrumentation and the design and testing of a new generation of instrumental techniques. We recommend that a vigorous program of instrument development, testing, and intercalibration be undertaken immediately and be continued throughout the Global Tropospheric Chemistry Program.

The results of our community survey of current chemical instrumentation are summarized in Tables 9.1 through 9.24. These tables list the *detection technique*; the *sampling mode* (continuous, real time (1); continuous, integrative (2); intermittent (3); and remote (4)); the *development status* (concept developed (1); benchtop instrument (2); laboratory prototype available for field testing (3); field tested or commercial instrument (4)); the *detection limit*; the *time resolution* (the time period over which the signal is observed or averaged in order to attain the stated detection limit); the estimated *accuracy* (the difference in the measured value versus the "true" value); *precision* (the repeatability under conditions of constant concentration); *calibration* in the field (the lowest concentrations at which test gas mixtures have been prepared for purposes of technique calibration); the *weight/power requirements*; the *platform usage* (techniques were considered as applicable for sampling on the ground, aboard ships, or on aircraft; an "all" entry implies that the technique is suitable for all three applications; a technique need not have been operated or designed for all three applications in order to be designated as "all"); and *interferences* or *constraints* (species, processes, or environmental conditions that cause artifacts, positive or negative, competing or output signals, and so on; one can never be confident that a technique is completely interference free; one can only evaluate the technique for known or suspect interferences; environmental constraints can present operational problems for a technique; they differ from interferences because they either limit or invalidate the applicability of the technique, or change the baseline sensitivity, calibration, accuracy, or precision of the technique).

Most of the above classification parameters are difficult to define and, in themselves, warrant discussion. We reviewed and, in general, accepted the classification provided by the respondents of our survey. Therefore, the tables presented here do not necessarily reflect a consensus among the authors; rather the tables are a first attempt to list—without ranking—a variety of techniques applicable to global tropospheric measurements.

REMOTE SENSING TECHNOLOGY

Spaceborne remote sensors could provide near-global measurements and thus offer the ultimate goal of obtaining a three-dimensional distribution of certain atmospheric trace constituents. We hope this approach will eventually provide the tropospheric chemistry community with the opportunity to iterate a variety of distribution measurements with evolving mathematical models of the troposphere. In assessing the capability of current remote sensor technology for performing measurements in the global troposphere, we reviewed three classes of remote sensors: imaging spectroradiometers, passive remote sensors, and active remote sensors. We found that significant technological advances, both relative to the species that can be detected and spatial resolution, are necessary to satisfy the long-term needs of the Global Tropospheric Chemistry Program. Because of the significant potential of remote sensing technologies for any future global measurement program, we elaborated extensively on their present status and future outlook, and this review is presented in Appendix B.

AIRCRAFT PLATFORMS

There is a wide variety of aircraft platforms currently available in the United States from government, university, and private operators. A detailed compilation of the specifications and characteristics of these aircraft is presently being developed by the National Center for Atmospheric Research (NCAR) for use in planning research programs requiring such platforms. A brief summary of research aircraft platforms in the United States is presented in Table 9.25. Much of the information in Table 9.25 was obtained from the questionnaire sent to research aircraft operators by the Working Group for Coordination of Research Aircraft (WGCRA) and collected by the NCAR Research Aviation Facility. NCAR has organized and stored the responses in a data bank that can be easily accessed through the NCAR computer from remote locations by those interested in further details.

Available platforms range from simple, two-engine aircraft with limited range and space for scientific equipment (over 20) to long-range, four-engine turbo-jet and turbo-prop transports. Currently, three jet aircraft and five turbo-prop aircraft are being utilized in some aspect of tropospheric chemistry research. In some cases, the aircraft platform is available as an unmodified vehicle, and in others, as a complete aircraft measuring system often dedicated for extensive periods of time to meteorological and atmospheric chemistry studies.

One present limitation is the shortage of large, well-instrumented aircraft with sufficient interior space, power, exterior mounting locations, and specialized meteorological instrumentation for extended long-range flights investigating tropospheric chemistry in the boundary layer or lower free troposphere. At present, the NCAR Electra and the NASA Electra fill that requirement. The NOAA P-3s are a possibility, although available space is limited. In general, however, we believe that the current aircraft fleet number and type are adequate to undertake the Global Tropospheric Chemistry Program, although improvements and modifications to some aircraft and the meteorological support equipment aboard them will undoubtedly be necessary.

OCEANOGRAPHIC PLATFORMS

The Commander, Naval Oceanography Command, with the assistance of the University-National Oceanographic Laboratory System (UNOLS), publishes annually the Oceanographic Ships Operating Schedules. If an effective national oceanographic program is to exist, efforts must be made to maximize use of existing oceanographic platforms, including "piggy backing" and coordinated scheduling. To accomplish this goal, the Naval Oceanography Command has developed the Oceanographic Management Information System. A subset computer file, the Research Vehicle Reference Service (RVRS), has also been established as a centralized source of information pertaining to ship characteristics, latest operating schedules, last known positions, and points of contact for the vessel operators. Table 9.26 was compiled from information obtained through RVRS and reflects the status of oceanographic ships at academic institutions as of 1982. UNOLS coordinates scheduling of oceanographic research vessels in the United States academic fleet.

In discussing the current status of the fleet of ships, the subpanel noted that there is no oceanographic vessel designed specifically to carry out tropospheric chemistry research or dedicated to this area of research. Therefore, oceanographic sampling platforms will be a compromise between the needs of the atmospheric chemistry community and the mission for which the ship was designed. Nevertheless, sufficient ship platforms are available to undertake the proposed field research at sea in the Global Tropospheric Chemistry Program, as there are currently over 30 active ships from academic institutions and 35 operated by the federal government. As with other platforms, however, local contamination can be a serious problem if careful planning is not undertaken. Because of these potential problems, the subpanel includes the following history and summary of the past experience of scientists who made air chemical observations over the oceans. Although the earth is predominantly covered with water and the seas are the primary source of heat and water vapor to the atmosphere, a vast majority of all air chemical observations occur over land. This is not only a result of the inaccessibility of many parts of the world ocean, but also a result of the difficulty of making accurate observations in the presence of a ship or platform (Roll, 1965). Atmospheric physics was incorporated into the magnetic studies of the *Galileo* and *Carnegie* from 1910 through 1927 (Wait, 1946), and the foundations of modern marine meteorology were derived from the geochemical cruise of the *Meteor* in 1926-1927 (Roll, 1965).

Many of the observations of aerosol concentrations over the seas during the first half of the twentieth century were performed from ships of opportunity as the investigator traveled to scientific meetings (Parkinson, 1952; Hess, 1948; Landsberg, 1934). Gunn (1964) replicated the atmospheric conductivity apparatus of the *Carnegie* aboard a U.S. freighter, and Östlund and Mason (1974) studied tritium exchange over a wide range of latitudes from USNS *Towle*, enroute from the United States to Antarctica.

During the 1960s and 1970s, ships of opportunity were used by Elliott (1976) and Hogan et al. (1972) to characterize surface aerosol concentrations over the world ocean. Elliott (1976) supplied portable aerosol detectors to the meteorological observer aboard the *Glomar Challenger* and obtained systematic aerosol observations at many oceanic locations not visited by scheduled ships. Hogan et al. (1973) provided similar instruments to ships' officers contacted through the meteorology and oceanography programs of the State University of New York Maritime College. Aerosol observations were accomplished synoptically by these officers, as a part of the six hourly synoptic weather observations.

Meteorology, aerosols, and air chemistry have been combined with oceanography on several oceanographic cruises. Cobb and Wells (1970) studied atmospheric conductivity from NOAA ships, and Jaenicke (1974) examined not only air chemistry, but air-sea exchange of several gases from the *Meteor* as part of the first GEOSECS cruise. Probably the largest shipborne meteorological programs, to date, were accomplished during GATE and FGGE, and similar experiments as part of MONEX.

Aerosol particle and atmospheric chemistry measurements at sea are difficult for several reasons. There is a large gradient in aerosol particle and gas concentration near the surface. This gradient is influenced by the presence of the ship, and as the ship is constantly moving with respect to the sea, precise, simultaneous measurements of both concentration and height are nearly impossible. The ship itself is a strong source of every conceivable vapor and particle, and the combination of operational wind screens and forced ventilation found on every ship makes it very difficult to find uncontaminated sampling positions. These positions also vary with wind and the ship's speed in frustrating ways. For example, Hogan (1981) found that, with a beam wind of greater than 30 knots, he was able to sample uncontaminated air only from the highest observation level on USCGC *Northwind*, and then only on the upwind portion of the roll.

Moyers et al. (1972) discuss the contamination problems associated with aerosol sampling on board ships. They describe a bow tower sampling system combined with control of the sampling by the relative wind direction. This system has been utilized on a number of research vessels for atmospheric chemistry studies and has proven to be quite satisfactory for collecting uncontaminated samples. (See, for example, Duce and Hoffman, 1976, and Graham and Duce, 1979.) Problems with contamination indicate that in general, however, air chemistry measurements should be limited to real-time measurements as much as possible.

We concluded that there appears to be no immediate need for new or additional oceanographic platforms to satisfy the needs of atmospheric chemists. The shortage of funds to operate and/or modify the platforms for dedicated use appears to be the main constraint for their use in tropospheric experiments.

ACKNOWLEDGMENTS

We gratefully acknowledge the help from our colleagues who responded to our instrument survey and to the additional requests for help and clarification. Several individuals have helped us in the preparation of this document. These include James Hoell of NASA-Langley Research Center, for detailed information on N_xO_y instruments; Jorg Mohnen of ASRC, for coordinating the ship survey; and Byron Phillips of NCAR, for coordinating the aircraft survey.

BIBLIOGRAPHY

Cobb, W. E., and H. J. Wells (1970). The electrical conductivity of oceanic air and its correlation to global atmospheric pollution. *J. Atmos. Sci. 27*:814–819.

Duce, R. A., and G. L. Hoffman (1976). Atmospheric vanadium transport to the ocean. *Atmos. Environ. 10*:989–996.

Elliott, W. R. (1976). Condensation nuclei concentrations over the Mediterranean Sea. *Atmos. Environ. 10*:1091–1094.

Graham, W. F., and R. A. Duce (1979). Atmospheric pathways of the phosphorus cycle. *Geochim. Cosmochim. Acta 43*:1195–1208.

Gunn, R. (1964). The secular increase of the worldwide fine particle pollution. *J. Atmos. Sci. 21*:168–181.

Hess, V. F. (1948). On the concentration of condensation nuclei in the air over the North Atlantic. *Terrestrial Magnetism and Atmospheric Electricity 53*:399–403.

Hess, V. J. (1951). Further determinations of the concentrations of condensation nuclei in the air over the North Atlantic. *J. Geophys. Res. 56*:553–556.

Hogan, A. W. (1981). Aerosol measurements over and near the South Pacific Ocean and Ross Sea. *J. Appl. Meteorol. 20*:1111–1118.

Hogan, A. W., A. L. Aymer, J. M. Bishop, B. W. Harlow, J. C. Klepper, and G. Lupo (1967). Aitken nuclei observations over the North Atlantic Ocean. *J. Atmos. Sci. 6*:726–727.

Hogan, A. W., M. H. Degani, and C. Thor (1972). Study of maritime aerosols. Report to the U.S. Environmental Protection Agency, Division of Meteorology. Contract 70-64, 42 pp.

Hogan, A. W., V. A. Mohnen, and V. J. Schaefer (1973). Comments on "Oceanic aerosol levels deduced from measurements of the electrical conductivity of the atmosphere." *J. Atmos. Sci. 30*:1455–1460.

Jaenicke, R. (1974). Size distribution of condensation nuclei in the NE trade regime of the African coast. *J. Rech. Atmos. 8*:723–733.

Keafer, L. S., Jr., ed. (1982). *Tropospheric Passive Remote Sensing*. NASA Publ. No. CP 2237, June 1982. Order No. N82-26637, 95 pp.

Kuhlbradt, E., and J. Refer (1935). Die meteoroligischen Methoden und das aerologische Beobachtungsmaterial wiss. *Ergebnisse der Atlantische Expedition Meteor 14*:1925–1927.

Landsberg, H. (1934). Observations of condensation nuclei in the atmosphere. *Mon. Weather Rev.*, Dec:442–445.

Landsberg, H. (1938). Atmospheric condensation nuclei. *Ergeb. Kosm. Phys. 3*:155–252.

Levine, J. S., and F. Allario (1982). The global troposphere: biogeochemical cycles, chemistry and remote sensing. *Environ. Monitoring Assessment 1*:263–306.

Mason, B. J. (1957). The ocean as a source of cloud-forming nuclei. *Geofis. Pura Appl. 36*:148.

Moyers, J. L., R. A. Duce, and G. L. Hoffman (1972). A note on the contamination of atmospheric particulate samples collected from ships. *Atmos. Environ. 6*:551–556.

Moyers, J. L., and R. A. Duce (1972). Gaseous and particulate iodine in the marine atmosphere. *J. Geophys. Res. 77*:5229–5238.

Östlund, H. G., and A. S. Mason (1974). Atmospheric HT and HTO, 1. Experimental procedures and tropospheric data 1968–1972. *Tellus 26*:91–102.

Parkinson, W. C. (1952). Note on the concentration of condensation nucleii over the western Atlantic. *J. Geophys. Res. 57*:314–315.

Roll, H. U. (1965). *Physics of the Marine Atmosphere*. Academic, New York, 426 pp.

Seinfeld, J. H. (1981). *Report of the NASA Working Group on Tropospheric Program Planning*. NASA RP 1062.

Wait, G. R. (1946). Some experiments relating to the electrical conductivity of the lower atmosphere. *J. Wash. Acad. 36*:321–343.

TABLE 9.1 Gaseous Sulfur

Technique	Development Status[a]	Sampling Mode[b]	Detection Limit	Time Resolution	Accuracy	Precision	Calibration	Weight/Power Requirements	Platform Usage	Interferences/ Constraints
Sulfur Dioxide										
Pulsed fluorescence	4	1	3 ppbv	1–2 min	15%	3%	5 ppbv	30 kg, 300 W	All	None identified
Gas chromatography/flame photometric detection	3	3	20 pptv	3–5 min	15%	10%	100 pptv	50 kg, 300 W	All	Adsorption, baseline shifts from H_2O
Continuous flame photometry with particle filter	4	1	0.3 ppbv	5 s	10%	5%	0.4 ppbv	30 kg, 300 W	All	Selective scrubbing of RSH, H_2S required, uses SF_6-doped H_2 fuel
Condensation collection/ion chromatography	3	2	10–50 pptv	30 min	TBD	TBD	NA	30 kg, 1 kW	Ground	H_2S, readily oxidizable sulfur
HiVol filter pack-CO_3^{2-} impregnated cellulose	4	2	100 pptv	30 min	10%	5%	1 ppbv	15 kg, 500 W	All	Lowered efficiency at very low RH
Automated West-Gaeke colorimetry	4	3	1 ppbv	60 min	10%	5%	1 ppbv	5 kg, 100 W	Ground, ship	Possible interferences: Cl_2, HCl, H_2S, thiols, aldehydes
KI wet chemical w/scrubber	3	1	3 ppbv	20 s	20%	5%	5 ppbv	5 kg, 100 W	Ground, ship	NO_2, H_2O reductants
Liquid chemiluminescence w/$KMnO_4$	2	3	10 pptv	15 min	20%	10%	NA	4 kg, 150 W	Ground, ship	None identified
Differential optical absorption spectroscopy	3	1	100 pptv	5 min (2 km)	20%	5%	NA	200 kg, 2 kW	Ground	None identified
Tunable diode laser	3	1	0.3 ppbv	0.1 s	10%	5%	5 ppbv	200 kg, 1.5 kW	All	None identified
Single-photon laser-induced fluorescence	4	1	3 pptv 30 pptv	10 min 3 min	15%	10%	40 pptv	700 kg, 3 kW	All	None identified
Hydrogen Sulfide										
Impregnated tape/ photometry	4	2 or 3	0.2 ppbv	60 min	20%	10%	5 ppbv	5 kg, 200 W	All	
Gas chromatography/flame photometric detection	4	3	20 pptv	3–5 min	15%	10%	0.2 ppbv	50 kg, 300 W	All	Severe adsorption problems
Methylene blue colorimetry	3	3	1 ppbv	120 min	10%	5%	5 ppbv	10 kg, 200 W	Ground, ship	Protect from light, high NO_2 interferes
Fluorescence quenching	3	3	<0.2 ppbv	60 min	TBD	20%	1 ppbv	10 kg, 200 W	All	NO_2 interferes
Ozone chemiluminescence	2	1	1 ppb	2 s	TBD	10%	5 ppbv	80 kg, 1.2 kW	All	Organosulfur compounds
Organosulfur Compounds										
Dimethyl sulfide (DMS) by chemisorption/gas chromatography/FPD	4	3	0.5 pptv	30 min	6%	2%	100 pptv	150 kg, 1.5 kW	All	High levels of SO_2
COS by cryogenic trapping/ GC/FPD	4	3	100 pptv	20 min	10%	5%	0.5 ppbv	150 kg, 1.5 kW	All	None identified
Thiols, DMS by O_3 chemiluminescence	2	1	100 pptv	5 s	TBD	10%	2 ppbv	80 kg, 1 kW	All	H_2S, CS_2 interference can be eliminated; high olefins interfere

[a] 1 = concept developed, 2 = bench-top instrument, 3 = laboratory prototype available for field testing, 4 = field-tested or commercial instrument.
[b] 1 = continuous, real time; 2 = continuous, integrative; 3 = intermittent; 4 = remote.
NOTE: TBD = to be determined; NA = not available.

TABLE 9.2 NO (Gas)

Technique	Development Status[a]	Sampling Mode[b]	Detection Limit	Time Resolution	Accuracy	Precision	Calibration	Weight/Power Requirements	Platform Usage	Interferences/ Constraints
Ozone chemiluminescence	4	1	5 ppt	1 s	10%	5%	7 ppt	80 kg, 1.2 kW	All	None known/detection limit dependent on water vapor concentration
Single-photon laser-induced fluorescence	3	1	30 ppt	1 min	15%	10%	20 ppt	700 kg, 3 kW	All	None known
Two-photon laser-induced fluorescence	2	1	5 ppt 350 ppt	1 min 30 min	15%	2%	400 ppt	700 kg, 3kW	All	None known
Lidar laser-induced fluorescence	3	4	5 ppt	1 s (1 km)	15%	2%	NA	NA	All	Solar flux, white fluorescent noise/clouds, high aerosol concentrations, temperature changes
Multiphoton ionization	1	1	1 ppb	1 s	TBD	5%	NA	700 kg, 3 kW	All	NO_2, RNO_2/temperature changes
Tunable diode laser	4	1	100 pptv	1 s	10%	5%	6 ppb	200 kg, 1.5 kW	All	None known
Resonant ion laser	2	1	10 ppb	1 s	TBD	TBD	NA	NA	All	NO_2, RNO_2/temperature changes, limited UV laser energy
Long-path UV absorption	4	4	100 ppt	5 min (1 km)	20%	5%	NA	200 kg, 2 kW	Ground	None known/high aerosol concentration

[a] 1 = concept developed, 2 = bench-top instrument, 3 = laboratory prototype available for field testing, 4 = field-tested or commercial instrument.
[b] 1 = continuous, real time; 2 = continuous, integrative; 3 = intermittent; 4 = remote.
NOTE: TBD = to be determined; NA = not available.

TABLE 9.3 NO_2 (Gas)

Technique	Development Status[a]	Sampling Mode[b]	Detection Limit	Time Resolution	Accuracy	Precision	Calibration	Weight/Power Requirements	Platform Usage	Interferences/ Constraints
Ozone chemiluminescence (NO_2 to NO converter)	4	1	10 ppt	1 s	30%	5%	25 ppt	80 kg, 1.2 kW	All	PAN, other organic nitrates/detection limit dependent on water vapor concentration
Chemiluminescence (luminol)	3	1	30 ppt	1 s	TBD	TBD	NA	10 kg, 100 W	Ground, ship	PAN, other organic nitrates/no NO interference
Gas chromatography	2	3	3 ppb	10 min	15%	5%	NA	50 kg, 2 kW	All	CO_2, other electron capture sensitive trace gases
Photofragmentation laser-induced fluorescence	2	1	100 ppt 5 ppt 0.5 ppt	1 s 1 min 30 min	15%	5%	4 ppt	700 kg, 3 kW	All	TBD/temperature changes, laser energy and respective rate
Multiphoton ionization	1	1	1 ppt	1 s	TBD	TBD	NA	NA	All	RNO_2/temperature changes, UV laser energy and respective rate
Tunable diode laser	4	1	100 ppt	1 s	10%	5%	NA	200 kg, 1.5 kW	All	None known
Resonant ion laser	1	1	1 ppb	NA	NA	NA	NA	NA	All	RNO_2/temperature changes, UV laser energy and respective rate
Long-path UV absorption	4	4	100 ppt	1 min	20%	5%	NA	200 kg, 2 kW	Ground	None known/high aerosol concentration, high O_3; path-length limited by visibility
Photothermal										
Photoacoustic	2	1	5 ppb	100 s	TBD	5%	1 ppb	NA	All	H_2O, aerosols/temperature changes
Resonant photoacoustic	2	1	2 ppb	3 min	TBD	5%	1 ppb	NA	Ground	NO_3, O_3, aerosols
Zeeman modulated	1	1	TBD	TBD	TBD	TBD	NA	NA	All	None known

[a] 1 = concept developed, 2 = bench-top instrument, 3 = laboratory prototype available for field testing, 4 = field-tested or commercial instrument.
[b] 1 = continuous, real time; 2 = continuous, integrative; 3 = intermittent; 4 = remote.
NOTE: TBD = to be determined; NA = not applicable.

TABLE 9.4 NO_3 (Gas)

Technique	Development Status[a]	Sampling Mode[b]	Detection Limit	Time Resolution	Accuracy	Precision	Calibration	Weight/Power Requirements	Platform Usage	Interferences/ Constraints
Photofragmentation laser-induced fluorescence	1	1	20 ppt 1 ppt	1 s 1 min	TBD	10%	NA	700 kg, 3 kW	All	TBD/temperature changes, laser rep rate
Long-path UV absorption	4	4	1 ppt	5 min (10 km)	30%	5%	NA	200 kg, 2 kW	Ground	None known/water vapor, high aerosol concentration; path length limited by visibility
Tunable diode laser absorption	2	1	100 ppt	NA	10%	5%	NA	200 kg, 1.5 kW	All	None known/TBD

[a]1 = concept developed, 2 = bench-top instrument, 3 = laboratory prototype available for field testing, 4 = field-tested or commercial instrument.
[b]1 = continuous, real time; 2 = continuous, integrative; 3 = intermittent; 4 = remote.
NOTE: TBD = to be determined; NA = not applicable.

TABLE 9.5 HNO_3 (Gas)

Technique	Development Status[a]	Sampling Mode[b]	Detection Limit	Time Resolution	Accuracy	Precision	Calibration	Weight/Power Requirements	Platform Usage	Interferences/ Constraints
Chemiluminescence with converter (two channel with nylon filter)	4	1	100 ppt	10 s	15%	5%	0.5 ppb	200 kg, 1500 W	All	None known/serious absorption problems
Tunable diode laser absorption	4	1	300 ppt	60 s	10%	5%	6 ppb	200 kg, 1500 W	All	None known
Direct denuder tube (tungstic acid) with C.L. detector	3	3	70 ppt	40 min	20%	10%	2 ppb	20 kg, 1000 W	All	TBD/must separate NH_3
Direct denuder (NaF) with C.L. detector	3	3	20 ppt	8 hr	20%	10%	NA	20 kg, 1000 W	All	None known
Denuder difference, IC analysis	4	3	100 ppt	30 min	20%	5%	0.2 ppb	60 kg, 1000 W	All	Coarse particle nitrate
Filter pack	4	3	10 ppt	60 min	20%	2%	15 ppt	20 kg, 500 W	All	NH_4NO_3, high aerosol loading
Condensation collection	4	3	10–90 ppt	30 min	TBD	5%	NA	30 kg, 1000 W	All	Hydrolysis of organic nitrates/temperature changes
Long-path FT-IR spectroscopy	4	1	5 ppb	5 min (1 km)	20%	5%	NA	500 kg, 3 kW	Ground	None identified

[a]1 = concept developed, 2 = bench-top instrument, 3 = laboratory prototype available for field testing, 4 = field-tested or commercial instrument.
[b]1 = continuous, real time; 2 = continuous, integrative; 3 = intermittent; 4 = remote.
NOTE: TBD = to be determined; NA = not applicable.

TABLE 9.6 HNO_2 (Gas)

Technique	Development Status[a]	Sampling Mode[b]	Detection Limit	Time Resolution	Accuracy	Precision	Calibration	Weight/Power Requirements	Platform Usage	Interferences/ Constraints
Photofragmentation laser-induced fluorescence	1	1	15 ppt	1 min	TBD	10%	NA	700 kg, 3 kW	All	TBD/temperature changes, laser rep rate
Long-path UV absorption	4	4	20 ppt	10 min (10 km)	20%	5%	NA	200 kg, 2 kW	Ground	None known/path length limited by visibility
Condensation	4	3	15 ppt	1 hr	TBD	5%	NA	30 kg, 1 kW	All	PAN/temperature changes, cannot distinguish hydrolysis products

[a]1 = concept developed, 2 = bench-top instrument, 3 = laboratory prototype available for field testing, 4 = field-tested or commercial instrument.
[b]1 = continuous, real time; 2 = continuous, integrative; 3 = intermittent; 4 = remote.
NOTE: TBD = to be determined; NA = not applicable.

TABLE 9.7 PAN (Gas)

Technique	Development Status[a]	Sampling Mode[b]	Detection Limit	Time Resolution	Accuracy	Precision	Calibration	Weight/Power Requirements	Platform Usage	Interferences/ Constraints
Chemiluminescence with converters	2	1	TBD	TBD	TBD	TBD	NA	80 kg, 1.2 kW	All	Possibly organic nitrates, H_2O_2/selectivity of PAN converter to be established
Long-path FT-IR spectroscopy	4	1	1 ppb	10 s (100 m)	25%	10%	NA	500 kg, 3 kW	Ground	TBD
Gas chromatography-electron capture detection	4	3	5 ppt	1 min	15%	15%	40 ppb	50 kg, 300 W	All	None known/residual water vapor on column

[a]1 = concept developed, 2 = bench-top instrument, 3 = laboratory prototype available for field testing, 4 = field-tested or commercial instrument.
[b]1 = continuous, real time; 2 = continuous, integrative; 3 = intermittent; 4 = remote.
NOTE: TBD = to be determined; NA = not applicable.

TABLE 9.8 NH_3 (Gas)

Technique	Development Status[a]	Sampling Mode[b]	Detection Limit	Time Resolution	Accuracy	Precision	Calibration	Weight/Power Requirements	Platform Usage	Interferences/ Constraints
Oxalic acid denuder	4	3	100 ppt	4 hr	15%	15%	NA	10 kg, 150 W	All	Evaporation of oxalic acid, NH_4HO_x
Citric acid denuder	2	3	100 ppt	4 hr	15%	5%	5 ppb	20 kg, 250 W	All	None known
Tungstic acid denuder, converter and C.L. detector	3	2	100 ppt	40 min	TBD	5%	5 ppb	20 kg, 100 W	All	TBD (Must separate HNO_3)
Acid-impregnated filter pack	4	3	0.2 ppb	3 hr	20%	2%	NA	10 kg, 200 W	All	Loss or gain of NH_3 on prefilter/blank variability
Fluorescence derivatization	4	1	0.2 ppb	3 min	20%	10%	4 ppb	20 kg, 150 W	All	Gaseous primary amines/ temperature dependence
Condensation collection	2	3	0.15 ppb	30 min	TBD	5%	NA	30 kg, 1 kW	All	Volatility of ammonium salts
Tunable diode laser absorption	3	1	100 ppt	1 min	10%	5%	5 ppb	20 kg, 1.5 kW	All	None known
Long-path FTIR spectroscopy	4	1	3 ppb	5 min (1 km)	20%	5%	10 ppb	500 kg, 3 kW	Ground	None known
Photofragmentation laser-induced fluorescence	2	1	0.01 ppt 1 ppt	30 min 0.5 min	TBD	TBD	NA	700 kg, 3 kW	All	None known/limited by laser rep rate
IR heterodyne radiometry	4	4	0.5 ppb	1 min	15%	10%	NA	NA	Ground	Limited in detecting small volume variability

[a] 1 = concept developed, 2 = bench-top instrument, 3 = laboratory prototype available for field testing, 4 = field-tested or commercial instrument.

[b] 1 = continuous, real time; 2 = continuous, integrative; 3 = intermittent; 4 = remote.

NOTE: TBD = to be determined; NA = not available.

TABLE 9.9 H_2

Technique	Detection Limit	Time Resolution	Precision	Weight/Power Requirements	Interferences/Constraints
Doped carrier ECD gas chromatography	10 ppbv	10 min	1%	30 kg, 0.5 kW	None identified
Mercuric oxide reduction and atomic absorption	10 ppbv	5 min	1%	20 kg, 0.5 kW	CO and other reducing species unless separated by GC or removed absorbents

TABLE 9.10 OH

Technique	Detection Limit	Time Resolution	Precision	Weight/Power Requirements	Interferences/Constraints
Radio carbon tracer (Campbell)	$10^5/cm^3$	100 s	20%	180 kg, 3 kW	Methoxyl radical
Short-pulse two-wavelength single-photon laser-induced fluorescence, in situ (Davis)[a]	$10^6/cm^3$	20 min		1000 kg, 9 kW	Laser-generated OH; aerosol fluorescence
Two-photon laser-induced fluorescence, in situ (Davis)[a]	$2.5 \times 10^5/cm^3$				None
Two-wavelength laser-induced fluorescence, lidar (Wang)[a]	$10^6/cm^3$	20 min		800 kg, 11 kW	Laser-generated OH; aerosol fluorescence
Low-pressure laser-induced fluorescence (Hard, Rateike)	$10^6/cm^3$	8 min			Wall loss of OH
308-nm laser-induced fluorescence (McDermid)	TBD				
High-rep rate laser-induced fluorescence (Anderson)[a]	$10^6/cm^3$	20 min			Aerosol fluorescence

[a]Detection limit cited is for moist conditions at 1 atmosphere pressure. At 500-mbar pressure, detection limits could be 3 to 5 times lower, depending on the LIS technique employed.

NOTE: TBD = to be determined.

TABLE 9.11 O_3 (Gas)

Technique	Development Status[a]	Sampling Mode[b]	Detection Limit	Time Resolution	Accuracy	Precision	Calibration	Weight/Power Requirements	Platform Usage	Interferences/ Constraints
UV absorption (commercial)	4	1	1 ppb	30 s	10%	2%	10 ppb	10 kg, 150 W	All	None identified
UV absorption (special)	3	1	0.5 ppb	1 s	3%	3%	NA	150 kg, 100 W	All	None identified
Ethylene chemiluminescence	4	1	1–2 ppb	1 s	10%	5%	10 ppb	10 kg, 50 W	All	None identified, accuracy dependent on ozone source
KI wet chemical	4	3	1 ppb	20 s	5%	2%	NA	5 kg, 100 W	Ground	NO_2, H_2O_2, and other reductants

[a]1 = concept developed, 2 = bench-top instrument, 3 = laboratory prototype available for field testing, 4 = field-tested or commercial instrument.
[b]1 = continuous, real time; 2 = continuous, integrative; 3 = intermittent; 4 = remote.
NOTE: NA = not applicable.

TABLE 9.12 H_2O_2

Technique	Development Status[a]	Sampling Mode[b]	Detection Limit	Time Resolution	Accuracy	Precision	Calibration	Weight/Power Requirements	Platform Usage	Interferences/ Constraints
Gas										
Bubbler/luminol chemiluminescence	4	2	0.5 ppbv	15 min	Environmental dependent	5%	NA	50 kg, 300 W	All	Metal ions (?), oxidants, in situ production of peroxide
Condensation collection/fluorescence	3	3	TBD (very low)	30 min	TBD	5%	NA	30 kg, 1 kW	Ground, ship	In situ production
Tunable diode laser absorption	3	1	300 ppt	1 min	10%	5%	NA	200 kg, 1.5 kW	All	None identified
Photofragmentation differential laser-induced fluorescence	2	1	30 pptv	1 min	15%	5%	TBD	700 kg, 3 kW	All	None identified
Liquid										
Luminol chemiluminescence	4	2 or 3	1–2 ppbm	Sampling limited	Environmental dependent	5%	4 ppbm	40 kg, 200 W	All	Oxidants/limited by blank variability
Catalase/peroxidase fluorimetry	4	3	1–2 ppbm	Sampling limited	10–15%	5%	5 ppbm	50 kg, 300 W	All	None identified/limited by blank variability
Chemiluminescence (peroxyoxalate)	2	3	TBD	NA	TBD	TBD	NA	NA	All	NA
Scopoletin fluorescence quenching	2	3	~0–1 ppbm	Sampling limited	TBD	2%	NA	40 kg, 200 W	All	Apparently no interference from oxidants, organic peroxides

[a]1 = concept developed, 2 = bench-top instrument, 3 = laboratory prototype available for field testing, 4 = field-tested or commercial instrument.
[b]1 = continuous, real time; 2 = continuous, integrative; 3 = intermittent; 4 = remote.
NOTE: TBD = to be determined; NA = not applicable.

TABLE 9.13 CO_2

Technique	Detection Limit	Time Resolution	Precision	Weight/Power Requirements	Interferences/Constraints
Nondispersive infrared	3 ppmv	0.2-20 s (limited by the physical transport of gas and calibration)	0.04%	30 kg, 0.5 kW	H_2O, O_2 pressure broadening
Dual catalyst flame ionization gas chromatography		8 min	0.04%	30 kg, 0.5 kW	None identified
Precision manometer		4 hr	0.02%		N_2O

TABLE 9.14 CO

Technique	Detection Limit	Time Resolution	Precision	Weight/Power Requirements	Interferences/Constraints
Catalyst flame ionization gas chromatography	1 ppb	10 min	0.4%	30 kg, 0.5 kW	None identified
Mercuric oxide reduction and atomic absorption detection	10 ppb	5 min	1.0%	20 kg, 0.5 kW	H_2 and other reducing species unless separated by GC or removed by absorbents
Tunable diode laser IR absorption	1 ppb	10 s	2.0%	250 kg, 2.5 kW	None identified

TABLE 9.15 CH_4, C_2H_6, C_2H_4, C_3H_8 (Gases)

Technique	Detection Limit	Time Resolution	Precision	Weight/Power Requirements	Interferences/Constraints
Flame ionization gas chromatography	1 ppb	10 min	0.2%	30 kg, 0.5 kW	None identified

TABLE 9.16 Aldehydes

Technique	Development Status[a]	Sampling Mode[b]	Detection Limit	Time Resolution	Accuracy	Precision	Calibration	Weight/Power Requirements	Platform Usage	Interferences/ Constraints
HCHO (gas)										
Colorimetric/DNPH derivatization and LC	4	3	1 ppbv	1 hr	15%	5%	5 ppbv	50 kg, 250 W (includes LC)	All	Interferences separated by LC
Differential optical absorption spectroscopy	2	3	0.6 ppbv	5 min (2 km)	20% TBD	5%	NA	20 kg, 2 kW	Ground	NO_2 at high concentrations
Tunable diode laser absorption	3	1	100 ppt	1 s	10%	5%	1 ppbv	200 kg, 1.5 kW	All	None known
HCHO (liquid)										
Chemiluminescence (gallic acid, H_2O_2, pH 12.5)	4	3	100 pphm	5 min	NA	NA	NA	20 kg, 1 kW	All	None identified at low concentrations
Colorimetric (pararosanoline-bisulfite)	4	3	1 ppbm	1 hr	NA	NA	NA	20 kg, 200 W	All	None identified at low concentrations
Colorimetric (DNPH derivatization and LC)	4	3	8 ppbm	1 hr	15%	5%	75 ppbm	50 kg, 500 W (includes LC)	All (?)	None identified
Colorimetric (MBTH)	4	3	~10 ppbm	1 hr	15%	10%	NA	20 kg, 200 W	All	Total aldehyde measurement
Other Aldehydes										
Colorimetric (DNPH derivatization and LC)	4	3	15 ppbm (for HCHO)	1 hr	15%	5%	75 ppbm	50 kg, 500 W (includes LC)	All (?)	Acetone interferences with C_2H_5CHO detection

[a] 1 = concept developed, 2 = bench-top instrument, 3 = laboratory prototype available for field testing, 4 = field-tested or commercial instrument.
[b] 1 = continuous, real time; 2 = continuous, integrative; 3 = intermittent; 4 = remote.
NOTE: TBD = to be determined; NA = not applicable.

TABLE 9.17 CCl_3F (F11)

Technique	Detection Limit	Time Resolution	Precision	Weight/Power Requirements	Interferences/Constraints
ECD gas chromatography	3 pptv	15 min	0.5%	30 kg, 0.5 kW	None identified
Gas chromatography NOAA porous cell A column	5 ppt (sensitivity 2 ppt)	10 min		50 kg, 2 kW	Other electron capture trace gas species

TABLE 9.18 CCl_2F_2 (F12)

Technique	Detection Limit	Time Resolution	Precision	Weight/Power Requirements	Interferences/Constraints
ECD gas chromatography	5 pptv	15 min	0.5%	30 kg, 0.5 kW	None identified
Gas chromatography NOAA porous cell A column	5 ppt (sensitivity 2 ppt)	10 min		50 kg, 2 kW	Other electron capture trace gas species

TABLE 9.19 Anions (Liquid)

Technique	Detection Limit	Time Resolution	Precision	Weight/Power Requirements	Interferences/Constraints
$SO_3^=$					
Chemiluminescence	5 ppbm			20 kg, 1 kW	None identified
Ion chromatography	30 ppbm			50 kg, 500 W	None identified
Automated colorimetric technique (auto analyzer)	1 ppm				
$SO_4^=$					
Ion chromatography	50 ppbm			50 kg, 500 W	None identified
Isotope dilution (IDA)	10^{-8} g			Liquid scintillation counter needed	$SO_3^=$, Sr^{++}, Ba^{++}
Automated colorimetric technique (auto analyzer)	1 ppm				
NO_3^-					
Ion chromatography	100 ppbm			50 kg, 500 W	SO_3^-, Br^-, can be avoided
Direct UV absorption	100 ppbm			20 kg, 100 W	Aromatics, Fe^{+++}, can be avoided
Automated colorimetric technique (auto analyzer)	100 ppbm				
Cl^-					
Ion chromatography	100 ppbm			50 kg, 500 W	None identified
Automated colorimetric technique (auto analyzer)	1 ppm				

TABLE 9.20 Cations (Liquid)

Technique	Detection Limit	Time Resolution	Precision	Weight/Power Requirements	Interferences/Constraints
Ion chromatography for NH_4^+, Na^+, K^+	50 ppbm	minutes		50 kg, 500 W for analysis	None identified
Atomic absorption for Ca^{++}, Mg^{++}	20 ppbm	minutes		50 kg, 500 W for analysis	None identified

TABLE 9.21 Trace Metal Vapors

Technique	Detection Limit	Time Resolution	Precision	Weight/Power Requirements	Interferences/Constraints
Activated charcoal column	~1 ng/m^3	1 hr		5 kg, 300 W (sampler)	Collection efficiency influenced by temperature and humidity, specific interferences are a function of analytical technique
Noble metal adsorber	<1 ng/m^3	1 hr		5 kg, 300 W (sampler)	Collection efficiency influenced by temperature and humidity, specific interferences are a function of analytical technique
Airborne mercury spectrometer	2-5 ng/m^3				SO_2

TABLE 9.22 Aerosol Composition

Technique	Development Status[a]	Sampling Mode[b]	Detection Limit	Time Resolution	Accuracy	Precision	Calibration	Weight/Power Requirements	Platform Usage	Interferences/ Constraints
Aerosol Sulfur										
Filter pack/IC or other wet chemistry	4	3	0.3 $\mu g/m^3$	30 min	10%	6%	NA	20 kg, 300 W	All	Requires artifact-free filter medium
Flame photometric detector with SO_2 denuder	4	1	1.5 $\mu g/m^3$	10 s	20%	5%	10 $\mu g/m^3$	30 kg, 300 W	All	Loss of H_2SO_4/mass flow control required
FPD with heated denuder for H_2SO_4	3–4	1	3 $\mu g/m^3$	1 min	20%	10%	10 $\mu g/m^3$	30 kg, 300 W	All	Mass flow required for airborne sampling
Heated denuder for H_2SO_4	3	3	0.4 $\mu g/m^3$	8 hr	10%	5%	NA	10 kg, 500 W	All	None known
Low-pressure impactor/flash volatilization-FPD	4	3	1 μg S/m^3 (6 ng S/ stage)	1 hr	10-20%	5%	15 ng S	20 kg, 500 W	All	Coarse refractory S excluded, volatiles lost in low-P stages
Aerosol Nitrate										
Filter pack/IC or other wet chemistry	4	3	0.2 $\mu g/m^3$	30 min	10%	5%	NA	20 kg, 300 W	All	Nitrate artifacts
Filter pack with pre-denuder for HNO_3	4	3	0.1 $\mu g/m^3$	30 min	15%	5%	20 $\mu g/m^3$	60 kg, 1 kW	Ground, ship	Coarse particle nitrate losses
Heated denuder with CL detector	3	2–3	0.1 $\mu g/m^3$	8 hr	15%	5%	NA	20 kg, 1 kW	Ground, ship	None known
Tungstic acid cartridge with CL detector	3	3	0.1 $\mu g/m^3$	40 min	TBD	5%	5 $\mu g/m^3$	20 kg, 100 W	All	TBD
Aerosol Strong Acid										
Filter extraction/pH	4	3	±0.1 pH	1 hr	15%	10%	NA	20 kg, 300 W	All	Coarse basic particles must be absent
Filter extraction/gran titration	4	3	5 neq/m^3	1 hr	5%	2%	0.2 μeq.	20 kg, 300 W	All	Coarse basic particles must be absent
Filter/^{14}C amine	3	3	<5 neq/m^3	3 hr	10%	5%	TBD	10 kg, 500 W	All	Coarse basic particles must be absent
Aerosol Ammonium										
Tungstic acid cartridge with CL detector	3	3	0.2 $\mu g/m^3$	40 min	TBD	5%	10 $\mu g/m^3$	20 kg, 100 W	All	TBD
Filter extraction/ion selective electrode	4	3	1 $\mu g/m^3$	30 min	10%	5%	10 $\mu g/m^3$	20 kg, 300 W	All	Loss or gain of NH_3 during sampling (NH_4NO_3)
Filter extraction/indophenol colorimetry	4	3	0.2 $\mu g/m^3$	30 min	10%	5%	10 $\mu g/m^3$	20 kg, 300 W	All	Loss or gain of NH_3 during sampling (NH_4NO_3)

Aerosol Carbon										
Thermoevolution/organic, elemental	4	3	0.5 $\mu g/m^3$	6–24 hr	15%	10%	10 $\mu g/m^3$	80 kg, 500 W	All	Contamination problems with most filter media
Solvent extraction (organic)	4	3	1 $\mu g/m^3$	6–24 hr	20%	10%	NA	20 kg, 300 W	All	Lack of specificity for organics
Soot carbon reflectance photometry	4	3	20 ng/m^3	0.5–24 hr (variable)	15%	5%	NA	5 kg, 200 W	All	Possible interference from inorganic aerosols
Proton inelastic scattering	3	3	~ 20 ng/m^3	0.5–24 hr (variable)	10%	5%	NA	5 kg, 300 W (for sampling)	All	Requires uncommon facilities
Elemental/Metals Analysis										
X-Ray fluorescence (XRF)	4	3	0.1-10 ng/m^3	0.5–24 hr (variable)	10%	2%	NA	5 kg, 300 W (for sampling)	All	Various line overlap interferences can be eliminated
Proton-induced X-ray emission (PIXE)	4	3	~ 0.5 ng/m^3	>0.5 hr	10%	5%	NA	5 kg, 300 W (for sampling)	All	Limited by elements in filter blank
Proton inelastic scattering (light elements)	3	3	10–50 ng/m^3	0.5–24 hr (variable)	10%	5%	NA	5 kg, 300 W (for sampling)	All	None known
Atomic absorption spectroscopy	4	3	~ 1 ng/m^3	>0.5 hr	10%	2%	NA	5 kg, 300 W (for sampling)	All	Matrix and interelement effects must be overcome
Inductively coupled plasma emission	3	3	~ 1 ng/m^3	>0.5 hr	15%	10%	NA	5 kg, 300 W (for sampling)	All	Matrix and interelement effects must be overcome
Neutron activation analysis	4	3	~ 0.5 ng/m^3	0.5–24 hr (variable)	10%	10%	NA	5 kg, 300 W (for sampling)	All	Limited by elements in filter blank

[a] 1 = concept developed, 2 = bench-top instrument, 3 = laboratory prototype available for field testing, 4 = field-tested or commercial instrument.
[b] 1 = continuous, real time; 2 = continuous, integrative; 3 = intermittent; 4 = remote.
NOTE: TBD = to be determined; NA = not applicable.

TABLE 9.23 Aerosols—Physical Measurements

Technique	Detection Limit	Time Resolution	Precision	Weight/Power Requirements	Interferences/Constraints
Condensation nuclei technique (pulsating)	$D > 50$ Angstroms Conc. $> 10\ cm^{-3}$	2 s		60 kg, 220 W	
Condensation nuclei technique (continuous)	$D > 500$ Angstroms Conc. $> 10^{-2}\ cm^{-3}$	5 s		15 kg, 200 W	
Electrical mobility	$D > 32$ Angstroms Conc.-size dependent	2 min		25 kg, 100 W	
Optical scattering (white light)	$D > 0.3\ \mu m$ size particle			25 kg, 250 W	Index of refraction particle shape
Optical scattering laser cavity	$0.1\ \mu m < D < 6\ \mu m$ single particle			20 kg, 150 W	Index of refraction particle shape
Laser scattering	$0.3 < D < 32\ \mu m$			20 kg, 250 W	Index of refraction particle shape
Cloud and Precipitation Particles					
Nephelometer (integrated optical)	$D > 0.1\ \mu m$ Conc. $> 100\ cm^{-3}$	5 s		25 kg, 150 W	
Transmissometers (long path)	$D > 0.1$ (path length dependent)	5 s		100 kg, 500 W	
Laser scattering	$0.5\ \mu m < D < 47\ \mu m$			20 kg, 150 W	Index of refraction particle shape
Laser image (one-dimensional shadow)	$10\ \mu m < D < 300 \mu m$			20 kg, 100 W	Index of refraction particle shape
Laser image (one-dimensional shadow)	$150\ \mu m < D < 12$ mm			20 kg, 100 W	Index of refraction particle shape

TABLE 9.24 Cloud Condensation Nuclei

Technique	Detection Limit	Time Resolution	Precision	Weight/Power Requirements	Interferences/Constraints
Static diffusion chamber integrated light scatter detector	$20\ cm^{-3}$ $0.25 < S(\%) < 2$	45 s		20 kg, 200 W	Large insoluble particles
Continuous flow diffusion chamber single-particle optical detector	$0.02\ cm^{-3}$ $0.1 < S(\%) < 2$	5 min		100 kg, 1500 W	Large insoluble particles
Isothermal haze chamber single-particle optical detector	$0.02\ cm^{-3}$ $0.01 < S(\%) < 0.25$	5 min		40 kg, 250 W	Large insoluble particles

TABLE 9.25 Aircraft Summary

Airplane	Maximum Altitude and Range	Research Missions	Maximum Gross Weight (lbs.)	Availability	Operator
Single Engine					
U-2/ER-2 (3)	65,000′ 2,400 nm	High-altitude photography, radiometry, air sampling	Contact for details	2-3 months notice	High Altitude Missions Branch Mail Stop 240-6 NASA Ames Research Center Moffett Field, CA 94035 John C. Arvesen (415) 694-5376
T-28 (armored)	20,000′ 100 nm	Severe storm penetration, cloud physics	9,000	3-12 months	Institute of Atmospheric Sciences South Dakota School of Mines and Technology Rapid City, SD 57701 Dr. Paul L. Smith, (605) 394-2291
Twin Engine					
DeHavilland Canada DHC-6-200 Twin Otter	20,000′ 800 nm	Flight dynamics, turbulence research, air quality, cloud physics, CO_2 flux measurement for Agriculture Canada	11,579	Prefer 3-months notice	National Research Council of Canada Flight Research Lab Ottawa, Canada K1A 0R6 Ian McPherson (613) 998-3594 or 998-3071
Convair C131A	22,000′ 1,650 nm	Cloud physics, aerosols, cloud chemistry/acid rain, air pollution, volcanic effluents, mesoscale studies	47,000	Undergoing development	Department of Atmospheric Sciences University of Washington Seattle, WA 98195 Dr. Peter Hobbs (206) 543-6027
T-29B	16,000′ 1,300 nm	Radar systems development, microwave systems development	43,575	Available	Georgia Institute of Technology, SEL/EES Engineering Experiments Station Atlanta, GA 30332 Dr. J. Lee Edwards (404) 424-9641
King Air B200T	35,000′ 1,800 nm	Cloud physics and dynamics, air motion and turbulence, air chemistry, radiation	14,000	Flexible, 3-15 months	Research Aviation Facility National Center for Atmospheric Research P.O. Box 3000 Boulder, CO 80307 Mr. Byron Phillips (303) 497-1032

Table continues on next page.

TABLE 9.25 Continued

Airplane	Maximum Altitude and Range	Research Missions	Maximum Gross Weight (lbs.)	Availability	Operator
Sabreliner	45,000′ 1,200 nm	Air chemistry, air motion, cloud physics, radiation	20,000	Flexible, 3-15 months	Research Aviation Facility National Center for Atmospheric Research P.O. Box 3000 Boulder, CO 80307 Mr. Byron Phillips (303) 497-1032
Queen Air 80A	25,000′ 765 nm	Air motion, cloud physics, air chemistry, radiation	8,800	Flexible, 3-15 months	Research Aviation Facility National Center for Atmospheric Research P.O. Box 3000 Boulder, CO 80307 Mr. Byron Phillips (303) 497-1032
King Air 200T	35,000′ 1,700 nm	Cloud physics and dynamics, air motion, boundary layer, and icing research	14,000	Same as NCAR aircraft May–October	Department of Atmospheric Sciences University of Wyoming P.O. Box 3038 University Station Laramie, WY 82071 Dr. Wayne R. Sand (307) 766-3245
Beechcraft Baron 56TC	32,800′ 1,600 nm	Atmospheric electrical measurements and supporting meteorological measurements, marine boundary layer, surface fluxes		Flexible	Airborne Research Associates 46 Kendal Common Road Weston, MA 02193 Dr. Ralph Markson (617) 899-1834
Cessna Conquest	35,000′ 2,000 nm	Configured for photography support of ballistic missile testing		Available on noninterference basis	Aeromet P.O. Box 571030 Tulsa, OK 74157 D. Ray Booker (918) 299-2621
Cessna Chancellor	31,000′ 1,000 nm	Configured for test and development of radar		Available on noninterference basis	Aeromet P.O. Box 571030 Tulsa, OK 74157 D. Ray Booker (918) 299-2621
Gates Learjet	41,000′ 1,000 nm	Infrared astronomy, meteorology, air sampling, radiation, clear air turbulence, wind shear, and boundary layer studies	13,500	3–12 months notice	NASA/Ames Research Center Mail Stop 211-12 Moffett Field, CA 94035 Dr. Curt Muehl (415) 694-6431
Beechcraft Queen Air		Air quality and meteorology		Available	Sonoma Technology, Inc. 3402 Mendocino Avenue Santa Rosa, CA 95401 Dr. Donald L. Blumenthal (707) 527-9372

Cessna 411	22,000′ 605 nm	Air chemistry, cloud physics, boundary layer, long-range transport, studies with tracers, solar radiation, urban source plume characterization	6,500	Heavy summer use; unavailable January and February; one month notice	Battelle NW Laboratory P.O. Box 999 Richland, WA 99352 P. M. Potter and F. O. Gladfelder (509) 375-3862/3863
DC-3	21,000′ 864 nm	Air chemistry, cloud physics, boundary layer, long-range transport, studies with tracers, solar radiation, urban source plume characterization	26,200	Heavy summer use; unavailable January and February; one month notice	Battelle NW Laboratory P.O. Box 999 Richland, WA 99352 P. M. Potter and F. O. Gladfelder (509) 375-3862/3863
Beechcraft 58TC		Cloud physics, aerosols, turbulence data			Particle Measuring Systems 1855 South 57th Court Boulder, CO 80301 Dr. Robert G. Knollenberg (303) 443-7100
Cessna Citation II	43,000′ 1,600 nm	Cloud physics, photographic, wind profile measurement	13,300	Available; 1–3 months notice	Center for Aerospace Sciences University of North Dakota P.O. Box 8216, University Station Grand Forks, ND 58202 Dr. Tony Grainger (701) 777-3170
Piper Cheyenne II		Cloud physics and dry ice instrumentation aircraft			Colorado International Corporation P.O. Box 3007 Boulder, CO 80307 Ralph Papania, Jr. (303) 443-0384
Beechcraft Queen Air A-80/8800		Air quality and meteorological measurements, lidar system		Normally limited to SRI; special arrangements are possible	SRI International 333 Ravenswood Avenue Menlo Park, CA 94025 Dr. Warren B. Johnson (415) 859-4755
Piper Navajo PA-31 (turbo charged)		Basic meteorological and cloud physics variables, cloud seeding			Atmospherics Incorporated 5652 East Dayton Avenue Fresno, CA 93727 Dr. Thomas J. Henderson (209) 291-5575
King Air C90	25,000′ 1,300 nm	Air pollution, cloud physics, cloud chemistry, and air chemistry	9,650	Available for cooperative studies	Air Quality Division of Air Resources Laboratory NOAA 325 Broadway Boulder, CO 80303 Dr. Rudolf Pueschel (303) 497-6181
Four Engine					
P-3 (2)	32,000′ 3,500 nm	Hurricane research, boundary layer, cloud physics, air-sea interaction (AXBT), air chemistry, radiation	135,000	Hurricane research first priority July 1–November 1; otherwise available; 6 months notice	NOAA Office of Aircraft Operations P.O. Box 520197 3401 N.W. 59th Avenue Miami, FL 33152 W.D. Moran (305) 526-2939

Table continues on next page.

TABLE 9.25 Continued

Airplane	Maximum Altitude and Range	Research Missions	Maximum Gross Weight (lbs.)	Availability	Operator
Lockheed Electra	30,000′ 2,450 nm	Cloud physics and dynamics, air motion and turbulence, air chemistry, radiation	116,000	Flexible, 6–18 months	Research Aviation Facility National Center for Atmospheric Research P.O. Box 3000 Boulder, CO 80307 Mr. Byron Phillips (303) 497-1032
NKC-135A (Boeing 707)	40,000′ 5,000 nm	Measurement of passive infrared emission from targets, background and atmosphere	240,000	Available for periods of up to two weeks with 9–12 months notice, 9 out of 12 months per year	Air Force Geophysics Laboratory AFGL/OPR Hanscom AFB, MA 01731 Mr. Brian Sandford (617) 861-3370
NKC-135	40,000′ 4,000 nm	Ionospheric physics, aeronomy	275,000	Should be requested by March 1 for 12 months beginning following October 1. Collaboration with existing AFGL mission plans welcomed	Air Force Geophysics Laboratory Hanscom AFB, MA 01731 Mr. James G. Moore (617) 861-3128
Lockheed NC130-B	24,000′ 1,850 nm	Remote sensing and spacecraft sensing development; geology, geobotany, agriculture, urban analysis	135,000	Available with 6–12 months notice, particularly October–March	Medium Altitude Missions Branch NASA/Ames Research Center Mail Stop 211-12 Moffett Field, CA 94035 Mr. R. H. Mason (415) 694-5348
Lockheed C141-A	45,000′ 3,000 nm	Infrared astronomical research	320,000	Available year-round with 8–9 months notice	NASA/Ames Research Center Mail Code 211-12 Moffett Field, CA 94035 Robert M. Cameron and Louis C. Haughney (415) 965-5338/5339
Convair 990	41,000′ 3,000 nm	Atmospheric and space science, earth resources, air chemistry; radar geology, aerodynamic oceanography, meterology	240,000	6–12 months notice	NASA/Ames Research Center Moffett Field, CA 94035 Mr. John O, Reller, Jr. (415) 694-5392
LC-130 (7)					National Science Foundation Antarctic Aviation Support
UH-IN (7)					National Science Foundation Antarctic Aviation Support

TABLE 9.26 Oceanographic Ships, Academic Institutions, 1982

Name of Ship	Home Port	Operator	Length (ft)	Displacement (tons)	Cruise Speed (kt)	Sea State (Beaufort)	Endurance (days)	Range (nm)	Scientific Party	Cost ($1000/day based on year as stated)
Robert D. Conrad AGOR-3	New York NY	Lamont-Doherty Geological Obs.	209	1428	10	5	45	19,000	23	6.8/1982
Melville AGOR-14	San Diego CA	Scripps Inst. Oceanogr.	244.8	2075	10	4	41	9,000	29	12.0/1982
Gyre AGOR-21	Galveston TX	Texas A&M Univ.	174	946	9.5	8	60	8,000	22	7.3/1982
Alpha Helix	Seward AK	Univ. of Alaska	132.8	554	10	3	30	5,800	15	7.0/1982
Cape Hatteras	Beauford NC	Duke Univ.	135	539	11	6	24	6,800	12	4.1/1982
Cape Henlopen	Lewes DE	Univ. of Delaware	120	165	12.5	5	14	2,390	12	3.7/1982
Kana Keoki	Honolulu HI	Univ. of Hawaii	156	1080	11	4	42	12,500	15	5.9/1982
Moana Wave	Honolulu HI	Univ. of Hawaii	174	1437	10	5	45	8,000	12	3.2/1982
Ridgely Warfield	Annapolis MD	Johns Hopkins Univ.	106	162	14	4	14	1,500	10	4.8/1982
Cape Florida	Miami FL	Univ. of Miami	135	539	11	7	21	7,680	12	3.9/1982
Columbus Iselin	Miami FL	Univ. of Miami	170	830	13	5	30	9,700	16	6.5/1982
Cayuse	Moss Landing CA	Moss Landing Marine Lab.	80	173	9	4	20	4,500	8	3.4/1982
Wecoma	Newport OR	Oregon State Univ.	177	1015	11.5	6	30	5,000	16	5.7/1982
Endeavor	Narragansett RI	Univ. of Rhode Island	177	972	12	5	30	5,540	16	5.5/1982
New Horizon	San Diego CA	Univ. of California, San Diego	170	1080	10	5	32	7,600	13	6.1/1982
Ellen B. Scripps	San Diego CA	Univ. of California, San Diego	95	234	9	4	14	6,480	8	3.4/1982
Thomas Washington	San Diego CA	Univ. of California, San Diego	209	1362	11.5	6	42	8,700	21	9.6/1982
Robert G. Sproul	San Diego CA	Univ. of California, San Diego	125	520	10		22		12	3.7/1985 estd.
Blue Fin	Savannah GA	Skidaway Inst. of Oceanography	72	86	10	4	12	3,000	8	0.7/1980
Velero IV	Wilmington CA	Univ. of Southern California	110	650	9.5	4	30	11,500	12	4.4/1982
Ida Green	Galveston TX	Univ. of Texas Inst. for Geophysics	135		10	7	21	5,000	12	4.4/1981
Fred H. Moore	Galveston TX	Univ. of Texas Inst. for Geophysics	165		10	7	26	6,000	20	7.2/1981
Longhorn	Port Aransas TX	Univ. of Texas, Port Aransas Marine Lab.	80	200	9	3	18	2,000	10	1.5/1981
Thomas G. Thompson	Seattle WA	Univ. of Washington	208.8	1401	9.5	5	40	8,500	22	8.2/1982
Onar	Seattle WA	Univ. of Washington	65	95	8.5	4	5	750	6	1.05/1980
Hoh	Seattle WA	Univ. of Washington	65	81	8	2	10	800	6	1.05/1980
Atlantis II	Woods Hole MA	Woods Hole Oceanographic Inst.	210.3	2300	12		30	10,000	25	10.0/1984 estd.
Knorr	Woods Hole MA	Woods Hole Oceanographic Inst.	244.8	1915	10		45	10,000	23	9.2/1982
Oceanus	Woods Hole MA	Woods Hole Oceanographic Inst.	177	960	12.5		25	7,000	12	5.7/1982
Johnson	Fort Pierce FL	Harbor Branch Foundation, Inc.	123.7	335	12			2,880		
Sea Diver	Fort Pierce FL	Harbor Branch Foundation, Inc.	98.6	234	10			5,000		
Westward	Woods Hole MA	Sea Education Assn.	125	250	7					

APPENDIXES

APPENDIX A

Current Tropospheric Chemistry Research in the United States

There are programs under way in the United States and elsewhere that are already addressing certain aspects of global tropospheric chemistry. In the United States, these programs, active in several agencies, are formally coordinated through the Subcommittee on Atmospheric Research of the Committee on Atmospheres and Oceans, which is established under the Federal Coordinating Council for Sciences, Engineering, and Technology. None of the current programs includes all of the Global Tropospheric Chemistry Program elements recommended in this report, and all of them together will not, without significant augmentation, achieve the goals of the recommended global program. Relatively few of the existing programs have global tropospheric chemistry issues as their focus. Most of the larger programs are focused on urban or regional problems such as air pollution or acid rain. Nevertheless, many of the research tasks encompassed by the existing federal programs will contribute substantially to solving tropospheric chemical problems encountered on the global scale. A brief summary of the various programs follows.

NSF's ATMOSPHERIC CHEMISTRY PROGRAM

The Atmospheric Chemistry Program of the Atmospheric Sciences Division of the National Science Foundation (NSF) supports a wide range of laboratory, field, and modeling investigations of the troposphere, the stratosphere, and planetary atmospheres. The ultimate goal of understanding the complex interactions of hundreds of chemical reactions with transport and radiation phenomena is supported by studies of the pathways and kinetics of molecular-level processes, of the global cycling of chemical elements, and of new approaches to the measurement of trace species, including free radicals. Trace gases and aerosols are investigated in both clean and polluted atmospheres.

The program lends support to advances in methodologies for the identification of individual aerosol particles and to the study of the interaction of these particles with their gaseous environment. The chemistry of the particle-size domain between molecular dimensions and filterable particles receives added emphasis.

The Atmospheric Sciences Division of NSF, through the University Corporation for Atmospheric Research, supports the National Center for Atmospheric Research (NCAR) to initiate, coordinate, and carry out atmospheric research that requires long-term cooperative efforts among scientists at NCAR and at universities and government laboratories and to provide and develop facilities and related services for the atmospheric research community. Research within the Atmospheric Chemistry and Aeronomy Division at NCAR has recently focused on the role of CO, NO_x, CH_4, and nonmethane hydrocarbons in the tropospheric O_3 budget, on biospheric processes as they influence the atmosphere, on biomass burning as a source of atmospheric trace gases, and on cloud chemistry and acid precipitation and its causes.

NASA's GLOBAL TROPOSPHERIC EXPERIMENT

A research effort named the Global Tropospheric Experiment (GTE) has been initiated by the National Aeronautics and Space Administration (NASA) to study the chemistry of the global troposphere and its interaction with the stratosphere. The first phase of the project, aimed at developing and validating measurement techniques for H_xO_y and NO_x trace species in tropospheric chemical cycles, is designed to lead to the development and implementation of a cooperative global tropospheric chemistry research program with the goal of understanding the chemical cycles that control the composition of the global troposphere and its changes.

The immediate emphasis of the GTE is on the development, testing, and evaluation of measurement techniques that can achieve, under field conditions, the extreme sensitivity required to determine accurately atmospheric concentrations of key chemical species. A later phase of the GTE will focus on widespread, systematic measurements supported by modeling and laboratory studies to understand the principal processes that govern key chemical cycles in the troposphere.

The role of the global troposphere as the source and sink for the stratosphere, the details of the troposphere-stratosphere interchange, the processes that control global tropospheric O_3, and the atmospheric role in biogeochemical cycles are of particular interest to NASA, as is the eventual development of an enhanced capability to study the troposphere and its composition from space.

Instrument development in the initial phase of the GTE will involve a three-step test and evaluation program comprising a ground-based intercomparison, an airborne intercomparison in the tropical troposphere with particular attention to the boundary layer over the ocean and over tropical forests, and an airborne intercomparison in the upper troposphere. This strategy will systematically expose the measurement systems under current development and evaluation to conditions that will be encountered in later global tropospheric chemistry field experiments.

NOAA's GEOPHYSICAL MONITORING FOR CLIMATIC CHANGE PROGRAM

The NOAA Air Resources Laboratory's Geophysical Monitoring for Climatic Change (GMCC) program operates four baseline observatories at which measurements are made of atmospheric trace constituents important for climatic change. These observatories are located in remote clean-air sites where the measured values are representative of background concentrations of trace gases and particles in the atmosphere. The observatories are located at Barrow, Alaska; Hilo (Mauna Loa), Hawaii; American Samoa; and South Pole, Antarctica; and are also components of the WMO Background Air Pollution Monitoring Network (BAPMoN) Program.

The objectives of the GMCC program are as follows:

1. To determine concentrations, their variations with time and space, and properties of atmospheric trace gases and aerosols that can potentially have an impact on climate;
2. To understand the sources, sinks, transport, modification, and budgets of those trace constituents; and
3. To apply (in collaboration with others) those measurement data to determine their effect on global weather and climate. The GMCC monitoring effort at the baseline observatories is primarily for long-term surveillance of atmospheric trace species concentrations. Additional flask sampling measurements of selected atmospheric trace constituents are made near Niwot Ridge, Colorado; at Boulder, Colorado; at a global network of CO_2 flask sampling stations; at a number of total O_3 monitoring stations located mainly in the contiguous United States; and occasionally at other locations in support of particular research objectives. GMCC research primarily focuses on analyses of these data sets.

Programs for the measurement of O_3 (surface, vertical profiles, and total column), atmospheric aerosols, halocarbons (Freons 11 and 12), N_2O, water vapor, and precipitation chemistry are currently under way by NOAA at one or more of the GMCC program sites.

A primary objective for the GMCC program is to make measurements of trace species in a monitoring mode to document long-term trends. As such, most programs are continuing, in contrast to expeditionary. In the near future, programs will begin in the automation of Dobson spectrometers, and on the measurement of radiatively active trace gases, and of gases and aerosols in the Arctic.

NSF's SEAREX PROGRAM

The Ocean Science Division of NSF has been sponsoring a coordinated research effort investigating the atmospheric transport of material from continental regions to the ocean. The Sea-Air Exchange (SEAREX) program was initiated in 1977 and directly involves 11 universities and laboratories from the United States, France, and England, and cooperative ancillary programs with investigators from New Zealand, Australia, Japan, and the People's Republic of China. The SEAREX program has concentrated its efforts on investigating air-sea exchange in the westerly and trade wind regimes of the North and South Pacific Ocean. The objectives of SEAREX are as follows:

1. The measurement of the rate of exchange of selected trace elements and organic compounds across the sea-air interface,
2. The investigation of the mechanisms of exchange of these substances, and
3. The identification of the sources for these substances in the marine atmosphere.

Chemical substances being investigated in SEAREX include selected heavy metals (e.g., Pb, Cd, Hg, Zn, Se, Fe, Mn, V, and Cr), soil dust, sea salt, halogens, ^{210}Pb and ^{210}Po, sulfate, nitrate, phosphate, particulate organic carbon, and such organic species as PCB, DDT, HCB, aliphatic hydrocarbons, phthalate plasticizers, fatty acids, fatty and polycyclic alcohols, and low-molecular-weight ketones and aldehydes. Extensive field programs have been carried out at Enewetak Atoll, Marshall Islands in 1979, at American Samoa at the NOAA GMCC station site in 1981, and in northern New Zealand in 1983. A final SEAREX field program is planned from an oceanographic research vessel located at about 40°N, 170°W in 1986.

THE NATIONAL ACID PRECIPITATION ASSESSMENT PROGRAM

The Acid Precipitation Act of 1980 established the Interagency Task Force on Acid Precipitation to develop and implement a comprehensive National Acid Precipitation Assessment Program. The act required the task force to produce a national plan for a 10-year research program. The purpose of the National Acid Precipitation Assessment Program is to increase understanding of the causes and effects of acid precipitation. The national program includes research, monitoring, and assessment activities that emphasize the timely development of a firmer scientific basis for decision making. The National Acid Precipitation Assessment Program is co-chaired by the Environmental Protection Agency (EPA), NOAA, and the Department of Agriculture, and it includes a major involvement by the national laboratories of the Department of Energy (DOE).

Research is proposed in the National Plan in nine categories. Each of the research tasks described focuses on a specific area and generally involves the coordinated participation of several agencies.

Focusing on research needs and tasks relevant to atmospheric sciences in general and atmospheric chemistry in particular, the plan pursues the following objectives:

1. Identify the natural emissions that can influence precipitation chemistry and develop an experimental data base that sufficiently characterizes the source strengths of these species.
2. Quantify pollutant emissions of interest from man-made sources with supporting energy-use, economic, and technical data for appropriate sources and regions for certain time periods.
3. Develop and maintain quality-assured emission models and methods to support other task groups assessing control strategies for acid deposition.
4. Conduct special research projects into economic, energy-use, technological, and other factors that affect pollutant emissions from major man-made sources.
5. Determine the important aspects of meteorological transport of acidic substances and their precursors on spatial scales ranging from local to global.
6. Determine the important overall physical and chemical pathways and specific reaction processes regulating the formation of acid substances in the atmosphere through laboratory and field measurements and theoretical interpretation.
7. Determine the relative importance of wet and dry removal processes for acid substances and their precursors within the atmosphere.
8. Develop state-of-the-science modeling frameworks using advanced products resulting from the atmospheric processes research program. The models will serve as the media for integrating the full spectrum of phenomenological research in acid deposition and as the primary assessment tools in identifying future control strategies for mitigation.
9. Develop a comprehensive data base for evaluation and verification of acidic deposition models and associated process component modules.
10. Determine the spatial and temporal variations in the composition of atmospheric deposition within the United States for a period measured in decades through a National Trends Network.

The time schedules contained in this federal plan call for preliminary assessment of the research results by 1985-1986 and a more complete assessment by 1988-1989.

OTHER RELATED PROGRAMS

In addition to the above programs that have regional- to global-scale tropospheric chemistry as their major thrust, several mission agencies support work that is applicable to the research goals of a global tropospheric chemistry program.

The DOE has supported for many years a program

designed to yield understanding of the relationship between energy production activities and effects on the atmosphere, and, from work of its predecessor agencies, to study the impact of nuclear explosions in the atmosphere. Chemistry program elements involve laboratory and field studies of the mechanisms and kinetics of the production and transformation of emissions related to use of energy. A major research area is the study of the physics and chemistry of those processes that control the removal from and reinsertion into the atmosphere of gases and aerosols. Recent emphasis has also been placed on problems related to emerging energy technologies. Since 1978, the DOE has been engaged in a major effort to analyze the causes and climatic consequences of CO_2 buildup in the atmosphere, an activity that requires measurement of fluxes and involves study of the carbon cycle as a whole in the biosphere, including the atmosphere.

The EPA, as the regulatory agency with responsibility for establishing and enforcing environmental standards within the limits of various statutory authorities, conducts a major atmospheric research and monitoring program. A major fraction of this activity is directed toward understanding and characterizing problems within urban and regional air sheds with focus on criteria pollutants, hazardous pollutants, long-range transport, transformations, particles, vehicular emissions, and large-scale and long-term effects of air pollution on the biosphere. An analysis of this program is beyond the scope of this report, but the program has produced, and continues to produce, results of immediate application to global tropospheric chemistry investigations. Particularly notable examples are the studies of the long-range transport and fate of atmospheric pollutants and the extensive transport and fate models designed for applications on scales up to the regional.

APPENDIX B

Remote Sensor Technology

The following overview will provide a perspective of the ultimate role of spaceborne remote sensor techniques in providing global measurements to improve understanding of the processes involved in the chemistry, dynamics, and transport phenomena in the global troposphere. To provide this perspective, this overview will highlight two spaceborne sensors that are currently providing measurements from space, provide a survey of potential instrument techniques that can provide key measurements important to tropospheric science, and project some technological developments that need to be implemented to realize the full potential of remote sensor techniques from space. Ultimately, remote sensor techniques should be utilized from orbiting satellites for long-duration missions to exploit the potential to measure and detect long-term trends related to changes in the global balance of the troposphere due to anthropogenic and nonanthropogenic processes.[1] However, the path to developing research instruments for long-duration satellites should capitalize on other airborne and spaceborne platforms, including high-flying aircraft, the Space Shuttle, and possibly spaceborne pallets or free flyers of shorter duration missions. This approach will provide the atmospheric science community with the opportunity to iterate a variety of measurements with evolving mathematical models of the troposphere, and provide the instrument scientific community with the opportunity to develop and evaluate advanced sensor technology to optimize the measurement base required to validate the mathematical models developed.

[1]*Report of the NASA Working Group on Tropospheric Program Planning*. J. H. Seinfeld. NASA RP 1062. 1981.

PRESENT SPACEBORNE SENSOR MEASUREMENTS

Three classes of remote sensors have demonstrated unique capabilities in meeting some of the measurement needs in the global troposphere. The first class includes imaging spectroradiometers currently being used in Earth Observation Satellite Systems for meteorological and earth resource measurements. These sensors have recently shown the ability to detect regions of elevated haze layers and aerosol loading in the troposphere. A second class of instruments includes passive remote sensors that measure spectral emission or absorption of atmospheric molecules with external sources of radiation. Vertical distributions of molecular species, pressure, and temperature can be inferred through the use of inversion algorithms. A third class of instruments includes active remote sensors in which lasers in the ultraviolet, visible, and infrared portion of the spectrum are used in a similar mode as an active radar system.

Through a combination of scattering by aerosols and molecules in the atmosphere and selective absorption by atmospheric molecules, these sensors can provide range-resolved measurements of tropospheric molecules and aerosol loading.

HAZY AIR MASSES

Current research objectives and approaches for spectroradiometer systems encompass the following elements. Overall, the current objective is to evaluate capabilities of existing satellite and aircraft systems in conjunction with existing image processing systems for monitoring air pollution episodes.

- Outline the extent of the pollution area as seen on the visible imagery.
- Attempt to measure the threshold quantitatively over land as well as water.
- Note the condition, if any, under which the area is seen on the infrared imagery.
- Track the motion of the pollution area by noting the motion of its "center of gravity" or the motion of any discrete edge or feature.
- Compare any apparent motion with wind information obtained from conventional meteorological sources.
- Delineate the radiance values associated with individual pixels within the area and draw isopleths of selected radiances.
- Measure the variation of the radiance, i.e., the change of the isopleth values, as a function of the time of day.
- Make a correction for wind-related changes in the apparent density of the haze to obtain the variation in radiance related primarily to the solar zenith angle.
- Follow up university efforts to calibrate the SMS/GOES satellite visible sensors, and continue this effort by comparing the radiances as seen by the GOES sensors over a region of uniform brightness, with radiances expected from that surface through a normal haze-free atmosphere, as predicted by the Fraser model. In this model, the solar relationship to the surface is included.
- Adjust the radiances measured by the SMS/GOES sensors according to the calibration, and with the adjusted values, calculate the optical depth and mass loading at various locations of the smog area by using the Fraser model.
- Correlate any computed smog density with various meteorological parameters, as well as with various ground-based measurements. The most interesting meteorological parameters for initial investigation appear to be wind at various levels, the existence and height of inversion layers, relative humidity, and ground visibility. The ground-based smog measurements are optical thickness and particulate count.
- Attempt to generate at least two sets of correlations, one based on the model-calculated mass loading, and another based on the basic SMS/GOES radiances.

Under the modeling phase of the study, the approach involves an investigation of the scattered sunlight radiance. The radiance of the sunlight scattered toward a satellite is being computed for models of the air pollution for two purposes: (1) To determine the response of the radiance to the important physical parameters; (2) To estimate the accuracy with which air pollution parameters such as aerosol optical thickness, sulfate mass, and visibility can be derived from satellite observations. The important variable parameters for study in specifying the model are the bidirectional reflectivity of the ground, the aerosol optical thickness, the relative humidity and amount of water vapor, the composition of the aerosols, and their size distribution.

The computed radiation characteristics will be compared with experimental results. Even more importantly, the aerosol parameters derived from satellite observations will be verified with experimental results. A large body of good experimental data on the physical and chemical characteristics of dense air pollution was obtained during the Persistent Elevated Pollution Episodes (PEPE/NEROS) Field Measurement Program during the summer of 1980. The basic program does not provide for measurements of the aerosol optical thickness, which is the most important aerosol parameter that will be derived from satellite observations. Solar transmission observations at about 12 stations obtained during the PEPE/NEROS experiment will be utilized. The transmission observations will be analyzed to derive values of the aerosol optical thickness.

The satellite observations expected during the PEPE/NEROS experiment include the visible and infrared spin and scan radiometer (VISSR) on GOES. In addition, more precise radiometer data will be used from the coastal zone color scanner (CZCS) on Nimbus 7.

MEASUREMENT OF AIR POLLUTION FROM SATELLITES (MAPS)

An earth-orbiting experiment flown on the Space Shuttle can provide scientists with data to accurately map changes within the earth's atmosphere. The experiment, Measurement of Air Pollution from Satellites (MAPS), charted concentrations of CO gas around the world over a range of latitudes extending from 30°S to 38°N during the second Space Shuttle flight. The performance of MAPS on this flight proved the system to be a faster, more efficient method of mapping trace gases in the earth's atmosphere than chromatograph devices.

MAPS used a gas filter radiometer to obtain measurements of the CO mixing ratio in the middle troposphere and stratosphere. The radiometer method is simpler and much less expensive than previous gas chromatograph devices. The experiment produced concentration maps of CO.

Early analysis of MAPS data concentrated on measurements performed during orbital Pass 15 on November 13, 1981. This orbital pass of the Shuttle began over Central America, continued east over the Mediterranean Sea, turned southeast over the Persian Gulf, the Arabian Sea, and extended to the southern tip of India. The concentration of CO within this extended area ranged from 70 ppb over the eastern Atlantic Ocean and 140 ppb in the Mediterranean area. These measurements were performed in less than 20 min from the vantage point of the Shuttle. Similar measurements from aircraft platforms would have required observations over many hours.

Analysis of experiment data so far indicates significant concentrations of middle troposphere CO mixing with both north-south and east-west variation over the North Atlantic and the Mediterranean Sea and the Middle East. Accuracy of the measurement has been determined to be within 15 percent with a repeatability of about 5 percent from orbit to orbit. CO gas is produced by natural processes (e.g., oxidation of CH_4 in wetland areas and forest fires) and man's activities (e.g., slash-and-burn agriculture and automobile emissions). Man's contribution to the global atmospheric budget of CO has grown significantly during this century, resulting in a large asymmetry between the CO concentration in the northern and southern hemispheres.

The major atmospheric sink for CO is a complex sequence of photochemical reactions that oxidize CO molecules into CO_2. One of the initial steps in this oxidation process is the combination of CO with the hydroxyl radical OH. OH plays a major role in a variety of other atmospheric processes, notably reactions involving SO_2, nitrogen oxides, and chlorofluoromethanes.

CO oxidation can potentially divert OH from reactions with these other gaseous species and alter the overall chemical balance within the earth's atmosphere.

NASA plans to refly MAPS on the seventh Shuttle mission, scheduled for summer 1984, to study seasonal variations in the total abundance and regional distribution of CO within the earth's atmosphere.

Although the second Shuttle flight was abbreviated, the experiment collected data for about 42 hr, and the investigators were able to corroborate the sampled areas with the instrument readings taken with underflying aircraft and with other surface information.

Reduction of MAPS data is continuing at the present time, and the first global map of CO concentrations at low-latitudes to midlatitudes is now available.

FUTURE INSTRUMENT TECHNIQUES

Passive Remote Sensors

In general, remote sensors for atmospheric applications can be developed to measure changes in the three basic properties of electromagnetic waves including energy (absorption or emission), wavelength (frequency shifts), and polarization. Passive spectroscopic remote sensors, summarized in this section, correspond to the class of sensors that measures one of these properties (absorption or emission) with an external source of radiation. For tropospheric measurements from space or aircraft observing platforms, the downward-viewing modes (nadir) provide the widest vertical and horizontal coverage. Solar occultation measurements from space and airborne platforms have been used extensively in stratospheric applications, but due to the extent of global cloud cover, geographical coverage of the troposphere is severely limited in this operational mode. External sources of radiation for nadir-viewing experiments include upwelling thermal radiance of the earth-atmosphere system, reflected radiation from the surface of the earth, and scattered radiation from molecules and aerosols in the atmosphere. Compared to detection of direct solar radiation through the atmosphere, as in stratospheric solar occultation measurements, these sources of radiation are relatively weak and require more sensitive detection instruments. However, in addressing some of the tropospheric scientific questions, sensors developed for stratospheric observations in solar occultation can be used from ground-based platforms to provide some vertical layering of tropospheric constituents, which are important in establishing global concentrations of well-mixed gases, or those gases showing seasonal and temporal variability. In addition, with improvements in sensitivity and instrument optimization to a 300°K source and broader spectral response functions, many sensors developed for stratospheric applications should become viable tropospheric sounders from space platforms.

Before addressing the question of the available passive remote sensing techniques, some basic properties of atmospheric radiation should be considered for the various external sources of radiation. In viewing the upwelling thermal radiance of the earth-atmosphere system, passive instruments detect radiation that is a composite of energy transmitted through the atmosphere, that is absorbed and reemitted by atmospheric molecules at all altitudes between the source and the sensor. The thermal emission of this radiation is primarily governed by the temperature lapse rate of the lower atmosphere, which ranges from approximately 300°K near the ground to 220°K in the lower stratosphere. Since the Planck function of a blackbody radiator peaks at approximately 10 μm for a 300°K blackbody, the desired wavelength region varies from 4.5 to 15.0 μm. Also, in this wavelength range, contributions to atmospheric radiance from scattered solar radiation are small and can be considered of a second order. Because of the pressure broadening of the spectroscopic absorption or emission lines, the energy received in the wings of the atmospheric emission lines reflects the presence of molecules at high pressures (i.e., lower altitudes), while energy received near line centers is representative of molecules at lower pressures (i.e., higher altitudes). This gives rise to the possibility that by selectively measuring radiation at various spectral regions from line center, one could, in principle, generate vertical profiles of gas concentrations. This is analogous to the inference of temperature profiles by remote sensing methods using thermal infrared wavelengths where wavelengths in various parts of the emission line of a uniformly mixed gas (e.g., CO_2 and N_2O) are used to obtain altitude discrimination. In order to invert radiances detected at the top of the atmosphere to concentration profiles, inversion algorithms must be developed that take into account emission of the earth-atmosphere system, governed by the temperature lapse rate. In inverting the radiance to obtain useful concentration accuracies, temperature profiles must be simultaneously measured with the radiance to a desired accuracy of approximately ± 2°K. For measurements of minor trace gases in the troposphere, thermal emissions are weak due to a combination of low concentrations and low reservoir temperatures. Also, the temperature lapse rate and the relatively low-pressure gradient in the atmosphere limit the degree of vertical discrimination possible, independent of the spectral resolving power of the instrument beyond a resolution of approximately 0.01 cm^{-1}. Furthermore, radiances emitted and absorbed near the ground (within the first 3 km) are faced with the limitation of a small temperature gradient between the earth and the layer of atmosphere near the earth, thus making fine discrimination of layers near the ground difficult.

In the free troposphere, however, broad layers of the atmosphere can be identified when simulating the inversion of the earth-atmosphere radiance in the thermal infrared, and weighted averages of gas concentrations at specific altitudes can be obtained. Therefore, these techniques provide the synoptic coverage and spatial resolution required to study the global transport of minor trace gases such as CO (refer to previous section). Such observations are required to address those scientific questions related to global budgets and distributions of gas concentrations and to the relative changes that occur over time of some of the well-mixed gases such as CH_4, N_2O, and CO_2, especially in the free troposphere.

The other two external sources of radiation that can be used in tropospheric sounding of atmospheric trace gases include reflected radiation from the earth's surface and scattered radiation from the earth's atmosphere. In the latter mode, scattered radiation from aerosols and molecules is predominant in the ultraviolet and visible portion of the spectrum and can be used to infer integrated molecular concentrations through direct absorption by atmospheric spectra. In the ultraviolet and visible portion of the spectrum, the dependence of line width on pressure has a smaller functional dependence than in the infrared, and vertical layering of the atmosphere using this physical process is more difficult to infer than in the thermal infrared. However, inversion of radiance data in the ultraviolet and visible is less sensitive than the thermal infrared to the knowledge of atmospheric temperature profiles. Relevant tropospheric molecules in this spectral region include O_3, SO_2, and NO_2.

In viewing reflected solar radiation from the earth's surface, one should restrict observations to approximately the 1.0- to 3.5- μm region, since in this range the radiance at the top of the atmosphere is primarily due to reflected solar radiation, and absorption can be used as the physical process to infer molecular concentrations. Some vertical layering of molecular concentrations can be achieved through pressure broadening of the molecular absorption lines in the atmosphere, and integrated measurements to the ground are possible. The accuracy of the retrieved concentrations has a smaller functional dependence on the temperature profile than in the upwelling thermal infrared region. In the intermediate spectral band (i.e., 3.5 to 4.5 μm), the radiance at the top of the atmosphere is composed of comparable values of the upwelling thermal radiance and reflected solar radiation. Although interesting absorption and emission lines of major atmospheric molecules lie in this region, inversion of the radiance measurement to molecular concentrations is complicated by the complexity of the radiative transfer equation, which makes it difficult to quantitatively invert radiances to obtain molecular concentrations.

A summary of the current passive remote sensors developed under the NASA remote sensing program has been reviewed by Levine and Allario[2] and will not be discussed further.

---

[2] "The global troposphere: biogeochemical cycles, chemistry, and remote sensing." J. S. Levine and F. Allario. *Environmental Monitoring and Assessment 1*:263-306. 1982.

Active Remote Sensors

The use of laser techniques for measuring range-resolved concentrations of aerosols, molecular constituents, and meteorological parameters in the stratosphere and troposphere has represented a major research activity in the NASA research and applications programs over the last decade. In considering some of the fundamental limitations in using external sources of radiation for measuring tropospheric molecules, it should be obvious that the use of powerful and tunable, monochromatic sources in the ultraviolet, visible, and infrared portions of the spectrum has the potential to remove some of the limitations of passive remote sensing imposed by the physical processes of the atmosphere. For example, in probing the troposphere from the top of the atmosphere in the nadir mode, the laser beam has the potential to probe down to the surface of the atmosphere, and through the process of molecular absorption and range gating (to be discussed later), the vertical distribution of molecular concentrations, aerosols, and meteorological parameters can, in principle, be obtained to a spatial resolution of ≤ 0.1 km. The ability to meet these measurement needs is dependent on the energy of the laser and the magnitude of the differential absorption cross section of the species to be measured. In general, the ability of lidar systems to obtain individual measurement parameters to a given accuracy depends on the magnitude of optical scattering coefficients and molecular absorption cross sections, if one assumes sufficient flexibility in system parameters, such as telescope size, detector quantum efficiency, and energy of the laser transmitter.

Research in the NASA program has evolved from initial studies with fixed-wavelength lasers to measure optical backscatter from aerosols, to the use of tunable and narrow bandwidth lidar systems for remote measurements of atmospheric gases and aerosols in the troposphere. Investigations have been conducted by using Raman scattering techniques for remote measurements of water vapor and SO_2, but this technique was found to be limited to high gas concentrations, short ranges, and nighttime operation as a result of the small Raman scattering coefficients. The differential absorption lidar (DIAL) technique can overcome some of these limitations because absorption cross sections can be 6 to 8 orders of magnitude larger than the Raman cross sections. This technique has recently been demonstrated as viable for detecting O_3 and aerosol layering in the troposphere from an aircraft platform. In general, successful demonstrations of the feasibility of DIAL techniques have evolved as the technology of the tunable laser sources has improved in energy output, spectral purity, and amplitude and frequency stability. DIAL techniques, to date, have been applied from fixed and mobile ground stations to detect aerosols and SO_2 emitted by stack plumes and to measure the vertical distribution of tropospheric aerosols, O_3, and water vapor from aircraft platforms.

The principle of the DIAL techniques is discussed below. In general, the DIAL technique depends on the existence of a molecular feature (absorption line) that is specific to a gas molecule, whose spectroscopic characteristics are well known in the atmosphere (i.e., line intensity, line position, and line broadening as a function of pressure), and which is relatively free of spectroscopic interference from other molecules in the atmosphere. In order to detect the molecular feature, the wavelength of the laser is tuned to overlap the absorption feature, preferably at the center of the line. A second laser wavelength is required whose wavelength is removed from the molecular feature but sufficiently close by in wavelength to detect the same atmospheric scattering properties of the atmosphere at the two wavelengths. The two laser wavelengths are pulsed and can be fired simultaneously or within a time spacing that essentially freezes atmospheric dynamics during the measurement period (i.e., ≤ 100 ns). The two laser pulses are backscattered to the receiver telescope by atmospheric aerosols and molecules, so that the return signal represents a time history of the atmospheric scattering and absorption properties of the atmosphere for each laser pulse. This time history is related to a spatial profile of the atmospheric scattering and absorption properties of the atmosphere, through the equation $(c\Delta t)/2 = \Delta R$, where Δt represents a time gating interval that can be selectable in time to correspond to a range gate interval, ΔR. In the data processing mode, the ratio of return signals at the "on" wavelength, P_{on}, to the "off" wavelength, P_{off}, is measured and, through the lidar equation, can be related to the molecular concentration of the gas in the atmosphere as a function of range from the transmitting telescope. Another useful mode for a DIAL experiment is to employ a continuous wave (CW) laser that uses reflection from the ground to return the two laser wavelengths back to the receiving telescope. In the infrared, where pressure broadening of atmospheric spectral lines as a function of altitude is larger than in the ultraviolet and visible portion of the spectrum, one can selectively probe various segments of the absorption line by tuning the wavelength of the transmitting laser from the central peak into the wings to obtain vertical layering of the selected tropospheric molecule. Conceptually, this is similar to the passive solar reflected technique discussed earlier. In this case, however, the source of radiation is monochromatic, allowing a single absorption line to be probed. Furthermore, the radiative transfer equation for the transmitted and reflected signals through the atmosphere can be described by straightforward atmospheric processes.

A summary of the current active remote sensors developed under the NASA remote sensing program is given by Levine and Allario[3] and will not be discussed further here.

FUTURE THRUSTS

A Workshop on Passive Remote Sensors for performing tropospheric measurements was conducted in Virginia Beach, Virginia, July 20-23, 1981. The purpose of this workshop was to define the long-range role of passive remote sensors in tropospheric research and to identify the technology advances necessary to implement that prescribed role. Recommendations and conclusions from the two panels of the workshop attendees are given below. Details leading to these conclusions are given in NASA CP 2237[4] and are abstracted below.

RECOMMENDATIONS AND CONCLUSIONS

Sensor Systems Panel Recommendations

Following are the conclusions and recommendations from a systems point of view for passive remote sensing of tropospheric constituents:

1. Passive remote systems exhibit promise and should be developed for two-layer measurement of some of the more abundant tropospheric species (e.g., O_3, CO, CH_4, CO_2, HNO_3, H_2O, and NO).
2. A measurement scenario consisting of a combination of nadir viewing and solar occultation should be considered for measurement of gases such as O_3 and HNO_3. Measurement of these gases in the troposphere presents a unique challenge in that well over 90 percent of the total burden of the gas resides in the stratosphere.
3. For multilayer (i.e., more than two) measurements of a wide range of species, a nadir-viewing instrument capable of obtaining continuous spectra in the 3- to 15- μm spectral region with a spectral resolution of less than 0.1 cm^{-1} is desired.
4. Gas filter radiometer instruments (e.g., MAPS and HALOE) should be developed concurrently with a scanning instrument (see previous recommendation). Such systems may provide near-term two-layer tropospheric measurements of gases such as CO and CH_4 with only modest improvements in system performance. For gases such as O_3 and NO, major problems may be encountered with the gas cell technology and resolving more than one atmospheric layer.
5. Further development of aerosol retrieval algorithms is required for obtaining aerosol thickness and size distributions on a global scale. Although the technology currently exists for obtaining aerosol information over water, additional channels extending from the visible to near infrared would be desirable for future measurements.
6. The prospect and the possibility of initiating a feasibility study to determine if polarization measurements of scattered solar radiation can yield the refractive index (i.e., composition) of aerosols should be reassessed.
7. Existing spectrally scanning radiometers (e.g., interferometers) and/or gas filter systems should be employed from balloon and Shuttle platforms to study the effects of instrument noise and background fluctuations on inversion techniques. Short-term nadir-viewing Shuttle missions should be coordinated with existing solar occultation missions to study the feasibility of utilizing simultaneous occultation and nadir-viewing data.
8. Feasibility studies for both gas filter and spectrally scanning instruments should be initiated to study (a) the extent to which nadir-viewing systems can obtain profiles of two or more layers within the troposphere, (b) the accuracy requirements on molecular line parameters, meteorological parameters (i.e., temperature and pressure), radiance data, and background effects, and (c) the extent to which solar scattering can be used to obtain lower-level tropospheric data.

Sensing Technology Panel Recommendations

The technology needs presented earlier are restated in Table B.1 as critical needs and are recommended as technology development thrusts for elements of several passive sensing techniques for tropospheric research. In one sense, this table indicates the needs once a system is chosen. In a larger sense, the technology requirements and their apparent difficulty and cost should be a critical part of the instrument evaluation studies. Certain technology thrusts, however, are needed regardless of the sensing system choice. Examples from Table B.1 are as follows: (1) detector arrays, (2) cryogenic cooling (of sensors and optics), (3) sophisticated optical elements, and (4) data processing as an integral part of the sensor. Although not explicit in Table B.1, calibration techniques and equipment should be added to the list of needed technology thrusts. In response to the continuing need for more sensitive and accurate measurements over the full globe for long periods of time, producing great volumes of data, the workshop participants felt that the application of technology advances in these five areas would yield the greatest benefit in passive remote sensing of the troposphere.

---

[3] *Ibid.*

[4] *Tropospheric Passive Remote Sensing*. L. S. Keafer, Jr. (ed.). NASA CP 2237.

TABLE B.1 Technology Thrusts

Sensor	Technology Needs
Gas filter radiometry	
All types	Gas filter test cells
	Linear, high dynamic range detectors
	Highly uniform optical elements
Broadband spectrometry	
All types	Onboard smart processing
	Cryogenics/cooling
Grating type	10^3 element arrays
	Large gratings
Interferometer	Mitigation of background fluctuations
	Multiaperture, multiband interferometer
	In-flight alignment verification
Narrow-band spectrometry	
Laser heterodyne type	Tunable lasers and heterodyne arrays
Fabry-Perot	Improved coatings at long wavelengths

SUMMARY

In the current state of development of remote sensors, two sensor systems are being used from spaceborne platforms, and several active and passive systems have progressed to the stage where they are being used or proposed in field measurement programs from airborne and ground-based platforms to measure gaseous species (e.g., CO, O_3, SO_2, H_2O, NH_3, and NO_2), tropospheric aerosols, mixing heights, and optical extinction coefficients. Future developments of remote sensing systems are aimed toward extending this capability to other molecular species with improved sensitivity and vertical resolution.

Passive remote sensors are categorized generically into four categories: (1) gas filter correlation techniques, (2) interferometry, (3) infrared heterodyne radiometry, and (4) spectroradiometers. Technological improvements in the sensitivity of passive remote sensors should expand the number of instruments within each generic class as potential payloads for airborne and space platforms. Despite the potential improvements in the sensitivity of passive remote sensors through technological improvements in systems and detector technology, several fundamental limitations in passive sensors will restrict the use of these instruments for some of the scientific experiments envisioned in tropospheric research. Passive sensors in the thermal infrared have difficulty measuring molecules in the biosphere and are limited to measuring in broad vertical layers in the middle and upper troposphere. The relatively long integration times required to measure low-concentration species will restrict the horizontal resolution of the measurement to broad geographic areas and will probably require the use of geostationary satellite platforms to investigate regional areas, such as the northeastern U.S. corridor. Instruments using earth-scattered sunlight are capable of measuring total integrated burdens of molecular species, but have inherently limited temporal and geographic coverage. Despite these limitations, passive remote sensors do exhibit several attractive characteristics for global measurements. Their systems and technology are the most advanced for satellite missions. Passive instruments using upwelling thermal radiance have potential for measuring a large number of interesting tropospheric molecules simultaneously and should be effective in studying the chemistry of selected chemical systems. Passive instruments using earth-scattered sunlight have the capability of measuring the total burden of several major tropospheric gases. A satellite mission incorporating both types of instruments should be investigated in light of the scientific requirements for an early dedicated satellite for the lower atmosphere.

Considerable effort has been expended by NASA and other organizations during the past decade to develop active remote sensing techniques using lasers. The active remote sensing systems in the NASA Air Quality Program can be classified into fixed and tunable wavelength systems. The former is important primarily in measurements of tropospheric and stratospheric aerosols, aerosol extinction, and measurements of inversion layer heights. Tunable wavelength lidar systems currently have the highest potential for measuring a variety of trace gases in the troposphere simultaneously, with vertical range extending to the ground and with vertical resolution approaching 1 km. Active remote sensors under development, however, have high potential to meet some of the major scientific requirements for measurements in the troposphere from an airborne platform, including vertical resolution of approximately 1 km for major species, vertical range extending to the ground, day/night operation, true column-content measurements, and inherent high spectral resolution and tunability to measure tropospheric species simultaneously and uniquely in a background of interfering gases. For applications to space platforms, however, major technological developments must be made in the sources themselves, including high power and efficiency, improved collimation and spectral purity, wider tunability, and higher frequency and amplitude stability. In order to perform many of the scientific investigations of the troposphere, it will be necessary to develop appropriate sources for active remote sensing systems. Further, existing instruments must be tested under flight conditions not only to demonstrate their operational reliability and sensitivities, but also to provide data that can be interpreted and analyzed under conditions as close to an actual mission as possible. The time scale for utilization of laser systems and research satellite investigations is currently difficult to estimate because of the rapidly emerging technology of laser systems. Therefore, in generating scientific requirements for satellite missions, the science requirements must be tempered somewhat by the status of laser technology, and, in some cases, priorities for missions must be dictated by availability of laser technology.

APPENDIX C

Element Cycle Matrices

A matrix approach was used in an effort to systematically, but simply, indicate what is currently known about the primary species involved in the sulfur, carbon, halogen, nitrogen, and trace element cycles, as well as the importance of the species in each cycle to an overall understanding of that cycle. Individual species were rated to indicate current knowledge and their importance in each cycle relative to their major sources, removal and transformation processes, and tropospheric distribution. In the "knowledge" category, ratings range from one (very low, or no, knowledge) to four (high knowledge level—almost all is known that it is necessary to know). In the "importance" category, ratings range from one (the factor is very important in understanding the complete element cycle) to four (the factor has little or no importance in understanding the element cycle).

The combination of the "knowledge" and "importance" factors for each component of each matrix leads to an "urgency" factor for that component. An urgency factor of A indicates a very important component of an element cycle about which very little is known. Low "urgency" factors (e.g., B or C) flag areas that require research emphasis in each element cycle. A "D" urgency factor indicates a relatively unimportant component of a cycle about which quite a lot is known. These areas would require relatively little research emphasis in that cycle.

Two matrices are presented for each element cycle in the following tables (Tables C.1 through C.5). The first is the "knowledge" and "importance" matrix. Both the "knowledge" and "importance" ratings are indicated in each matrix element, with the "importance" rating in parentheses. The second matrix indicates the "urgency" factor ratings for each cycle.

TABLE C.1 Sulfur Cycle

	H_2S	SO_2	DMS	COS	Sulfate	Other
Knowledge and (Importance)[a]						
Source						
Anthropogenic	4 (2)	4 (1)	2 (2)	2 (2)	3 (1)	2 (2)
Biological	1 (2)	1 (1)	1 (2)	1 (2)	2 (1)	1 (2)
Other natural	1 (3)	1 (2)	1 (3)	1 (3)	1 (2)	1 (3)
Stratospheric	—	—	—	—	1 (2)	1 (3)
Removal						
Wet (land)	2 (2)	3 (1)	1 (2)	1 (2)	3 (1)	3 (2)
Wet (ocean)	1 (2)	1 (1)	1 (2)	1 (2)	1 (1)	1 (2)
Dry (land)	2 (3)	2 (1)	1 (2)	1 (2)	2 (1)	1 (2)
Dry (ocean)	1 (2)	2 (1)	1 (2)	1 (2)	1 (1)	1 (2)
Transformation						
Homogeneous aq. phase	2 (3)	2 (2)	1 (3)	1 (3)	3 (2)	1 (3)
Homogeneous gas phase	2 (3)	2 (1)	2 (2)	2 (2)	—	1 (2)
Heterogeneous	2 (2)	2 (1)	1 (2)	1 (2)	3 (1)	1 (2)
Distribution	2 (2)	2 (2)	1 (2)	1 (2)	2 (2)	2 (3)
Urgency Factor[b]						
Source						
Anthropogenic	C	B	B	B	B	B
Biological	A	A	A	A	A	A
Other natural	B	A	B	B	A	B
Stratospheric	—	—	—	—	A	B
Removal						
Wet (land)	B	B	A	A	B	C
Wet (ocean)	A	A	A	A	A	A
Dry (land)	B	A	A	A	A	A
Dry (ocean)	A	A	A	A	A	A
Transformation						
Homogeneous aq. phase	C	B	B	B	C	B
Homogeneous gas phase	B	A	B	B	—	A
Heterogeneous	B	A	A	A	B	A
Distribution	B	B	A	A	B	C
Interaction with other cycles	H_xO_y	C N H_xO_y Aerosol	C N H_xO_y	C H_xO_y	C H_xO_y Aerosol	C N H_xO_y

[a]Knowledge: 1 = low, 4 = high; Importance: 1 = high, 4 = low.

[b]Urgency factor: A = extremely urgent for that cycle, B = considerably urgent, C = moderately urgent, D = of little urgency.

TABLE C.2 Carbon Cycle

	Carbonaceous Aerosol	Methane and Its Reaction Products	Gas Phase Organic Compounds	
			C_2-C_5	C_5
Knowledge and (Importance)[a]				
Source				
Anthropogenic	2 (3)	2 (3)	2 (3)	2 (3)
Biological	1 (1)	1 (1)	1 (1)	1 (1)
Other natural	—	—	1 (2)	1 (2)
Stratospheric	—	—	—	—
Removal				
Wet (land)	1 (2)	—	1 (2)	1 (2)
Wet (ocean)	1 (3)	—	1 (3)	1 (3)
Dry (land)	1 (2)	—	1 (2)	1 (2)
Dry (ocean)	1 (3)	—	1 (3)	1 (3)
Transformation				
Homogeneous aq. phase	—	1 (2)	1 (2)	1 (2)
Homogeneous gas phase	—	2 (1)	1 (1)	2 (1)
Heterogeneous	1 (1)	—	1 (1)	1 (1)
Distribution	1 (1)	3 (1)	1 (1)	1 (1)
Urgency Factor[b]				
Source				
Anthropogenic	C	C	C	C
Biological	A	A	A	A
Other natural	—	—	A	A
Stratospheric	—	—	—	—
Removal				
Wet (land)	A	—	A	A
Wet (ocean)	B	—	B	B
Dry (land)	A	—	A	A
Dry (ocean)	B	—	B	B
Transformation				
Homogeneous aq. phase	—	A	A	A
Homogeneous gas phase	—	A	A	A
Heterogeneous	A	—	A	A
Distribution	A	B	A	A
Interaction with other cycles	S N C Aerosol	H_xO_y N C	H_xO_y N	H_xO_y N Aerosol

[a]Knowledge: 1 = low, 4 = high; Importance: 1 = high, 4 = low.
[b]Urgency factor: A = extremely urgent for that cycle, B = considerably urgent, C = moderately urgent, D = of little urgency.

TABLE C.3 Trace Element Cycles

	Hg	Se	As	B	P	Pb
Knowledge and (Importance)[a]						
Source						
Anthropogenic	3 (1)	3 (1)	3 (1)	2 (2)	3 (2)	4 (1)
Biological	2 (1)	2 (1)	2 (2)	2 (1)	1 (2)	1 (2)
Other natural	2 (2)	1 (2)	2 (2)	2 (1)	2 (2)	2 (2)
Stratospheric	—	—	—	—	—	—
Removal						
Wet (land)	2 (1)	2 (1)	2 (1)	2 (1)	3 (1)	3 (1)
Wet (ocean)	2 (1)	1 (1)	2 (1)	2 (1)	2 (1)	3 (1)
Dry (land)	1 (3)	1 (2)	1 (2)	1 (4)	2 (2)	2 (3)
Dry (ocean)	1 (2)	1 (3)	1 (3)	1 (3)	2 (2)	2 (4)
Transformation						
Homogeneous aq. phase	1 (3)	2 (3)	2 (3)	2 (2)	2 (4)	1 (4)
Homogeneous gas phase	2 (1)	1 (3)	1 (3)	1 (2)	1 (4)	1 (4)
Heterogeneous	1 (2)	1 (1)	1 (1)	1 (2)	1 (4)	1 (3)
Distribution	2 (1)	1 (1)	2 (1)	1 (1)	2 (1)	3 (1)
Urgency Factor[b]						
Source						
Anthropogenic	B	B	B	B	C	B
Biological	A	A	B	A	B	B
Other natural	B	A	B	A	B	B
Stratospheric	—	—	—	—	—	—
Removal						
Wet (land)	A	A	A	A	B	B
Wet (ocean)	A	A	A	A	A	B
Dry (land)	B	A	A	B	B	C
Dry (ocean)	A	B	B	B	B	C
Transformation						
Homogeneous aq. phase	B	C	C	B	C	B
Homogeneous gas phase	A	B	B	A	B	B
Heterogeneous	A	A	A	A	B	B
Distribution	A	A	A	A	A	B
Interactions with other cycles	H_xO_y C	H_xO_y C	H_xO_y C	H_xO_y		

[a]Knowledge: 1 = low, 4 = high; Importance: 1 = high, 4 = low.

[b]Urgency factor: A = extremely urgent for that cycle, B = considerably urgent, C = moderately urgent, D = of little urgency.

TABLE C.4 Halogen Cycles

	F_o	F_i	F^-	Cl_o	Cl_i	Cl^-	Br_o	Br_i	Br^-	I_o	I_i	I^-
Knowledge and (Importance)[a]												
Source												
Anthropogenic	3(2)	2(2)	2(2)	3(1)	3(2)	3(2)	2(1)	2(2)	2(2)	2(2)	2(2)	2(2)
Biological	2(2)	1(3)	2(2)	2(1)	1(3)	2(2)	1(1)	1(3)	2(2)	2(1)	1(3)	2(2)
Other natural	—	3(3)	3(3)	2(3)	3(3)	3(3)	1(3)	2(3)	2(3)	2(3)	2(3)	2(3)
Stratospheric	3(4)	4(4)	4(4)	3(4)	3(4)	3(4)	2(4)	2(4)	2(4)	2(4)	2(4)	2(4)
Removal												
Wet (land)	3(2)	2(2)	3(2)	2(2)	3(2)	3(2)	2(2)	2(2)	2(2)	2(2)	2(2)	2(2)
Wet (ocean)	2(2)	2(2)	2(2)	2(2)	3(2)	3(2)	2(2)	2(2)	3(2)	2(2)	2(2)	2(2)
Dry (land)	2(3)	3(3)	3(3)	2(3)	2(3)	2(3)	2(3)	2(3)	2(3)	2(3)	2(3)	2(3)
Dry (ocean)	2(3)	2(3)	2(3)	2(3)	2(3)	2(3)	2(3)	2(3)	2(3)	2(3)	2(3)	2(3)
Transformation												
Homogeneous aq. phase	2(2)	3(2)	—	2(2)	3(2)	—	2(2)	2(2)	—	2(2)	2(2)	—
Homogeneous gas phase	2(2)	2(2)	—	2(2)	3(2)	—	2(1)	2(2)	—	2(1)	2(2)	—
Heterogeneous	1(2)	1(2)	—	2(2)	2(1)	—	1(2)	1(1)	—	1(2)	2(1)	—
Distribution	3(2)	2(2)	2(2)	3(2)	2(2)	3(2)	1(2)	1(2)	1(2)	2(2)	1(2)	1(2)
Urgency Factor[b]												
Source												
Anthropogenic	C	B	B	B	C	C	A	B	B	B	B	B
Biological	B	B	B	A	B	B	A	B	B	A	B	B
Other natural	—	C	C	C	C	C	B	C	C	C	C	C
Stratospheric	D	D	D	D	D	D	C	C	C	C	C	C
Removal												
Wet (land)	C	B	C	B	C	C	B	B	C	B	B	B
Wet (ocean)	B	B	B	B	C	C	B	B	B	B	B	B
Dry (land)	C	C	C	C	C	C	C	C	C	C	C	C
Dry (ocean)	C	C	C	C	C	C	C	C	C	C	C	C
Transformation												
Homogeneous aq. phase	B	C	—	B	C	—	B	B	—	B	B	—
Homogeneous gas phase	B	B	—	B	C	—	A	B	—	A	B	—
Heterogeneous	A	A	—	B	A	—	A	A	—	A	A	—
Distribution	C	B	B	C	B	C	A	A	A	B	A	A
Interactions with other cycles		Aerosol			S N H_xO_y Aerosol			H_xO_y Aerosol			H_xO_y Aerosol	

[a]Knowledge: 1 = low, 4 = high; Importance: 1 = high, 4 = low.
[b]Urgency factor: A = extremely urgent for that cycle, B = considerably urgent, C = moderately urgent, D = of little urgency.

NOTES: X_o = organic gaseous compounds; X_i = inorganic gaseous compounds; X^- = halides in droplets and aerosols; I^- includes IO_3^-.

TABLE C.5 Nitrogen Cycle

	Odd Nitrogen Species	NH_3/NH_4^+	HCN	N_2O
Knowledge and (Importance)[a]				
Source				
Anthropogenic	2 (2)	3 (3)	1 (2)	2 (1)
Biological	1 (1)	1 (1)	1 (3)	2 (1)
Other natural	2 (4)	2 (3)	—	2 (3)
Stratospheric	3 (3)	3 (4)	1 (3)	3 (4)
Removal				
Wet (land)	2 (1)	2 (1)	2 (2)	2 (4)
Wet (ocean)	1 (1)	1 (1)	2 (2)	2 (4)
Dry (land)	1 (1)	1 (1)	2 (3)	3 (4)
Dry (ocean)	1 (1)	1 (1)	2 (3)	3 (4)
Transformation				
Homogeneous aq. phase	2 (2)	2 (3)	2 (2)	—
Homogeneous gas phase	3 (1)	2 (2)	2 (2)	4 (2)
Heterogeneous	2 (2)	2 (3)	2 (2)	—
Distribution	2 (1)	1 (1)	2 (3)	4 (2)
Urgency Factor[b]				
Source				
Anthropogenic	B	C	A	A
Biological	A	A	B	A
Other natural	C	C	B	C
Stratospheric	C	D	B	D
Removal				
Wet (land)	A	A	B	C
Wet (ocean)	A	A	B	C
Dry (land)	A	A	C	D
Dry (ocean)	A	A	C	D
Transformation				
Homogeneous aq. phase	B	C	B	—
Homogeneous gas phase	B	B	B	C
Heterogeneous	B	C	B	—
Distribution	A	A	C	C
Interactions with other cycles	H_xO_y, S, C, Aerosol	S, Aerosol		H_xO_y

[a]Knowledge: 1 = low, 4 = high; Importance: 1 = high, 4 = low.
[b]Urgency factor: A = extremely urgent for that cycle, B = considerably urgent, C = moderately urgent, D = of little urgency.

Index

V

W